Phase-Locked LOOPS
Theory and Applications

John L. Stensby, Ph.D.
Department of Electrical and Computer Engineering
University of Alabama in Huntsville
Huntsville, Alabama

CRC Press
Boca Raton   New York

Library of Congress Cataloging-in-Publication Data

Stensby, John L.
    Phase-locked loops : theory and applications / John L. Stensby
        p.   cm.
    Includes bibliographical references and index.
    ISBN 0-8493-9471-6 (alk. paper)
    1. Phase-locked loops.   I. Title.
TK7872.P38S74   1997
621.3815'364--dc21
                                                                                                    96-50402
                                                                                                        CIP

   This book contains information obtained from authentic and highly regarded sources. Reprinted material is quoted with permission, and sources are indicated. A wide variety of references are listed. Reasonable efforts have been made to publish reliable data and information, but the author and the publisher cannot assume responsibility for the validity of all materials or for the consequences of their use.
   Neither this book nor any part may be reproduced or transmitted in any form or by any means, electronic or mechanical, including photocopying, microfilming, and recording, or by any information storage or retrieval system, without prior permission in writing from the publisher.
   The consent of CRC Press LLC does not extend to copying for general distribution, for promotion, for creating new works, or for resale. Specific permission must be obtained in writing from CRC Press LLC for such copying.
   Direct all inquiries to CRC Press LLC, 2000 Corporate Blvd., N.W., Boca Raton, Florida 33431.

© 1997 by CRC Press LLC

No claim to original U.S. Government works
International Standard Book Number 0-8493-9471-6
Library of Congress Card Number 96-50402
Printed in the United States of America 1 2 3 4 5 6 7 8 9 0
Printed on acid-free paper

PREFACE

This book presents basic theory and applications of phase-locked loops (PLLs). First, in a qualitative manner designed to motivate the reader, the subject is introduced, and several key applications are discussed. Next, basic models are developed for the components that comprise a PLL, and these are used to develop a model for the PLL. Then, both linear and nonlinear methods are employed to analyze the basic PLL model. Because PLL technology plays such an important role in modern electronic systems, the book includes an introduction to PLL component technologies that are used in system implementation.

The book is written in two parts. Part I provides coverage of elementary PLL theory and technology. It contains material that applications-oriented practitioners will find most helpful — material that is consulted most often during the development of systems that employ the phase-lock principle. This first part contains an analysis of the classical linear PLL model, and it includes coverage of basic PLL component technologies. Part II discusses the basic nonlinear PLL model. It describes basic phenomena that cannot be analyzed by using linear models and techniques. Furthermore, it describes theory and methods that are useful in establishing limits (and understanding these limits) on the ability of a PLL to do its job. That is, the material in Part II is useful for analyzing the ability of a PLL to achieve phase lock and extract phase information from a reference signal embedded in noise.

Difficult choices had to be made while writing this book. The subject of phase-locked loops (and synchronization theory in general) is wide and diverse. In a comprehensive manner, the subject cannot be covered in a single book. The adept student of PLLs will discover some topics in traditional PLL theory that are not discussed in this book. And he will find several topics covered here that are relatively new to the field. Part I covers most of the classical model development and linear PLL theory, without taking into account the chronology of topic development. Part II covers the most important topics from the classical nonlinear theory, but its coverage is weighted towards more recently developed (in the last 25 years) topics. To accommodate the nonspecialist in the field of PLLs, the pace of subject introduction is moderate, and the exposition is not terse. The subject matter is presented in a detailed manner with the aid of a significant number of technical figures and plots.

Powerful desktop computers have become ubiquitous in the work environment. These machines have made a tremendous impact on the way engineers and scientists interact, work, and learn. This is fortuitous for the study of nonlinear phenomena in PLLs and other systems. Classical qualitative theory and methods can be used to explain nonlinear phenomena in the PLL, but engineers need more than this. In PLL analysis and design, engineers can get much needed quantitative results by using numerical methods implemented on inexpensive desktop computers.

This philosophy is put to practice in this book. Numerical results are used extensively to describe the nonlinear phenomena that govern PLL behavior. Also, a considerable effort is devoted to the development of specialized numerical methods for PLL analysis. Powerful numerical algorithms are described for analyzing global, nonlinear behavior in second-order PLL models. Also, numerical methods are described for the analysis of noise in the nonlinear PLL model.

People interested in telecommunication, control, and instrumentation systems should find this book helpful. Only modest prerequisites are required to comprehend Part I. A good undergraduate training in communication and control systems should suffice as prerequisites for Part I. Normally, such training includes coverage of classical linear systems, applied random processes, and transform theory. In the appendices, extensive review material is given on bandpass processes and systems. Also, references are cited frequently in an effort to encourage the reader to consult other sources of information. Many of the main ideas in Part II should be comprehensible to those readers in command of the above-mentioned prerequisites; however, in most electrical engineering programs the topics in Part II are introduced at the first-year graduate level. Much of the nonlinear theory is described in a qualitative manner with the aid of figures and plots. Also, several appendices are devoted to coverage of basic nonlinear theory essential to the analysis of second-order PLLs. However, as is often the case when studying complex phenomena, a modest investment in time and effort can yield substantial dividends. Readers who consult the references cited in Part II will be rewarded by a more comprehensive understanding of PLL behavior.

Several approaches are possible for the study of this book. First, Part I is mandatory reading since it covers elementary linear PLL theory, technology, and applications. Readers can stop after Chapter 4 if they are not interested in nonlinear PLL theory; Part II should be studied by readers who are interested in the nonlinear nature of the PLL. Depending on specific interests, the study of Part II can be approached in several ways. Readers can stop after Chapter 5 if they are not interested in stochastic behavior and the influence of noise. On the other hand, readers could concentrate on Chapters 6 through 8, and consult Chapter 5 only when needed, if they are primarily interested in the treatment of noise.

The support of several people and organizations made writing this book possible. First, much of the material in this book was "student tested" in graduate-level courses at the University of Alabama in Huntsville. The author is indebted to numerous students who took his PLL courses and tolerated both his handwritten notes and early typewritten manuscript drafts. Dr. Merv Budge reviewed several versions of the manuscript; his patience while reviewing early versions of the manuscript is appreciated. Dr. Jia Li reviewed Chapter 5 and provided several valuable comments. Andrea Demby, Project Editor of CRC Press LLC, deserves much credit for her excellent editing and production work. Finally, Birgit Stensby, the author's wife, deserves extensive credit for proofreading much of the manuscript. Over the years, her instruction in writing style and grammar have been of great value. Without her tolerance and support, this book could not have been written.

John Stensby

University of Alabama in Huntsville
Huntsville, AL

ABOUT THE AUTHOR

John Stensby has worked in the field of phase-coherent communication systems for over 25 years. During his career, he has written extensively, including many articles on PLL theory, design, and implementation. Also, on a regular basis over the last 15 years, he has lectured on the subject of phase-locked loops, both in dedicated PLL courses and in general communication system courses. To industry and government, he is an active consultant involved in the design and implementation of phase-locked systems, radio frequency (RF) circuits, and microcomputer-based control systems. He worked for Rockwell International, Anaheim, CA, as an RF circuit design engineer; also, he was employed by Texas Instruments, Dallas, TX, as a communication systems engineer involved in the development of charge coupled devices for digital signal processing applications.

Dr. Stensby received his Ph.D. degree from Texas A&M University, College Station, TX. After receiving the degree, he started his teaching career as a member of the faculty at the University of Kansas, Lawrence, KS. For the last 13 years, he has been a member of the Electrical and Computer Engineering faculty at the University of Alabama in Huntsville.

Dr. Stensby enjoys running as a means of staying in shape and controlling stress. In his spare time he enjoys the hobby of amateur radio.

Important Symbols

Symbol	Definition
t	Real-time variable
≡	Is defined by (denotes a definition)
$\theta_i(t)$	Phase of input reference
$\theta_v(t)$	Phase of VCO
K_m	Phase detector gain
A	Input signal amplitude in RMS volts
K_1	VCO amplitude in RMS volts
ϕ	Closed-loop phase error
F(s)	Loop filter transfer function
e(t)	VCO control voltage
x(t)	Phase detector output
f(t)	Loop filter impulse response
L_1, L_2	Linear operators used to define loop filter
$a_0 \ldots a_m$	Numerator coefficients for loop filter
$b_0 \ldots b_n$	Denominator coefficients for loop filter
K_v	VCO gain constant
ω_0	VCO center frequency
$\theta_1(t)$	Relative phase of reference
$\theta_2(t)$	Relative phase of VCO
$\Theta_1(s)$	Relative phase of reference in Laplace domain
$\Theta_2(s)$	Relative phase of VCO in Laplace domain
G	Closed-loop gain term
s	Laplace variable
E(s)	VCO control voltage in Laplace domain
$G_0(s)$	Open-loop transfer function
$\Phi(s)$	Phase error in Laplace domain

$H(s)$	Closed-loop transfer function
$h(t)$	Closed-loop impulse response
ω_Δ	PLL detuning parameter
ω_i	Radian frequency of sinusoidal reference
K_p	Phase modulation index
β	Modulation index
K_f	Frequency modulation index
ω_m	Frequency of tone message
$|H(j\omega)|$	Magnitude of a transfer function
$\angle H(j\omega)$	Phase of a tranfer function
$r(t)$	Reference signal in additive noise
$\eta_{bp}(t)$	Bandpass Gaussian noise
$\eta_c(t), \eta_s(t)$	Quadrature components of bandpass Gaussian noise
$R_{\eta c}(\tau)$	Autocorrelation of quadrature component $\eta_c(t)$
$R_{\eta s}(\tau)$	Autocorrelation of quadrature component $\eta_s(t)$
$R_{\eta s \eta c}(\tau)$	Cross correlation of quadrature components
$S_{bp}(\omega)$	Power spectrum of bandpass noise
$S_{\eta c}(\omega)$	Power spectrum of component noise
N_0	White noise parameter
$\eta(t;\theta_2)$	Noise internal to PLL model
τ_c	Correlation time of $\eta_c(t)$ and $\eta_s(t)$
$\eta(t)$	White Gaussian noise
ρ	Signal-to-noise ratio in the loop
B_L	Noise-equivalent bandwidth
$S_\phi(\omega)$	Power spectrum of phase error
σ_ϕ^2	Variance of phase error
N_c, N_s	Quadrature components of noise process
K_D	Limiter phase detector gain constant
θ_n	Phase noise in limiter phase detector
g	Limiter phase detector characteristic, noiseless case
$\tilde{g}$	Control component in output of limiter phase detector
SNR_i	Signal-to-noise ratio on input of limiter phase detector
ω_{if}	IF frequency in long loop
ψ_{if}	IF signal phase in long loop
ω_c	Carrier frequency
$\Gamma_{bp}(t)$	Envelope of bandpass signal

$\psi_{bp}(t)$	Phase of bandpass signal
$x_{bp}(t)$	Bandpass signal
$x_c(t), x_s(t)$	Quadrature components
$x_{lp}(t)$	Lowpass equivalent
$\hat{x}(t)$	Hilbert transform of $x(t)$
t_p	Phase delay of bandpass filter
t_g	Group delay of bandpass filter
$U(t)$	Unit step function
θ_Δ	Phase step applied to reference
R	Frequency ramp applied to reference
α	Integrator gain
ω_n	Natural frequency
ζ	Damping factor
a, b	Imperfect integrator parameters
$v_u(t)$	Up-output on sequential phase detector
$v_d(t)$	Down-output on sequential phase detector
τ	Slow-time variable
Ω_h	Hold-in range
Ω_p	Pull-in range
T_p	Phase acquisition time for first-order PLL
$\|\vec{X}\|$	Euclidean norm of vector $\vec{X}$
φ_s	Saddle point
Ω_{po}	Pull-out range
$\dot{\phi}(\tau)$	Gain-normalized phase derivative
T_d	Pull-in time, second-order PLL
Ω_2	Half-plane pull-in range
Γ_s	Stable limit cycle
Γ_u	Unstable limit cycle
$s(\phi_{sp})$	Separatrix cycle stability test function
ϕ_{sp}	Phase value at saddle point
ω_f	Apparent frequency error in false locked loop
$X_d(N)$	Random walk displacement (in steps) after taking N steps
τ_s	Time required to take a step in random walk process
ℓ	Distance taken for every step of random walk process
R_{nm}	Number of random steps to the right for random walk
L_{nm}	Number of random steps to the left for random walk

$G(x)$	Probability distribution function for zero mean, unit variance Gaussian random variable
$X(t)$	Wiener process displacement at time t
D	Diffusion constant in Wiener process
$\mathfrak{J}$	Probability current
ΔX_t	Increment of random process X_t
Θ	Conditional characteristic function
$K^{(1)}$	Drift coefficient, one-dimensional Fokker–Planck equation
$K^{(2)}$	Diffusion coefficient, one-dimensional Fokker–Planck equation
$\mathfrak{J}_{ss}$	Steady-state probability current
t_a	Absorption time random variable
$\mathbb{C}_0$	Constant in absorbing boundary theory
$K_i^{(1)}$	Drift coefficient, general nth-order Fokker–Planck equation
$K_{ij}^{(2)}$	Diffusion coefficient, general nth-order Fokker–Planck equation
L_{FP}	Fokker–Planck operator
T_x	Relaxation time constant
R_k	Domain of attraction centered at kth stable state
φ_0	Phase at equilibrium point
φ_k	Phase $\varphi_k = \varphi_0 - 2\pi k$ at equilibrium point
ϕ_m	Modulo-$2\pi m$ phase error
$p_m(\phi_m, \vec{Y}, t)$	Density function for m-attractor model
$p_m(\phi_m)$	Marginal density function m-attractor model
$[\phi_m]_{2\pi}$	Modulo-2π value defined so that $-\pi \leq [\phi_m]_{2\pi} < \pi$
$U(\phi)$	Potential function
β_Δ	Product of SNR and detuning variables
N_+	Slip rate in forward direction
N_-	Slip rate in backwards direction
$e(t)$	State variable in second-order PLL model
ρ_2	Parameter in Fokker–Planck equation for second-order PLL
$E_s[\cdot]$	Ensemble average using steady-state density function describing bracketed quantity
$E_{\phi 1}[\cdot]$	Ensemble average using steady-state density $p_1(\phi_1)$
E_{dc}	DC component in approximation of $E_s[e \mid \phi_1]$
E_f	Sinusoidal component in approximation of $E_s[e \mid \phi_1]$

τ_a	Normalized first passage time
H^+, H^-	Hyperplanes at absorbing boundaries in state space
$\mathfrak{J}_{\phi ss}$	Steady-state probability current in ϕ_1 direction
ΔU	Change in steady-state potential over 2π in phase
$\{z_n(\phi_1)\}$	Basis of function space for 2π-periodic boundary conditions
$\{h_n\}$	Basis of function space for natural boundary conditions
$\psi_k(y;c)$	Weighted Hermite polynomial
$p_k(\phi_1)$	Periodic weight functions in density expansion
Q_k	Coefficients in expansion of $L_{FP}[p]$
g_k, f_k	Used in algebraic system for unknown initial conditions
$p_k^{(f)}$	Superscript (f) denotes a forward-direction numerical integration
$p_k^{(b)}$	Superscript (b) denotes a backward-direction numerical integration
$J(t)$	Integer-valued state index
a_{ik}	Transition rate
$\pi_k(t)$	Occupancy probability
$\mathbf{A}$	Transition rate matrix
$\vec{\Pi}(t)$	Occupancy probability vector
t_d	Time difference variable
$p_{ik}(t_d)$	Transition function
$P(t_d)$	Transition function matrix
λ_k	Eigenvalue of Fokker–Planck operator
$e_k(\phi_m, \vec{Y})$	Eigenfunction of Fokker–Planck operator
$U_{R_i}(\phi_m, \vec{Y})$	State space windowing functions
q_{ki}	Domain scale factor
Λ	Diagonal matrix containing Eigenvalues
$\vec{e}_k$	Eigenvector of transition rate matrix $\mathbf{A}$
$\text{mod}_m(i)$	Least positive residue function
a_k	Transition rate for cyclic jump model
r_j	Transition rate in unrestricted jump model
K_{min}, K_{max}	Bounds on cycle slip length
κ	Domain scale factor
R	Frequency of undesirable events
$\mathscr{L}$	Vector space of $2\pi m$-periodic functions
$\mathscr{L}^{(k)}$	Subspace of $\mathscr{L}$

$\oplus$	Direct sum
$e_i^{(k)}(\phi_m, \vec{Y})$	Eigenfunction in $\mathscr{L}^{(k)}$
$\lambda_i^{(k)}$	Eigenvalue corresponding to $e_i^{(k)}(\phi_m, \vec{Y})$
σ_y^2	Variance of modulation y
D	Bandwidth ratio
$H_n(y)$	Hermite polynomial
b, b^+	Linear operators

TABLE OF CONTENTS

PART I   ELEMENTARY THEORY AND APPLICATIONS 1

Chapter 1   Introduction 3
1.1   The Phase and Frequency of Signal Relative to a Reference 3
1.2   A Generic Problem 4
1.3   The Phase-Locked Loop 5
1.4   Basic Applications 6
    1.4.1   Coherent Demodulation of Amplitude-Modulated Signals 6
    1.4.2   Frequency Synthesis 7
    1.4.3   Demodulation of BPSK Signals 8
    1.4.4   Phase-Locked Receivers 9
1.5   Phase-Locked Loop Literature 11
1.6   Topical Outline of the Text 11

Chapter 2   Modeling the Phase-Locked Loop 15
2.1   Modeling PLL Components 16
    2.1.1   Modeling the Analog Phase Detector 16
    2.1.2   Modeling the Loop Filter 17
    2.1.3   Modeling the Voltage-Controlled Oscillator 18
2.2   Modeling the Nonlinear PLL 19
    2.2.1   PLL Description Based on an Ordinary Differential Equation 20
    2.2.2   PLL Description Based on a First-Order Nonlinear System 21
    2.2.3   PLL Description Based on an Integral Differential Equation 22
2.3   Modeling the Linear PLL 22
2.4   Modeling a PLL with an Angle-Modulated Reference Source 24
2.5   Modeling a PLL with a Noisy Reference Source 27
    2.5.1   Reference Signal Corrupted by Additive Noise 28
    2.5.2   Output of the Sinusoidal Phase Detector 29
    2.5.3   Nonlinear Model of the PLL with a Noisy Reference 31
    2.5.4   Linear Model of the PLL with a Noisy Reference 33
2.6   Modeling the Limiter Phase Detector 37
    2.6.1   Hard Limiting the Input and VCO Signals 39
    2.6.2   The Output of the Limiter Phase Detector 39
    2.6.3   Splitting the Detector Output into Control and Noise Components 40

    2.6.4   Practical Application of the Limiter Phase Detector Model..42
    2.6.5   Example: Limiter Phase Detector with Sinusoidal $g(\phi)$..........43
    2.6.6   Example: Limiter Phase Detector with Triangular $g(\phi)$..........44
2.7 Modeling the Long Loop 46
    2.7.1   Baseband Model of the Long Loop 47
Appendix 2.5.1: Narrowband Signals and Systems 51
    2.5.1.1 Modeling Bandpass Signals and Systems...................................51
    2.5.1.2 Lowpass Equivalent Signals and Systems..................................52
    2.5.1.3 Symmetrical Bandpass Filter...54
    2.5.1.4 Phase and Group Delays of a Bandpass System 55
    2.5.1.5 Bandpass Input/Output 57
Appendix 2.5.2 Narrowband Noise 60
    2.5.2.1 Relationships Between Correlations R_η, $R_{\eta c}$, $R_{\eta s}$, and $R_{\eta c \eta s}$ 62
    2.5.2.2 Symmetrical Bandpass Processes..65
Appendix 2.6.1 Evaluation of $E[\cos\theta_n]$ and $E[\sin\theta_n]$ for the Gaussian Noise Case 67

Chapter 3   **Linear Analysis of Common First- and Second-Order PLL** 71
3.1 The First-Order PLL..72
    3.1.1   Closed-Loop Transfer Function 72
    3.1.2   Transient and Steady-State Tracking Errors..............................72
    3.1.3   Noise-Equivalent Bandwidth..75
    3.1.4   Summary of the First-Order PLL Linear Model.......................75
3.2 The Second-Order PLL with a Perfect Integrator...............................76
    3.2.1   Transfer Functions 76
    3.2.2   Loop Stability..77
    3.2.3   Transient and Steady-State Tracking Errors..............................77
    3.2.4   Noise-Equivalent Bandwidth..82
    3.2.5   Summary for the Second-Order PLL Containing a Perfect Integrator Loop Filter 83
3.3 The Second-Order PLL with Imperfect Integrator 84
    3.3.1   Transfer Functions 85
    3.3.2   Loop Stability..86
    3.3.3   Transient and Steady-State Tracking Errors..............................86
    3.3.4   Noise-Equivalent Bandwidth..88
    3.3.5   Summary for the Second-Order PLL Based on the Imperfect Integrator Loop Filter..88

Chapter 4   **Phase-Locked Loop Components and Technologies** 91
4.1 Phase Detectors — Analog and Digital...92
    4.1.1   Integrated Circuit Four-Quadrant Analog Multipliers 93
    4.1.2   Diode Ring Mixer 96
    4.1.3   Exclusive OR Gate 98
    4.1.4   RS Flip-Flop 99
    4.1.5   Sequential Phase/Frequency Detectors...................................101
4.2 Loop Filters 108
4.3 Voltage-Controlled Oscillators...110

4.3.1 Voltage-Controlled Crystal Oscillators 111
4.3.2 RC Multivibrators 113
4.4 Lock Detection 115

PART II NONLINEAR PLL ANALYSIS 117

Chapter 5 Nonlinear PLL Behavior in the Absence of Noise 119
5.1 First-Order PLL with Constant Frequency Reference 121
    5.1.1 Phase Plane Analysis of a First-Order PLL 121
    5.1.2 Phase Acquisition in a First-Order PLL 123
5.2 A Second-Order PLL Using a Perfect Integrator 124
    5.2.1 Stable and Unstable Equilibrium Points 126
    5.2.2 A Phase Plane Analysis of the Perfect Integrator Case 128
    5.2.3 Pull-In Properties of a Second-Order Type II PLL 134
5.3 A Second-Order PLL Containing an Imperfect Integrator Loop Filter 135
    5.3.1 Stable and Unstable Equilibrium Points 137
    5.3.2 Phase Plane Structure Dependent on ω_Δ' — the Values Ω_p', Ω_2', and Ω_h' 139
    5.3.3 The High-Gain Case 142
    5.3.4 The Low-Gain Case 149
    5.3.5 General Phase Plane Characteristics for the Low-Gain Case 152
    5.3.6 Computing the Separatrix Cycle and Determining its Stability 154
5.4 Effects of IF Filtering on the Long Loop 156
Appendix 5.2.1 Pull-In Time For a Second-Order PLL 161
Appendix 5.3.1 Pull-In Range of a Second-Order PLL 165
Appendix 5.3.2 Computation of Separatrices for a Second-Order PLL 168
Appendix 5.3.3 The Separatrix Cycle of a Second-Order PLL 171
    5.3.3.1 Computing the Separatrix Cycle 173
    5.3.3.2 Stability of the Separatrix Cycle 174
    5.3.3.3 Bifurcation of a Periodic Limit Cycle from the Separatrix Cycle 175

Chapter 6 Stochastic Methods for the Nonlinear PLL Model 177
6.1 The Random Walk — A Simple Markov Process 178
    6.1.1 The Wiener Process as a Limit of the Random Walk 181
    6.1.2 The Diffusion Equation for the Transition Density Function 183
    6.1.3 An Absorbing Boundary on the Random Walk 185
    6.1.4 An Absorbing Boundary on the Wiener Process 187
    6.1.5 Gaussian White Noise as the Formal Derivative of the Wiener Process 188
6.2 The First-Order Markov Process 190
    6.2.1 An Important Application of Markov Processes 192
    6.2.2 The Chapman–Kolmogorov Equation 193

- 6.2.3 The One-Dimensional Kramers–Moyal Expansion............194
- 6.2.4 The One-Dimensional Fokker–Planck Equation..................198
- 6.2.5 Transition Density Function..202
- 6.2.6 Natural, Periodic, and Absorbing Boundary Conditions....203
- 6.2.7 Steady-State Solution to the Fokker–Planck Equation........206
- 6.2.8 The One-Dimensional First-Passage Time Problem............208
- 6.2.9 The Distribution and Density of the First-Passage Time Random Variable 210
- 6.2.10 The Expected Value of the First-Passage Time Random Variable 211
- 6.2.11 Ratio of Boundary Absorption Rates....................................214
- 6.3 The Vector Markov Process 216
  - 6.3.1 The n-Dimensional Kramers–Moyal Expansion 217
  - 6.3.2 The n-Dimensional Fokker–Planck Equation.......................223
  - 6.3.3 A Simple Example..225

Chapter 7 Noise in the Nonlinear PLL Model..............................229

- 7.1 Qualitative Nature of and Models for the Phase Error..................229
  - 7.1.1 Modulo-2π Phase Error Model...232
  - 7.1.2 Bistable Cyclic Model 236
  - 7.1.3 Generalizations: the Multistable/m-Attractor Cyclic Model..238
  - 7.1.4 Qualitative Properties of the Multistable Model for the Moderate Noise Case...240
  - 7.1.5 Phase Error Model Containing Absorbing Boundaries 241
- 7.2 Noise in the First-Order PLL...242
  - 7.2.1 Modeling the First-Order PLL with a Noisy Reference.......243
  - 7.2.2 Fokker–Planck Equation for the Time-Normalized, First-Order PLL Model 245
  - 7.2.3 Steady-State Density for the Modulo-2π Phase Error..........248
  - 7.2.4 Steady-State Distribution for the Modulo-2π Phase Error 253
  - 7.2.5 Mean of the Steady-State, Modulo-2π Phase Error..............253
  - 7.2.6 Variance of the Steady-State, Modulo-2π Phase Error........254
  - 7.2.7 Noise-Induced Cycle Slips 256
- 7.3 Noise in Second-Order PLLs 261
  - 7.3.1 Nonlinear Model for a Second-Order PLL with a Noisy Reference 261
  - 7.3.2 Fokker–Planck Equation for the Second-Order PLL............263
  - 7.3.3 A Simple Example: a PLL with an RC Lowpass Loop Filter 266
  - 7.3.4 Equation of Flow in the ϕ_1 Direction 267
  - 7.3.5 Conditional Expectation Method for Approximating $p_1(\phi_1)$...269
  - 7.3.6 Approximating the Conditional Expectation........................270
  - 7.3.7 Approximating $p_1(\phi_1)$ for the Perfect Integrator Case 279
  - 7.3.8 Noise-Induced Cycle Slips in Second-Order PLLs 280
  - 7.3.9 Average Cycle Slip Rate in the Forward Direction..............282
  - 7.3.10 Ratio of Cycle Slip Rates for the Second-Order PLL 285

Chapter 8 Numerical Methods for Noise Analysis in the Nonlinear PLL Model 289

8.1 Computing an Approximation to Steady-State $p_1(\phi_1, \vec{Y})$ 290
    8.1.1 Expansion of the Joint Density in a Complete Orthonormal Set 291
    8.1.2 A Coupled System of Differential Equations for the p_k 292
    8.1.3 Computing a 2π-Periodic Solution of the Coupled System 294
    8.1.4 The Case $n \geq 2$ 297

8.2 Approximating $p_1(\phi_1)$ for a PLL Containing an RC Lowpass Loop Filter 297
    8.2.1 Transformation of the Fokker–Planck Operator 298
    8.2.2 Form of Series Expansion 301
    8.2.3 Truncation of the Infinite Series 302
    8.2.4 Computing the Periodic Solution and Approximating $p_1(\phi_1)$ 304
    8.2.5 Practical Limitations of the Algorithm 306

8.3 Approximating $p_1(\phi_1)$ for a PLL Containing a Perfect Integrator Loop Filter 307
    8.3.1 Transformation of the Fokker–Planck Operator 308
    8.3.2 Development of the Coupled System of Differential Equations 308
    8.3.3 Computing an Approximation to the Marginal Density $p_1(\phi_1)$ 310

8.4 Modeling Cycle Slips as State Transitions of a Markov Jump Process 313
    8.4.1 Modeling the Cycle Slip Problem as a Finite State Markov Jump Process 314
    8.4.2 The Transition Rate Matrix 316
    8.4.3 Physical Interpretation and Properties of the Transition Rates 318
    8.4.4 The Dynamics of State Occupancy 320
    8.4.5 Eigenvalues and Eigenfunctions of the Fokker–Planck Operator 323
    8.4.6 Eigenvalues and Eigenfunctions Under Moderate Noise Conditions 325
    8.4.7 Calculation of the Transition Rates a_{ij} 327
    8.4.8 An m-Attractor Markov Jump Model with a Cyclic Structure 332
    8.4.9 Relationship of the Cyclic Jump Model to an Unrestricted Jump Model 334
    8.4.10 Computing the Slip Rates in the "Coarse-Grained", m-Attractor, Cyclic State Model 337

8.5 The First-Order PLL as a Simple Example 341
    8.5.1 The Eigenvalue Problem Formulated in Terms of a First-Order System of Differential Equations 343
    8.5.2 Decomposition of the Vector Space of $2\pi m$-Periodic Functions 344

	8.5.3	An Algorithm for Computing the $\lambda_0^{(k)}$	347
	8.5.4	Numerical Results	349
8.6		Eigenvalue Method Applied to a Second-Order System	354
	8.6.1	Development of the System Model and Fokker–Planck Equation	354
	8.6.2	Development of the Coupled System of Differential Equations	355
	8.6.3	Algorithm for Computing the p_n, h_n, and λ_k	357
	8.6.4	Numerical Results	360
Appendix 8.1.1		Hermite Polynomials	363
	8.1.1.1	Definitions and Elementary Properties	364
	8.1.1.2	Weighted Hermite Polynomials as Eigenfunctions of a Differential Operator	366
Appendix 8.4.1		Least Positive Residue Function	367
Appendix 8.5.1		Computing Eigenvalues of the Fokker–Planck Operator	367
References			373
Index			377

Part I
Elementary Theory and Applications

Chapter 1

INTRODUCTION

Rapid advances in the field of electronic components have made the use of phase-locked technology commonplace in military, aerospace, and consumer electronic systems. The use of this technology has added new functionality to these systems; also, it has enhanced the performance of many existing functions traditionally implemented without the benefits of phase-locked technology.

This chapter serves to introduce the notion of phase lock and the classical analog phase-locked loop (PLL), the main topics of this book. The analog PLL introduced here has found widespread use in applications, and it must be understood by those wishing to fully exploit the technology. To motivate such an understanding, this chapter summarizes several common applications that utilize PLLs.

A good understanding of the classical analog PLL is very helpful, if not necessary, in pursuing related topics in the broad area of synchronization in communication and control. Such topics as automatic frequency control loops, early–late gate data synchronizers, digital PLLs, and microprocessor-based software PLLs become easier to assimilate once the reader gains a basic understanding of the analog PLL. Finally, in this text a broad range of analytical techniques are employed. Many of these techniques can be applied to a broad selection of problems in engineering and the physical sciences.

1.1 THE PHASE AND FREQUENCY OF SIGNAL RELATIVE TO A REFERENCE

The phase of a bandpass signal is a relative quantity. In quantitative terms, it can be specified relative to a fixed reference signal. For example, consider $s_1(t) = \cos(\omega_0 t + \theta_1(t))$ as a reference signal that is supplied to a system by an external source; radian frequency ω_0 is a fixed constant, and $\theta_1(t)$ is a time-varying phase angle. As depicted by Figure 1.1, reference s_1 can be represented abstractly by a fixed unit-length phasor drawn along the horizontal axis. Now, consider a second signal $s_2(t) = \cos(\omega_0 t + \theta_2(t))$ that is generated by the system. Signal s_2 is represented

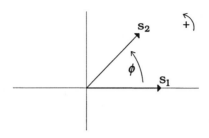

FIGURE 1.1
Phasor diagram displaying signals s_1 and s_2.

on Figure 1.1 as a unit length phasor displaced in angle from $s_1(t)$ by the amount $\phi(t) \equiv \theta_1(t) - \theta_2(t)$, where positive ϕ is measured on Figure 1.1 in a counter clockwise sense. Once ϕ is known, the phase of s_2 can be expressed in terms of (or relative to) the phase of s_1.

The phase difference $\phi(t)$ and its derivative are of interest here. In the present context where a system is supplied with the reference signal s_1, the quantity $\phi(t)$ is known as the *phase error*, and its control is a prime goal of the system. Roughly speaking, the system-generated signal s_2 is said to be *phase coherent* with the external reference s_1 if $\phi(t)$ remains near some constant ϕ_0 over some user-specified time interval. The derivative $d\phi/dt$ is a measure of the instantaneous *frequency error* between s_2 and reference s_1. Note that phase coherence does not require that the instantaneous frequency error remain small. It is possible to have arbitrarily large values of instantaneous frequency error even though $|\phi(t) - \phi_0|$ remains small over the time interval of interest.

The second derivative of phase is referred to as *frequency rate*. This quantity is of interest in many applications where a signal is shifted by a time-varying Doppler or modulated in frequency. For example, sinusoids that are ramped in frequency are used in radar signal design and simple ranging systems.

1.2 A GENERIC PROBLEM

A common requirement in modern electronic systems is that they *phase lock* an internal oscillator to a reference signal. That is, the internal oscillator of the system must be forced to produce a signal that is phase coherent with the supplied reference. The reference may be generated within the system, or it may come from some external source. As discussed in the applications section below, phase locking in this manner is a common requirement in systems that demodulate and synthesize signals.

INTRODUCTION

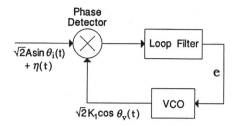

FIGURE 1.2
A simple baseband model of the PLL.

In many cases, the reference signal is buried in noise. This noise makes difficult the above-mentioned process of generating a phase-coherent local reference signal. Invariably, the amplitude and/or phase of the locally generated reference signal is corrupted by the external noise. This causes a degradation in the performance of the system that uses the locally generated reference. This noise-induced degradation must be understood and minimized by the system designer.

The phase-lock concept is fundamental in many problems outside of purely electronic systems. There are many situations where some form of feedback is used to synchronize some local periodic event with some observable external event. For example, feedback is used in many applications where it is necessary to control the speed of motors (both electric and fuel-burning types). In a more general sense, many biological processes employ some form of feedback to control the time occurrence of events. For example, most people schedule their daily activities to synchronize with timing information supplied by a watch or clock.

1.3 THE PHASE-LOCKED LOOP

Electronic circuits employing the phase-lock principle can provide practical solutions to the problem outlined above. Figure 1.2 depicts a block diagram of a *phase-locked loop* (PLL) often used as a starting point in the study of this class of electronic circuits. A considerable number of variations of this basic architecture have been proposed and used. However, many of these variants are mathematically equivalent to the loop described here. For reasons that will become clear as the topic is pursued, Figure 1.2 depicts what is often referred to as a *short loop* or a *baseband model* of the PLL.

The PLL depicted by Figure 1.2 contains (1) a phase comparator, (2) a loop filter, and (3) a voltage-controlled oscillator (VCO). The phase comparator may be a simple analog multiplier. A different, usually more complicated, phase comparator may be used to exploit attributes (like a high-input signal-to-noise ratio) of a given application. When present, the loop filter is low pass in nature, and it may be active or passive. The VCO oscillates with an instantaneous frequency that is functionally related to its control voltage e.

The external reference signal supplied to the loop is modeled as the sum of a desired signal $\sqrt{2}A\sin\theta_i(t)$ and an undesired additive noise component $\eta(t)$. In addition to random noise, the output of the phase comparator contains a component that quantifies the phase error $\phi \equiv \theta_i - \theta_v$. This component is processed by the loop filter, and the results are applied to the input of the VCO. Hopefully, this controls the phase of the VCO and results in a small value of phase error variance.

Under phase-locked conditions in a properly designed and operated PLL, the instantaneous phase θ_v of the VCO remains close to the phase θ_i of the desired input signal, and the phase error $\phi(t)$ is small in absolute value. This implies that the VCO leads the reference in phase by approximately $\pi/2$ radians. Of course, the presence of input noise η insures that θ_v and ϕ are random processes. However, in many practical applications, the PLL can be designed so that the variance of ϕ is acceptably small, and the PLL tracks well the phase θ_i of the desired input signal. In these applications, the PLL can be thought of as a filter that eliminates the bulk of the noise appearing at its input.

Phase lock is the end result of a process known as *pull in*. That is, pull in is the natural mechanism of the PLL by which phase lock is achieved after starting from an out-of-lock condition. Pull in may or may not be possible when the loop is closed. If possible, it may not be achieved, or it may not happen in a reasonable amount of time. Often, the pull-in mechanism must be aided by circuitry added to the basic PLL.

1.4 BASIC APPLICATIONS

The subject of phase lock first generated major interest with the advent of the space program in the early 1960s. Major advances in integrated circuits since this time have brought down the cost and trouble of using PLL technology. In addition to being used in military and space applications, PLLs have been incorporated into consumer electronic products which are taken for granted by the public. The remainder of this chapter describes some of these applications.

1.4.1 Coherent Demodulation of Amplitude-Modulated Signals

Amplitude-modulated (AM) signals can be demodulated coherently by using the system depicted by Figure 1.3. When this PLL is

INTRODUCTION

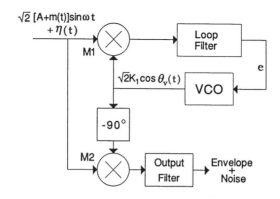

FIGURE 1.3
Coherent demodulation of AM.

phase locked to the carrier component of the AM signal, its VCO produces a sinusoid that is in phase quadrature with the received carrier. This implies that the output of the −90° phase shifter will be in phase with the carrier component of the signal and that an estimate of the envelope $\sqrt{2}[A + m(t)]$ appears at the terminals of the output filter.

This method of AM demodulation can produce excellent results. In a noisy environment it can produce performance which far exceeds the classical envelope detector. Of course, along with the additional performance of coherent demodulation come the additional complexity and cost. Phase-locked AM demodulators are far more complicated and expensive than simple envelope detectors. For this reason, coherent demodulation of AM is not used in most commercial broadcast receivers (AM radios).

1.4.2 Frequency Synthesis

The PLL plays a major role in the frequency synthesis of spectrally pure signals. Figure 1.4 depicts a block diagram of a simple PLL-based *frequency synthesizer*. As discussed below, this system is capable of generating a sinusoid at a frequency of Nf_o Hz, where integer N is user programmable. The frequency stability of the generated sinusoid is dependent on the stability of the reference, and it can be excellent.

In many applications, the reference signal at f_o Hz is obtained from a digital counter (frequency divider) circuit driven by a highly stable crystal oscillator. Hence, the phase comparator is fed digital signals in this application. Specialized phase comparators are used which exploit the noise-free nature of the signals involved. Often the phase comparator is augmented by a frequency comparator that helps the loop achieve a stable phase-lock state.

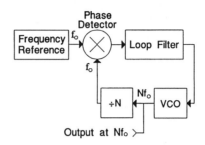

FIGURE 1.4
A simple PLL-based frequency synthesizer.

The VCO oscillates at Nf_o Hz when the PLL of the synthesizer is phase locked. Since N is user programmable, this system can generate a wide range of frequencies at a basic resolution of f_0 Hz. In many applications, it is possible to do this while maintaining adequate frequency stability, spectral purity, and switching times of the synthesized output signal.

1.4.3 Demodulation of BPSK Signals

Binary phase shift keying (BPSK) is a popular form of modulation which is used in the transmission of digital information. Let d(t) denote a binary (±A volts) waveform representing digital information. The BPSK modulator forms the signal $d(t)\cos\omega_0 t$; in applications, carrier frequency ω_0 is much higher than the rate of the clock used to generate d(t) (often, ω_0 is an integer multiple of the data clock frequency).

In many applications, data signal d(t) is constructed to have an average value of zero. Usually, this is accomplished by constructing the data signal from specially formulated binary symbols that have an average value of zero. A zero average value for d(t) is desirable from an efficiency standpoint; it ensures that the transmitted signal has no carrier component at ω_0, and all of the transmitted power appears in the sidebands where it is used to convey information.

At the receiver, efficient demodulation of the BPSK signal requires the use of a sinusoid that is phase coherent with the suppressed carrier of the received signal. This sinusoid can be generated by the *squaring loop* depicted by Figure 1.5. First, the received signal is squared to produce a second signal that has a strong component at $2\omega_0$. The bandpass filter (BPF) passes this $2\omega_0$ component, and it attenuates all other frequency components. The PLL locks on the $2\omega_0$ frequency component found in the output of the BPF. In this role, the PLL acts like a narrowband filter centered at $2\omega_0$. A digital counter is used to divide by two the frequency of the VCO. This division operation produces a

INTRODUCTION

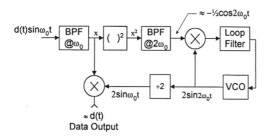

FIGURE 1.5
Squaring loop data demodulator.

signal that is phase coherent with the suppressed carrier of the received signal. Finally, this locally generated reference signal is used to demodulate the received signal and produce an estimate of data $d(t)$.

1.4.4 Phase-Locked Receivers

There are many applications which require the reception of a Doppler-shifted signal. Often, the amount of Doppler shift is changing with time, and there is a large amount of uncertainty in the frequency of the received signal. One possible solution to the problem of receiving this signal is to use a noncoherent approach based on a wide bandwidth receiver that can "hear" the signal regardless of the Doppler shift. However, this simple approach may bear a significant noise penalty since received noise power is proportional to receiver bandwidth. Large Doppler shifts would require large receiver bandwidths, and large bandwidths would lead to large amounts of received noise power and poor system performance.

A *phase-locked receiver* is a better solution to the problem of receiving a Doppler-shifted signal. The receiver electronically tunes itself so that it tracks out Doppler on the received signal. Hence, such a receiver can have a bandwidth that is comparable to that of the signal, so no noise penalty has to be paid to accommodate the unknown carrier frequency of the signal. In cases where the transmitted bandwidth is small compared to the amount of Doppler shift, a solution based on a tracking receiver will be substantially better than that afforded by the above-mentioned scheme that utilizes a wide-bandwidth, fixed-frequency receiver.

Figure 1.6 depicts a simplified block diagram of a tracking phase-locked receiver. Note that most of the receiver is in the feedback path. In the PLL literature, this architecture is known as a *long loop*. As discussed in Section 2.7, it reduces practical problems that can result from noise rectification in an imperfect phase comparator. Practical phase detectors are limited in dynamic range, and this limitation

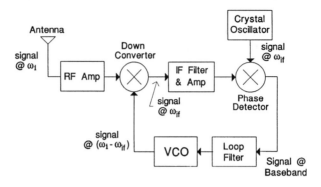

FIGURE 1.6
A simple phase-locked receiver.

(known as the *phase comparator threshold problem*) is a major reason for using an intermediate frequency (IF) path within the loop. In Section 2.7, this architecture is shown to be mathematically equivalent to a baseband PLL model similar to that depicted by Figure 1.2.

Normally, the receiver locks to the carrier frequency of the received signal. As the carrier frequency changes due to Doppler, the VCO is tuned automatically by loop dynamics so that the output of the down converter remains within the IF passband. In many practical applications, the *rate* of Doppler change is small (even if the total amount of Doppler is large), and the overall bandwidth of the closed loop may be on the order of a few tens of Hertz.

For example, consider an S-band (2.4 GHz) satellite in low earth orbit. Suppose that the satellite is transmitting a carrier that a ground station must recover. Depending on the particular overhead pass, the downlink signal of the satellite may contain as much as ±70 kHz of Doppler as seen by the ground station. Without Doppler tracking abilities, the wideband receiver of the ground station would need a bandwidth of 140 kHz. However, in this application, the rate of change in Doppler is moderate to small, and the signal of the satellite could be received by a PLL-based Doppler tracking receiver that utilizes a bandwidth on the order of 10 Hz.

Received noise power is directly proportional to receiver bandwidth. For the example outlined above, it turns out that the fixed-frequency, wideband receiver approach would pay a signal-to-noise ratio penalty of approximately 42 dB as compared to the phase-locked receiver. Penalties of this magnitude cannot stand; this is why narrow-band, phase-locked tracking receivers are used in almost all applications involving earth-orbiting satellites.

1.5 PHASE-LOCKED LOOP LITERATURE

The literature base dealing with PLLs is large and extensive, and a comprehensive review of it will not be attempted here. Instead, a sampling is given of some significant sources of PLL information; these sources are relatively easy to obtain since they have been distributed widely. Fortunately, computerized data bases are available which can be searched quickly to find individual articles of interest.

A number of important papers have been reprinted and bound in book form. The texts edited by Lindsey and Simon[1] and Lindsey and Chie[2] provide a survey of PLL theory as it existed up until 1985. In addition to providing a good background in PLL theory and practice, the papers reprinted here provide reference lists of additional works.

A number of books have appeared on the subject of phase lock. The work of Viterbi[3] is an early classic in the field. Lindsey's[4,5] books are very extensive, and they should be consulted by those students seeking a broad training on the subject matter. A very readable introduction to the topic is given by Gardner.[6] This work is aimed at the practicing engineer whose primary concern is methods that can be put to work solving practical problems. Stiffler[7] is concerned with the general theory of synchronous communication systems and the problem of optimum phase estimation. The work by Meyr and Ascheid[8] is extensive, and it is among the first texts that apply eigenfunction techniques to the Fokker–Planck equation describing the state transition density function of a PLL. Finally, the texts by Wolaver,[9] Best,[10] and Encinas[11] are excellent sources of practical information for the design engineer.

A number of books have appeared that apply PLLs to specific tasks. The work of Blanchard[12] focuses on applying phase-lock techniques to the coherent receiver. In their book, Klapper and Frankle[13] apply phase-lock and frequency-feedback techniques to receiver design. The classical book by Van Trees[14] shows how the PLL can be used to approximate an optimal phase estimator. Finally, books that cover the important area of PLL-based frequency synthesizer design include those by Rohde,[15] Egan,[16] Manassewitsch,[17] Stirling,[18] and Jerzy Gorski–Popiel.[19] PLL-based frequency synthesizer design has been revolutionized by large-scale integrated circuit technology, and advancements in this area are covered best by the application notes and literature published by the major semiconductor manufacturers in the frequency synthesizer market.

1.6 TOPICAL OUTLINE OF THE TEXT

The technology and science of phase-locked systems have witnessed explosive growth over the last few decades. A basic understanding of

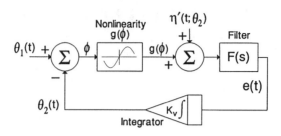

FIGURE 1.7
Baseband model of a phase tracking system.

these areas is essential for anyone working in the fields of communications, radar, and instrumentation. This text was written to help its readers achieve such an understanding.

This goal is made difficult by the wide diversity of the subject matter. Instead of trying to provide an encyclopedic-like coverage of the subject, this text concentrates on a model, illustrated by Figure 1.7, that has been the focus of much study. In this model, θ_1 (θ_2) denotes an input (output) phase angle, and the goal is to keep the phase error $\phi \equiv \theta_1 - \theta_2$ small. The quantity η' denotes an interference component that is modeled as a random process. It turns out that many phase-locked systems can be reduced to the point where they are described by this model.

This model is nonlinear, and a general analysis of it is made difficult by this fact. On the positive side, there are sophisticated theories and techniques for dealing with nonlinear models of this type. Unfortunately, most of these yield only qualitative and/or numerical results which are often hard or impossible to integrate into the design of practical systems. However, the model can be linearized about an operating point, and this linear model is sufficient for many engineering applications. In such cases, the practitioner can apply standard linear control system design techniques.

The elementary theory of PLLs is covered in Part I (Chapters 1 through 4) of the text. Design engineers will find this material useful in their work. Part II (Chapters 5 through 8) is devoted to the nonlinear theory. This material should be of interest to those who wish to gain a more complete knowledge of the PLL. It can serve as the starting point for those desiring to tailor their research efforts towards the effects of nonlinearities and noise on synchronization systems. Finally, most chapters contain appendices devoted to reference material referred to in the text.

In Chapter 2, mathematical models are given for the components that make up the PLL. Also, these component descriptions are integrated into a nonlinear model for the PLL with a noisy reference signal.

This model is linearized for the important case of small phase error. Some PLL applications contain an IF signal path in their feedback loops. In Chapter 2, these are handled by using a lowpass equivalent model of the IF filter.

Many practical system designs incorporate second-order PLL models that have been linearized. First, such models are well understood, and they are easy to work with. Second, many applications are served well by linear, second-order PLL models. Hence, Chapter 3 is devoted to the linear analysis of common first- and second-order PLLs. Both time and Laplace domain techniques are employed in this analysis, and they are used to analyze both transient and steady-state behavior.

Chapter 4 is devoted to descriptions of PLL components. Commonly used phase detectors are discussed in this chapter. These range from simple-to-understand analog multipliers to a relatively complicated sequential digital logic circuit that acts as both a phase and frequency detector. This digital phase/frequency detector plays a major role in digital PLLs employed in advanced frequency synthesizers. Voltage-controlled oscillators are discussed; these are naturally divided into two groups. A member of the first (second) group depends on a resonant condition (relaxation time) to determine its frequency of oscillation. Finally, the chapter contains descriptions of some commonly used loop filters.

Chapter 5 describes the nonlinear behavior of common first- and second-order PLLs in the absence of noise. The analysis employs the phase plane and modern bifurcation theory to produce a global analysis of PLL behavior. Some of the bifurcation theory-based results are relatively new to the PLL literature. Their importance stems from the fact that they explain fundamental behavior easily observed in practical applications.

The noise analysis techniques introduced in Chapter 2 are based on the linear model of the PLL, and they are useful for a large number of practical applications where the linear model applies. However, they are not appropriate for the important low signal-to-noise ratio case where the reference sinusoid is buried in noise, and the variance of the phase error is large enough so that the nonlinear model of the PLL must be used.

Specialized theory and techniques must be used to handle the low signal-to-noise ratio case where the nonlinear model of the PLL must be used. Chapter 6 serves to introduce this material, and it is the basis for the remainder of the book. The Fokker–Planck equation is the main development in Chapter 6. It governs the time evolution of the probability density function that describes the state of the PLL. This equation, and the theory and techniques associated with it, are developed for analyzing the PLL, but this material can be applied to many other nonlinear and/or time-varying systems.

Chapter 7 builds upon and applies the material introduced in Chapter 6. First, a class of models is developed in order to exploit the periodicity inherent in the phase detector of the PLL. Models from the class describe the phase error in a modulo-$2\pi m$ sense, where integer m, m ≥ 1, serves to index the class. The models facilitate greatly the development of theory and techniques that describe the steady-state probabilistic nature of the state of the PLL. Next, noise analysis techniques from Chapter 6 are applied to the first-order PLL, and they yield several practical results. Many of these results are quantitative in nature; often, they can be expressed in closed form, or they can be approximated accurately by simple formulas. Unfortunately, for second- and higher-order PLLs, this is not the case; for these PLLs, simple, closed-form quantitative results are not available in general. However, for second-order PLLs, qualitative results are given in Chapter 7, and they provide insight into the nonlinear phenomenon that they describe. Finally, for second-order PLLs operating under restricted conditions, Chapter 7 contains the development of an approximation to the steady-state density function that describes the modulo-2π phase error in the closed loop.

Chapter 8 is devoted to numerical methods for the analysis of noise in nonlinear PLL models. At least in principle, the numerical methods introduced in the chapter can be applied to PLLs of any order. And, for low-order PLLs, they can be implemented on an inexpensive personal computer. First, an algorithm is described which can be used to numerically approximate the steady-state probability density that describes the modulo-2π phase error in first- and second-order PLLs. It can provide accurate results over a wide range of phase error variances. To show the utility of this algorithm, it is applied to two different second-order PLLs. Next, a numerical algorithm is given for approximating the cycle slip rates in a PLL. It utilizes an m-state Markov jump process to model the state of the PLL; jumps between the states of the model are related to cycle slips in the PLL. It is useful under moderate noise conditions where noise-induced cycle slips are rare events. Finally, the cycle slip algorithm is applied to two simple examples, and numerical results are tabulated and displayed in graphical form.

Chapter 2

MODELING THE PHASE-LOCKED LOOP

The development of a model is an important part of the characterization of any electronic system or component. The model must be accurate over a realistic set of operating parameters. Also, the user of the model must possess a good understanding of the limitations imposed by simplifying assumptions made during the modeling process. Finally, the model must be tractable; from using the model, the user must be able to obtain qualitative and/or quantitative information.

Fortunately, good models for the PLL exist, and the standard ones are described in this chapter. In Section 2.1, a block diagram of the basic PLL is provided, and equations are given that model its components. A PLL with a noise-free reference is considered in Section 2.2, and its nonlinear equation of operation is developed. The PLL model is linearized in Section 2.3, and this linear model is analyzed by using classical system theory. An angle-modulated reference signal is modeled in Section 2.4, and its influence on the closed loop is analyzed. In Section 2.5 the PLL with a noisy reference signal is modeled as a nonlinear system driven by white Gaussian noise. Also, this model is linearized for the large signal-to-noise case, and this result is used to derive the statistics of the closed-loop phase error. In many practical applications, the noisy reference and voltage-controlled oscillator (VCO) signals are hard limited before they are compared in phase. A general phase detector model that includes hard limiting is discussed in Section 2.6. Finally, a simplified model of a phase-locked receiver is discussed in Section 2.7. This model has part of its error signal path centered at an intermediate frequency. In addition to the material contained in this chapter, the first dozen references listed in the bibliography and the tutorial paper by Gupta[49] should be consulted for information on modeling the PLL.

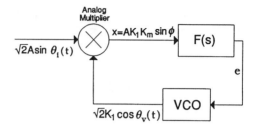

FIGURE 2.1
PLL with a noiseless reference signal.

2.1 MODELING PLL COMPONENTS

An analog multiplier phase comparator, loop filter, and VCO comprise the components of the baseband PLL depicted by Figure 2.1. Simple models of these components are discussed in this section.

2.1.1 Modeling the Analog Phase Detector

A noiseless reference signal is used to drive the PLL depicted by Figure 2.1. The phase comparator used here is a simple analog multiplier with output

$$x \equiv K_m \left[\sqrt{2} A \sin \theta_i \right] \left[\sqrt{2} K_1 \cos \theta_v \right] \\ = A K_1 K_m \left[\sin(\theta_i - \theta_v) + \sin(\theta_i + \theta_v) \right]. \quad (2.1\text{-}1)$$

The quantity K_m is the gain of the phase comparator; as discussed in what follows, it has units of volts/radians.

The sum frequency term $AK_1K_m \sin(\theta_i + \theta_v)$ in Equation 2.1-1 is considered to be a bandpass process centered around twice the input reference frequency. In most applications, it is filtered out by the combination of loop filter and VCO. Hence, this sum frequency term is discarded in what follows, and the phase comparator output is approximated as

$$x = AK_1 K_m \sin \phi, \quad (2.1\text{-}2)$$

where

$$\phi \equiv \theta_i - \theta_v. \quad (2.1\text{-}3)$$

MODELING THE PHASE-LOCKED LOOP

(a) $$\mathcal{L}[e] = \frac{a_m s^m + a_{m-1} s^{m-1} + \ldots + a_0}{b_n s^n + b_{n-1} s^{n-1} + \ldots + b_0} \mathcal{L}[x]$$

(b) $$\left[b_n \frac{d^n}{dt^n} + b_{n-1} \frac{d^{n-1}}{dt^{n-1}} + \ldots + b_0 \right] e = \left[a_m \frac{d^m}{dt^m} + a_{m-1} \frac{d^{m-1}}{dt^{m-1}} + \ldots + a_0 \right] x$$

(c) $$e(t) = \int_0^t f(u) x(t-u) du + e_0(t)$$

FIGURE 2.2
Laplace domain description of loop filter is given by (a). Time domain description is given by (b) and (c).

The quantity ϕ plays a prominent role in PLL analysis; it is known as the *closed-loop phase error*.

2.1.2 Modeling the Loop Filter

The error signal $x = AK_1 K_m \sin\phi$ shown on Figure 2.1 drives the linear, time-invariant loop filter to produce the VCO control voltage e. The relationship between the error signal and the control voltage can be given in both the Laplace and time domains as is summarized by Figure 2.2.

In the Laplace domain, the filter can be described by the transfer function

$$F(s) = \frac{a_m s^m + a_{m-1} s^{m-1} + \ldots + a_0}{b_n s^n + b_{n-1} s^{n-1} + \ldots + b_0}, \quad m \leq n, \qquad (2.1\text{-}4)$$

where integer n denotes the order of the filter. This rational function of s is the ratio of filter output to input in the Laplace domain. In practical PLLs, Equation 2.1-4 has no poles in the right half of the s-plane.

In the time domain, the nth-order differential equation

$$L_1[e] = L_2[x]$$

$$L_1 \equiv b_n \frac{d^n}{dt^n} + b_{n-1} \frac{d^{n-1}}{dt^{n-1}} + \ldots + b_0 \qquad (2.1\text{-}5)$$

$$L_2 \equiv a_m \frac{d^m}{dt^m} + a_{m-1} \frac{d^{m-1}}{dt^{m-1}} + \ldots + a_0$$

defines the loop filter. This equation is linear, and it has constant coefficients. Furthermore, it has a unique solution e(t), t ≥ 0, once initial conditions

$$e(0) = \epsilon_0$$

$$\left.\frac{d^k e}{dt^k}\right|_{t=0} = \epsilon_k, \quad 1 \leq k \leq n-1, \tag{2.1-6}$$

are specified.

A second time–domain relationship can be specified between the input and output of the loop filter. For t ≥ 0 this relationship is

$$\begin{aligned} e(t) &= \int_0^t x(\tau) f(t-\tau) d\tau + e_0(t) \\ &= \int_0^t x(t-\tau) f(\tau) d\tau + e_0(t), \end{aligned} \tag{2.1-7}$$

where

$$f(\tau) \equiv \mathcal{L}^{-1}[F(s)]$$

is the impulse response of the filter. In Equation 2.1-7, $e_0(t)$ is the zero-input response which depends only on initial conditions existing in the circuit at t = 0.

2.1.3 Modeling the Voltage-Controlled Oscillator

The VCO of the loop accepts as input the error-control voltage e and produces as output the sinusoidal signal $\sqrt{2}K_1\cos\theta_v$. The commonly used VCO model relates variables θ_v and e by

$$\frac{d\theta_v}{dt} = \omega_0 + K_v e, \tag{2.1-8}$$

where ω_0 and K_v are known constants. As can be seen from inspection of Equation 2.1-8, the VCO oscillates at frequency ω_0 radians per second when its input control voltage e is set to zero. Hence, ω_0 is known as the *center*, or *quiescent frequency* of the VCO. The frequency of oscillation of the VCO changes by K_v radians per second for every volt of control

MODELING THE PHASE-LOCKED LOOP

signal e applied as input. Hence, K_v is known as the *VCO gain parameter*, and it has units of radians per second-volt.

The analysis of the PLL is simplified by defining input reference and VCO phase variables which are relative to the quiescent phase $\omega_0 t$ of the VCO. The quantities

$$\theta_1(t) \equiv \theta_i(t) - \omega_0 t \qquad (2.1\text{-}9)$$

$$\theta_2(t) \equiv \theta_v(t) - \omega_0 t \qquad (2.1\text{-}10)$$

are relative input reference and VCO phases, respectively. From a modeling standpoint, relative phase θ_2 is considered as the output of the VCO even though a sinusoid is the physical output of this oscillator. Finally, note that closed-loop phase error can be expressed in terms of these relative phase variables as $\phi = \theta_1 - \theta_2$.

With the aid of Equations 2.1-8 and 2.1-10, it is easy to model the VCO as an integrator. Substitute Equation 2.1-10 into Equation 2.1-8 and obtain

$$\frac{d\theta_2}{dt} = K_v e, \qquad (2.1\text{-}11)$$

a result depicted by Figure 2.3. Note from Equation 2.1-11 that the response θ_2 of the VCO falls off at a rate that is inversely proportional to the frequency of forcing function e. This, and the lowpass nature of F(s), imply that undesired high-frequency signals in the phase detector output tend to have minimal influence on the PLL.

2.2 MODELING THE NONLINEAR PLL

The component descriptions developed in Section 2.1 are used below in a block diagram depicting a nonlinear model of the PLL. Three approaches are discussed for mathematically describing this model. First, the model is described by an (n + 1)th-order ordinary differential equation. Next, an n + 1 dimensional first-order system is used to describe the model. These descriptions of the PLL are used in Chapter 5. Finally, an integral–differential equation is given that describes the

FIGURE 2.3
Model of the VCO as an integrator.

model. This approach is utilized in Section 2.5 and Chapter 7 to model the effects of noise on the PLL.

2.2.1 PLL Description Based on an Ordinary Differential Equation

The PLL loop filter processes the phase comparator output $x = AK_1K_m\sin\phi$ to produce voltage e that drives the VCO. The differential equation

$$L_1[e] = AK_1K_m L_2[\sin\phi], \qquad (2.2\text{-}1)$$

obtained by combining Equations 2.1-2 and 2.1-5, describes this processing function.

Equation 2.2-1 contains the dependent variables ϕ and e. A second independent relationship must be obtained between these variables to complete the nonlinear dynamical model of the PLL. This required second equation can be derived by combining Equation 2.1-11 with Equation 2.1-3. This effort produces the results

$$\begin{aligned}\frac{d\phi}{dt} &= \frac{d\theta_1}{dt} - K_v e \\ &= \left[\frac{d\theta_i}{dt} - \omega_0\right] - K_v e.\end{aligned} \qquad (2.2\text{-}2)$$

Equations 2.2-1 and 2.2-2 describe the closed-loop dynamics of the PLL under consideration. They can be combined to produce

$$L_1\left[\frac{d\phi}{dt} - \frac{d\theta_1}{dt}\right] = -GL_2[\sin\phi], \qquad (2.2\text{-}3)$$

where

$$G \equiv AK_1 K_m K_v \qquad (2.2\text{-}4)$$

is a closed-loop gain constant. Constant G is assumed to be positive (if $G < 0$, perform the transformation $\phi = \pi + \hat{\phi}$, and take G as positive). Because of the dependence on $\sin\phi$, Equation 2.2-3 is a nonlinear differential equation. It is autonomous (time invariant) since only multiplicative constants appear in the equation.

The block diagram depicted by Figure 2.4 describes the nonlinear PLL model under consideration in this section. Equations 2.1-2 and

MODELING THE PHASE-LOCKED LOOP

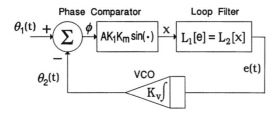

FIGURE 2.4
Nonlinear time-domain model depicting the function of each component.

2.1-3, which describe the phase detector, are implemented by the summing junction and the block containing the nonlinear operation $AK_1K_m\sin(\cdot)$. The loop filter block implements the linear transformation described by Figure 2.2. Finally, with a gain of K_v, the VCO block integrates the control voltage e to form the relative phase angle θ_2 defined by Equation 2.1-10. In Chapter 5, this model will be used to analyze the nonlinear behavior of common PLLs.

2.2.2 PLL Description Based on a First-Order Nonlinear System

Equation 2.2-3 can be written as a first-order system. Define the n + 1 dimensional vector

$$\vec{X} = \begin{bmatrix} x_1 & x_2 & \cdots & x_{n+1} \end{bmatrix}^T$$
$$= \begin{bmatrix} \phi & \dfrac{d\phi}{dt} & \dfrac{d^2\phi}{dt^2} & \cdots & \dfrac{d^n\phi}{dt^n} \end{bmatrix}^T. \qquad (2.2\text{-}5)$$

Then Equation 2.2-3 can be used to write

$$\frac{d\vec{X}}{dt} = \mathbb{C}\vec{X} - G\vec{F}(\vec{X}) + \vec{G}, \qquad (2.2\text{-}6)$$

where $\mathbb{C}$ is the (n + 1) × (n + 1) constant matrix

$$\mathbb{C} \equiv \begin{bmatrix} 0 & 1 & 0 & \cdots & 0 \\ 0 & 0 & 1 & \cdots & 0 \\ \vdots & \vdots & \vdots & & \vdots \\ 0 & 0 & 0 & & 1 \\ 0 & -b_0/b_n & -b_1/b_n & \cdots & -b_{n-1}/b_n \end{bmatrix} \qquad (2.2\text{-}7)$$

and $\vec{F}, \vec{G}$ are n + 1 vectors

$$\vec{F}(\vec{X}) = \begin{bmatrix} 0 & 0 & 0 & \ldots & 0 & b_n^{-1}L_2[\sin x_1] \end{bmatrix}^T$$

$$\vec{G} = \begin{bmatrix} 0 & 0 & 0 & \ldots & 0 & b_n^{-1}L_1\left[\dfrac{d\theta_i}{dt}\right] \end{bmatrix}^T.$$

(2.2-8)

Equation 2.2-6 is an n + 1 dimensional, nonlinear system that describes the PLL. Along with Equation 2.2-3, it is used in Chapter 5 to analyze the nonlinear behavior of some common PLLs.

2.2.3 PLL Description Based on an Integral Differential Equation

This approach to PLL modeling utilizes the impulse response f(t) instead of a differential equation to represent the loop filter in the model. Start with Equations 2.2-2, 2.1-7, and 2.1-2, and write the integral–differential equation

$$\frac{d\phi}{dt} = \left[\frac{d\theta_i}{dt} - \omega_0\right] - G\int_0^t f(\tau)\sin\phi(t-\tau)d\tau, \quad t \geq 0,$$

(2.2-9)

for the case of zero initial conditions. This equation is used in Section 2.5 and Chapter 7 to model the effects of noise on the PLL.

2.3 MODELING THE LINEAR PLL

Consider the case when the PLL is phase locked, and phase error ϕ is small in absolute value. Under this condition, the approximation $\sin\phi \approx \phi$ can be made, and the PLL model can be linearized. The linear loop can be described by the Laplace domain model depicted by Figure 2.5; the variables

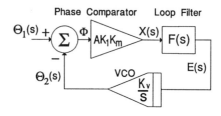

FIGURE 2.5
Linear Laplace domain model showing the function of each component.

MODELING THE PHASE-LOCKED LOOP

$$\Theta_1(s) \equiv \mathscr{L}[\theta_1(t)]$$
$$\Theta_2(s) \equiv \mathscr{L}[\theta_2(t)]$$
$$\Phi(s) \equiv \mathscr{L}[\phi(t)]$$
$$E(s) \equiv \mathscr{L}[e(t)]$$
(2.3-1)

are employed in the Laplace domain model.

A transfer function H(s) can be obtained that relates output Θ_2 to input Θ_1. From inspection of Figure 2.5, the open-loop transfer function is

$$G_0(s) \equiv \frac{\Theta_2(s)}{\Phi(s)} = G\left[\frac{F(s)}{s}\right], \qquad (2.3\text{-}2)$$

where the closed-loop gain constant G is given by Equation 2.2-4. This open-loop transfer function can be used to express Θ_2 as

$$\Theta_2(s) = G_0(s)\Phi(s). \qquad (2.3\text{-}3)$$

Now, substitute $\Phi = \Theta_1 - \Theta_2$ into Equation 2.3-3 and solve for

$$H(s) = \frac{\Theta_2(s)}{\Theta_1(s)} = \frac{G_0(s)}{1+G_0(s)}$$
$$= \frac{GF(s)}{s+GF(s)}$$
(2.3-4)

As given by Equation 2.3-4, H(s) is the closed-loop transfer function of the linearized PLL. This transfer function is used extensively in the practical design of PLL circuitry.

The poles of H are the roots of the polynomial

$$s + GF(s) = 0. \qquad (2.3\text{-}5)$$

Stability of the linear PLL model requires that these poles be in the left half of the complex plane. The stability issue can be studied by examining the loci of poles as G varies from zero to infinity. Examples of these *root locus* plots are considered in Sections 3.2.2 and 3.3.2 in Chapter 3.

As long as the phase error remains small, transfer function H(s) and standard linear system theory can be used to determine the response of the PLL to any input. This is accomplished by using the relationship

$$\Theta_2(s) = H(s)\Theta_1(s) \qquad (2.3\text{-}6)$$

to find the Laplace transform of the output given an input Θ_1. Then, time–domain output $\theta_2(t)$ can be calculated by computing the inverse transform of Θ_2.

The linear theory can be used to determine the closed-loop phase error. This is accomplished by using Equation 2.3-6 with $\Phi = \Theta_1 - \Theta_2$ to obtain

$$\begin{aligned}\Phi &= \Theta_1 - H\Theta_1 \\ &= (1-H)\Theta_1.\end{aligned} \qquad (2.3\text{-}7)$$

Alternatively, the linear analysis can be carried out in the time domain with the use of

$$h(t) \equiv \mathcal{L}^{-1}[H(s)], \qquad (2.3\text{-}8)$$

the impulse response of the closed loop. For $t \geq 0$, the output for a given input $\theta_1(t)$ can be expressed by the convolution

$$\theta_2(t) = \int_0^t h(t-\tau)\theta_1(\tau)d\tau, \qquad (2.3\text{-}9)$$

and the closed-loop phase error can be expressed as

$$\phi(t) = \theta_1(t) - \int_0^t h(t-\tau)\theta_1(\tau)d\tau. \qquad (2.3\text{-}10)$$

These results assume that input $\theta_1(t)$ is applied at $t = 0$, and all initial conditions are zero (θ_2 and its derivatives are zero at $t = 0^-$).

2.4 MODELING A PLL WITH AN ANGLE-MODULATED REFERENCE SOURCE

Consider the PLL depicted by Figure 2.1 with an angle-modulated reference centered at a frequency of ω_1. This reference signal is described

MODELING THE PHASE-LOCKED LOOP

in this section, and it can represent a frequency-modulated (FM) or phase-modulated (PM) signal. Assume that the loop is phase locked and that the phase error remains small for all time so that the linear model can be used.

For the case under consideration the phase of the reference can be expressed as

$$\theta_i(t) = \omega_i t + \chi(t), \qquad (2.4\text{-}1)$$

where χ depends linearly on message m(t). This implies that

$$\theta_1(t) = \omega_\Delta t + \chi(t), \qquad (2.4\text{-}2)$$

where $\omega_\Delta \equiv \omega_i - \omega_0$ is known as the *loop detuning parameter*. For the case of PM, message-dependent function χ is given by

$$\chi(t) = K_p m(t), \qquad (2.4\text{-}3)$$

where K_p is the modulation index with units of radians per volt. For the case of FM, function χ is

$$\chi(t) = K_f \int_0^t m(\tau) d\tau, \qquad (2.4\text{-}4)$$

where modulation index K_f has units of radians per second-volt.

A simple case of practical importance has m(t) as a sinusoid of frequency ω_m, and the sinusoidal steady-state loop response is desired. To unify the treatment of the two types of modulation during the solution of this problem, assume that

$$m(t) = \begin{cases} A_m \sin \omega_m t & \text{for PM} \\ A_m \cos \omega_m t & \text{for FM} \end{cases}. \qquad (2.4\text{-}5)$$

Then, for both cases

$$\chi(t) = \beta \sin \omega_m t, \qquad (2.4\text{-}6)$$

where

$$\beta = \begin{cases} A_m K_p & \text{for PM} \\ \dfrac{A_m K_f}{\omega_m} & \text{for FM} \end{cases} \qquad (2.4\text{-}7)$$

so that

$$\theta_1(t) = \omega_\Delta t + \beta \sin \omega_m t. \qquad (2.4\text{-}8)$$

The steady-state loop response can be obtained by substituting Equation 2.4-8 into Equation 2.3-9 and considering the limiting form of the results as time becomes large. The first of these steps yields

$$\theta_2 = \int_0^t h(t-\tau)\omega_\Delta \tau \, d\tau + \int_0^t h(t-\tau)\beta \sin \omega_m \tau \, d\tau. \qquad (2.4\text{-}9)$$

The second integral on the right-hand side is the response of a linear system to a sinusoid at frequency ω_m. The steady-state form of this response is

$$\beta|H(j\omega_m)|\sin(\omega_m t + \angle H(j\omega_m)), \qquad (2.4\text{-}10)$$

where $|H(j\omega_m)|$ and $\angle H(j\omega_m)$ represent the magnitude and phase, respectively, of the system at ω_m. Through a simple change of variable, the first integral on the right-hand side of Equation 2.4-9 can be expressed as

$$\omega_\Delta \int_0^t (t-\tau)h(\tau)\,d\tau$$

$$= \omega_\Delta \left[t\int_0^t h(\tau)\,d\tau - \int_0^t \tau h(\tau)\,d\tau \right]. \qquad (2.4\text{-}11)$$

Now, as time becomes large, the limiting form of Equation 2.4-11 follows from the observation that

$$\lim_{t \to \infty} \int_0^t h(\tau)\,d\tau = H(s)\big|_{s=0} = 1$$

$$\lim_{t \to \infty} \int_0^t \tau h(\tau)\,d\tau = -\frac{dH(s)}{ds}\bigg|_{s=0} = [GF(0)]^{-1}. \qquad (2.4\text{-}12)$$

Hence, the first integral on the right-hand side of Equation 2.4-9 produces the component

$$\omega_\Delta t - \frac{\omega_\Delta}{GF(0)} \qquad (2.4\text{-}13)$$

MODELING THE PHASE-LOCKED LOOP

in the steady state. Finally, combine Equations 2.4-10 and 2.4-13 to produce

$$\theta_2(t) = \omega_\Delta t - \frac{\omega_\Delta}{GF(0)} + \beta |H(j\omega_m)| \sin(\omega_m t + \angle H(j\omega_m)) \quad (2.4\text{-}14)$$

as the steady-state response of the linear model to a sinusoidally modulated reference.

The sinusoidal steady-state phase error can be found by using Equation 2.4-14 with Equation 2.4-8. The result of this combination is

$$\phi(t) = \frac{\omega_\Delta}{GF(0)} + \beta \left[\sin\omega_m t - |H(j\omega_m)| \sin(\omega_m t + \angle H(j\omega_m)) \right]. \quad (2.4\text{-}15)$$

Note that the average value of the phase error is inversely proportional to $GF(0)$, the open-loop DC gain of the PLL. Also, note from Equation 2.3-4 that $|H(j\omega_m)|$ and $\angle H(j\omega_m)$ approach unity and zero, respectively, as ω_m approaches zero. This implies that, as ω_m decreases, the sinusoidal component in Equation 2.4-15 becomes small, and the PLL tracks the modulation better. Conversely, as ω_m becomes large, the sinusoidal component of the phase error approaches $\beta\sin\omega_m t$, and the PLL tracks the modulation poorly.

2.5 MODELING A PLL WITH A NOISY REFERENCE SOURCE

This section describes a common method of including additive reference noise in PLL analysis. The input noise is modeled as a zero-mean, stationary, narrowband Gaussian process, and its statistical characteristics are discussed in Appendix 2.5.2. The input noise causes both the closed-loop phase error and the VCO phase to be random processes.

A modeling assumption is made that the bandwidth of the additive reference noise is much larger than the bandwidth of the closed loop. This implies that the VCO phase noise is slowly varying relative to the input noise, and it is nearly constant over a time interval that is large compared to the correlation time of the input noise. This assumption is valid in most practical applications where the input noise bandwidth is established by IF filtering, and the PLL is designed to act like a narrowband filter to eliminate most of the receiver IF noise. The assumption simplifies the task of describing the statistical properties of noise within the closed loop.

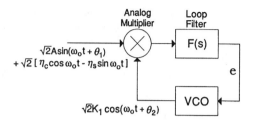

FIGURE 2.6
PLL with reference source containing additive noise.

2.5.1 Reference Signal Corrupted by Additive Noise

Figure 2.6 depicts a PLL supplied with the signal

$$r(t) \equiv \sqrt{2}A\sin(\omega_0 t + \theta_1) + \eta_{bp}(t), \quad (2.5\text{-}1)$$

where

$$\eta_{bp}(t) \equiv \sqrt{2}[\eta_c(t)\cos\omega_0 t - \eta_s(t)\sin\omega_0 t] \quad (2.5\text{-}2)$$

is a bandpass noise signal, and $\theta_1 = \theta_i - \omega_0 t$. It is assumed in what follows that η_c and η_s are zero-mean, stationary, lowpass Gaussian processes that have identical autocorrelation functions

$$\begin{aligned} R_{\eta_c}(\tau) &= E[\eta_c(t+\tau)\eta_c(t)] \\ &= E[\eta_s(t+\tau)\eta_s(t)] \quad (2.5\text{-}3) \\ &= R_{\eta_s}(\tau). \end{aligned}$$

Furthermore, it is assumed that η_c and η_s are uncorrelated so that

$$R_{\eta_c\eta_s}(\tau) = E[\eta_c(t+\tau)\eta_s(t)] = 0 \quad (2.5\text{-}4)$$

for all τ.

As discussed in Appendix 2.5.2, η_{bp} has a spectrum $S_{bp}(\omega)$ that has local symmetry around $\pm\omega_0$, and this spectrum can be constructed by translating the spectrum of η_c (or η_s) up to ω_0 and down to $-\omega_0$. In what follows, S_{bp} is modeled as having a constant amplitude of $N_0/2$ across a bandwidth of 2B centered at $\pm\omega_0$; Figure 2.7 depicts example spectral density plots for η_c and η_{bp}.

MODELING THE PHASE-LOCKED LOOP

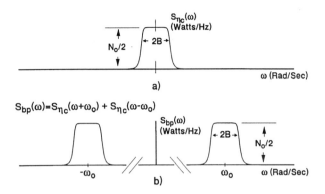

FIGURE 2.7
Example noise spectrums for (a) η_c and (b) η_{bp} where $\eta_{bp} = \sqrt{2}\,(\eta_c\cos\omega_0 t - \eta_s\sin\omega_0 t)$.

2.5.2 Output of the Sinusoidal Phase Detector

Initially, assume that the VCO has been disconnected from the loop filter, and its control voltage is zero. This implies that the VCO signal is

$$\sqrt{2}\,K_1\cos(\omega_0 t + \theta_2), \qquad (2.5\text{-}5)$$

where θ_2 is a random constant. The VCO will be reconnected to the loop filter after some initial analysis.

Under the open-loop conditions being considered, the phase detector forms the product of Equations 2.5-1 and 2.5-5 to produce

$$\sqrt{2}\,AK_m \sin(\omega_0 t + \theta_1)\!\left[\sqrt{2}\,K_1\cos(\omega_0 t + \theta_2)\right]$$
$$+\sqrt{2}\,K_m\!\left[\eta_c(t)\cos\omega_0 t - \eta_s(t)\sin\omega_0 t\right]\!\left[\sqrt{2}\,K_1\cos(\omega_0 t + \theta_2)\right] \qquad (2.5\text{-}6)$$

The terms at $2\omega_0$ produced by this product can be discarded since they will be eliminated by the combination of loop filter and VCO. For analysis purposes, this leaves the phase detector output as

$$x(t) = AK_1 K_m \sin\phi + K_1 K_m\,\eta(t;\theta_2), \qquad (2.5\text{-}7)$$

where

$$\eta(t;\theta_2) \equiv \eta_c(t)\cos\theta_2 + \eta_s(t)\sin\theta_2, \qquad (2.5\text{-}8)$$

and $\phi \equiv \theta_1 - \theta_2$.

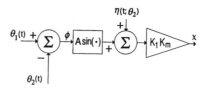

FIGURE 2.8
Baseband model of phase detector under open-loop conditions.

Figure 2.8 depicts an implementation of Equation 2.5-7. The first term on the right-hand side of this equation represents the nominal control signal used in loop operation. The term η in Equation 2.5-7 is an undesirable lowpass noise process. Note the important fact that noise η is introduced *after* the nonlinearity.

This noise process is Gaussian under the current assumptions that the loop filter and VCO are disconnected and θ_2 is a random constant. Under these conditions, calculation of the autocorrelation of η is straightforward. This autocorrelation can be calculated by using

$$R_\eta(\tau) = E\big[\eta(t+\tau;\theta_2)\eta(t;\theta_2)\big]$$
$$= E_{\theta_2}\big[E[\eta(t+\tau;\theta_2)\eta(t;\theta_2)\,|\,\theta_2]\big], \quad (2.5\text{-}9)$$

where the inner expectation is conditioned on θ_2, and the outer expectation is a statistical average over θ_2. With the aid of Equation 2.5-8, the conditional autocorrelation on the right-hand side of Equation 2.5-9 can be calculated as

$$E[\eta(t+\tau;\theta_2)\eta(t;\theta_2)\,|\,\theta_2]$$
$$= E\big[[\eta_c(t+\tau)\cos\theta_2 + \eta_s(t+\tau)\sin\theta_2]$$
$$\cdot [\eta_c(t)\cos\theta_2 + \eta_s(t)\sin\theta_2]\,|\,\theta_2\big]$$
$$= E\big[\eta_c(t+\tau)\eta_c(t)\cos^2\theta_2 + \eta_s(t+\tau)\eta_s(t)\sin^2\theta_2 \quad (2.5\text{-}10)$$
$$+ [\eta_s(t+\tau)\eta_c(t) + \eta_c(t+\tau)\eta_s(t)]\cos\theta_2\sin\theta_2\,|\,\theta_2\big]$$
$$= R_{\eta_c}(\tau)\big[\cos^2\theta_2 + \sin^2\theta_2\big]$$
$$= R_{\eta_c}(\tau)$$

MODELING THE PHASE-LOCKED LOOP

since $R_{\eta_c} = R_{\eta_s}$ as described by Equation 2.5-3. From Equation 2.5-9, it follows immediately that

$$R_\eta(\tau) = R_{\eta_c}(\tau). \qquad (2.5\text{-}11)$$

Note that the conditional expectation Equation 2.5-10 and the autocorrelation $R_\eta(\tau)$ do not depend on the random constant θ_2. An intuitive explanation of this follows from the observation that

$$\eta(t;\theta_2) = \text{Re}\left[(\eta_c + j\eta_s)e^{-j\theta_2}\right], \qquad (2.5\text{-}12)$$

where $\eta_c + j\eta_s$ is proportional to the lowpass equivalent of $\eta_{bp}(t)$ given by Equation 2.5-2. Hence, lowpass $\eta(t; \theta_2)$ is proportional to the cosine quadrature component of the bandpass process which results from subjecting $\eta_{bp}(t)$ to a carrier phase shift of $-\theta_2$ radians. A constant phase shift of the carrier of η_{bp} does not alter the autocorrelation of the lowpass quadrature components. For this reason, it is not surprising that the conditional expectation Equation 2.5-10 and the autocorrelation R_η do not depend on the random constant θ_2.

2.5.3 Nonlinear Model of the PLL with a Noisy Reference

To obtain the nonlinear PLL model for the noisy reference case, all one has to do is close the loop and replace the model of the phase detector in Figure 2.4 with the one given by Figure 2.8. The result of this operation is depicted by Figure 2.9. To complete the modeling process, a statistical characterization of $\eta(t; \theta_2)$ must be given.

In general, the task of determining the statistical properties of $\eta(t; \theta_2)$ is very difficult once the loop is closed so that e is a random process. The reason for this is that θ_2 is no longer constant; as can be seen from Figure 2.9, it is the output of a nonlinear system driven by a noisy

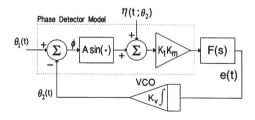

FIGURE 2.9
A nonlinear model which describes the PLL with a noisy reference.

signal. In general, the simple analysis that produced Equation 2.5-11 is no longer valid.

In many applications, $\theta_2(t)$ varies slowly compared to both $\eta_c(t)$ and $\eta_s(t)$. It can be considered constant over any interval that is equal to or less than (say) $4\tau_c$, where τ_c is the correlation time of η_c (or η_s). In other words, $\theta_2(t)$ varies slowly enough that it can be considered constant over any time interval of length τ, where $R_{\eta_c}(\tau)$ (or $R_{\eta_s}(\tau)$) is significantly different from zero. This key assumption is valid in most cases where the bandwidth $2B$ of the additive noise that appears on the PLL input is much greater than the PLL closed-loop bandwidth. Often, PLLs are designed with this as a goal in order to limit the effects of the input noise on the VCO.

Fortunately, this assumption on the nature of $\theta_2(t)$ simplifies the statistical description of $\eta(t; \theta_2)$, and it is utilized in what follows. As can be seen from Equations 2.5-9 and 2.5-10, $R_{\eta_c}(\tau)$ is the autocorrelation of $\eta(t; \theta_2)$ under closed-loop conditions when the above-mentioned assumptions on θ_2 are in effect. This same result can be obtained in a more terse manner by noting that

$$R_\eta(\tau) = E\big[[\eta_c(t+\tau)\cos\theta_2(t+\tau) + \eta_s(t+\tau)\sin\theta_2(t+\tau)]$$

$$\cdot [\eta_c(t)\cos\theta_2(t) + \eta_s(t)\sin\theta_2(t)]\big]$$

$$= \frac{1}{2}E\big[\eta_c(t+\tau)\eta_c(t)\cos(\theta_2(t+\tau)-\theta_2(t))\big]$$

$$+ \frac{1}{2}E\big[\eta_s(t+\tau)\eta_s(t)\cos(\theta_2(t+\tau)-\theta_2(t))\big] \quad (2.5\text{-}13)$$

$$+ \frac{1}{2}E\big[\eta_s(t+\tau)\eta_c(t)\sin(\theta_2(t+\tau)-\theta_2(t))\big]$$

$$- \frac{1}{2}E\big[\eta_c(t+\tau)\eta_s(t)\sin(\theta_2(t+\tau)-\theta_2(t))\big]$$

if the sum–frequency terms are neglected. Now, the assumption that $\theta_2(t+\tau) \approx \theta_2(t)$ for values of τ over which $R_{\eta_c}(\tau)$ is significant leads to the approximation

$$R_\eta(\tau) \approx \frac{1}{2}E\big[\eta_c(t+\tau)\eta_c(t) + \eta_s(t+\tau)\eta_s(t)\big]$$

$$= \frac{1}{2}R_{\eta_c}(\tau) + \frac{1}{2}R_{\eta_s}(\tau) \quad (2.5\text{-}14)$$

$$= R_{\eta_c}(\tau).$$

MODELING THE PHASE-LOCKED LOOP

This result implies that η has a flat spectrum S_η of height $N_o/2$, and that it has a bandwidth of $2B$ centered around $\omega = 0$ (see Figure 2.7a). The assumption that $2B$ is large compared to the closed-loop bandwidth implies that $S_\eta(\omega)$ is constant over the loop bandwidth, and that noise η "looks" to the PLL like white Gaussian noise with a double-sided spectral density of $N_o/2$ watts/Hz. Hence, in what follows, the dependence on θ_2 is dropped, and $\eta(t; \theta_2)$ is written as $\eta(t)$. Finally, with $\eta(t)$ used in place of $\eta(t; \theta_2)$, Figure 2.9 depicts a model of the nonlinear PLL driven by a reference embedded in additive white Gaussian noise.

An equation is obtained easily that describes the PLL with a noisy reference. From Equation 2.1-7 and Figure 2.9, it is readily seen that

$$e(t) = K_1 K_m \int_0^t [A \sin \phi(u) + \eta(u)] f(t-u) du, \quad t \geq 0, \quad (2.5\text{-}15)$$

where zero initial conditions are assumed, and the filter input is applied at $t = 0$. Finally, substitute this last result into Equation 2.2-2 and obtain

$$\frac{d\phi}{dt} = \frac{d\theta_1}{dt} - \int_0^t [G \sin \phi(u) + K_1 K_m K_v \eta(u)] f(t-u) du, \quad (2.5\text{-}16)$$

an integral–differential equation that describes the PLL with a noisy reference.

2.5.4 Linear Model of the PLL with a Noisy Reference

The approximation $\sin\phi \approx \phi$ can be used if the closed-loop phase error remains small. Under this condition, the PLL with a noisy reference can be modeled as depicted by Figure 2.10; that is, it can be modeled as a linear system with a zero-mean, white Gaussian noise input. In what follows, the zero-mean Gaussian response of this system

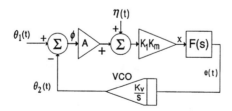

FIGURE 2.10
A linear model which describes the PLL with a noisy reference.

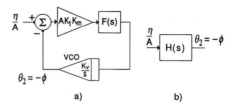

FIGURE 2.11
(a) Model used in computing phase error variance. (b) Block diagram equivalent to system.

is characterized, and an expression is obtained for the variance of the closed-loop phase error. Finally, commonly used parameters are discussed which characterize the noise performance of a PLL. They are known as the *noise-equivalent bandwidth* B_L and the *signal-to-noise ratio in the loop* ρ.

Since superposition applies to the linear model depicted by Figure 2.10, the effects of signal and noise can be analyzed separately. As can be seen from inspection of Figure 2.10, the effects of noise on the phase error can be determined by setting $\theta_1 = 0$, and analyzing the effects of η on output $\phi = -\theta_2$. Equivalently, use η/A in place of θ_1 in the linear analysis outlined in Section 2.3 (apply only noise to the linear model). Then, the output is $\phi = -\theta_2$, and this process can be modeled as a stationary, zero-mean Gaussian process.

This analysis procedure is illustrated by Figure 2.11. Under the condition $\theta_1 \equiv 0$ (noise only), the quantities η/A and $\phi = -\theta_2$ become the input and output, respectively, and the linear model reduces to that shown on Figure 2.11a. From an input–output standpoint, Figure 2.11a is equivalent to the system depicted by Figure 2.11b, where H(s) is given by Equation 2.3-4. Hence, the output power density spectrum can be expressed as

$$S_\phi(\omega) = \frac{S_\eta(\omega)}{A^2} |H(j\omega)|^2 \qquad (2.5\text{-}17)$$
$$= \frac{N_o/2}{A^2} |H(j\omega)|^2.$$

The variance of the phase error in the linear model is simply the AC power in ϕ. This quantity can be expressed as

$$\sigma_\phi^2 = \frac{1}{2\pi} \int_{-\infty}^{\infty} S_\phi(\omega) d\omega. \qquad (2.5\text{-}18)$$

MODELING THE PHASE-LOCKED LOOP

Substitute S_ϕ given by Equation 2.5-17 into Equation 2.5-18 and obtain

$$\sigma_\phi^2 = \frac{\frac{1}{2\pi}\int_{-\infty}^{\infty}\frac{N_o}{2}|H(j\omega)|^2 d\omega}{A^2} \qquad (2.5\text{-}19)$$

as the variance of the closed-loop phase error.

As given by Equation 2.5-19, variance σ_ϕ^2 has an intuitive interpretation. Note that Equation 2.5-19 is the ratio of two powers since A^2 is the power of the PLL reference signal. For this reason, the *signal-to-noise ratio in the loop* (also called *in-loop SNR*) is given the definition

$$\rho \equiv \frac{1}{\sigma_\phi^2}. \qquad (2.5\text{-}20)$$

The quantity ρ may be defined intuitively, but it has no direct physical meaning. Inside the loop there are no signal and additive noise components that have a power ratio given by ρ. However, in the PLL literature, the concept embodied by parameter ρ (or a similarly defined parameter) is in widespread use.

A second commonly used parameter in the PLL literature is the *loop-noise bandwidth* B_L (similar to the noise equivalent bandwidth parameter used in linear system theory). Parameter B_L is the one-sided bandwidth in hertz of a unity-gain, ideal lowpass filter that passes as much noise power as does the linear model depicted by Figure 2.11b. That is, the ideal filter is specified with bandwidth B_L in hertz, and it has an output noise power of σ_ϕ^2 when it is driven by a white process with a one-sided spectral density of N_o/A^2 watts/Hz. Hence, the loop-noise bandwidth B_L must satisfy

$$B_L \frac{N_o}{A^2} = \sigma_\phi^2. \qquad (2.5\text{-}21)$$

This equation can be combined with Equation 2.5-19 to obtain

$$\begin{aligned}B_L &= \frac{1}{4\pi}\int_{-\infty}^{\infty}|H(j\omega)|^2 d\omega \\ &= \frac{1}{2\pi}\int_{0}^{\infty}|H(j\omega)|^2 d\omega.\end{aligned} \qquad (2.5\text{-}22)$$

The fact that $|H(j\omega)|$ is an even function of ω has been used in the derivation of this result.

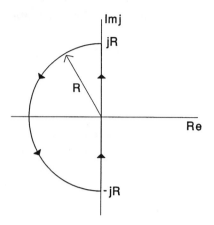

FIGURE 2.12
Contour of integration used to compute B_L.

A PLL loop-noise bandwidth can be calculated easily using the theory of residues. First, note that $H(j\omega)H^*(j\omega) = H(j\omega)H(-j\omega)$, so Equation 2.5-22 can be written as

$$B_L = \frac{1}{4\pi}\int_{-\infty}^{\infty} H(j\omega)H(-j\omega)\,d\omega. \tag{2.5-23}$$

Next, make the substitution $s = j\omega$, $d\omega = ds/j$ in this last equation to obtain

$$B_L = \frac{1}{4\pi j}\int_{-j\infty}^{j\infty} H(s)H(-s)\,ds. \tag{2.5-24}$$

Since the integrand $H(s)H(-s) \to 0$ as $s \to \infty$, this last equation can be expressed as

$$B_L = \lim_{R\to\infty}\frac{1}{4\pi j}\oint_C H(s)H(-s)\,ds. \tag{2.5-25}$$

where $\mathbb{C}$ denotes the semicircular contour depicted by Figure 2.12. However, the contour integral in Equation 2.5-25 can be evaluated as

$$\oint_C H(s)H(-s)\,ds = 2\pi j\sum_k r_k, \tag{2.5-26}$$

where r_k denotes the residue of $H(s)H(-s)$ at the kth left-half plane pole, and the sum is over all residues associated with poles that lie within the contour $\mathbb{C}$. Finally, the last two equations can be combined to yield

$$B_L = \frac{1}{2}\sum_k \left(\text{Residue of } H(s)H(-s) \text{ at Left Half Plane Pole } p_k\right), \quad (2.5\text{-}27)$$

where the sum is over all residues associated with poles that lie within the left half of the s-plane. This last formula provides a practical method of computing B_L; in Sections 3.1.3, 3.2.4, and 3.3.4 in Chapter 3, this method is applied to commonly used PLLs.

2.6 MODELING THE LIMITER PHASE DETECTOR

In many applications, the signal from which the PLL reference is derived can vary over a wide amplitude range. In order to maintain the PLL near its nominal operating design, these signal variations must not show up on the PLL input reference. That is, some form of amplitude control subsystem must be used to limit signal variations that reach the PLL phase detector. For example, consider the phase-locked receiver discussed in Section 1.4.4 in Chapter 1. In this application, the received signal may vary over a range of 100 dB or more. In addition to causing problems with the general circuitry of the receiver, this variation in signal strength would cause unacceptable extremes in the receiver closed-loop dynamics. For these reasons, phase-locked receivers contain one or more subsystems that control signal amplitude variations and maintain the phase comparator reference signal at a nearly constant level.

The hard limiter is a relatively simple and effective method of amplitude control. In most PLL modeling exercises, hard limiting is incorporated as part of the phase detector. In these cases, the limiter/detector combination is referred to as a *limiter phase detector*. This section is devoted to a discussion of this type of phase detector. The model development discussed in this section is based on the work of Rosenkranz,[50] Springett and Simon,[51] and Davenport.[52]

For the model developed in this section, the PLL is supplied with the bandpass signal

$$r(t) = \sqrt{2}\,A\sin\left(\omega_0 t + \theta_1(t)\right) + \eta_{bp}(t). \quad (2.6\text{-}1)$$

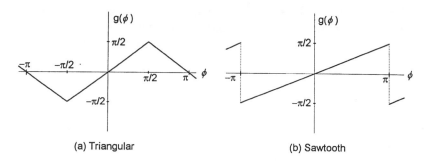

FIGURE 2.13
Commonly used nonsinusoidal phase error functions.

The quantity A denotes the reference signal amplitude (in RMS volts), and

$$\eta_{bp}(t) = \sqrt{2}\left[\eta_c(t)\cos\omega_0 t - \eta_s(t)\sin\omega_0 t\right] \quad (2.6\text{-}2)$$

is bandpass, zero-mean Gaussian noise. Note that this input signal and noise combination is identical to that used in Section 2.5. However, in this section, signal r(t) is passed through a hard limiter that strips off any amplitude variations and forms a ±1 binary signal. In a similar manner, the VCO output is hard limited to form a ±1 binary signal.

The limiter phase detector output signal is proportional to a function g of the phase difference between the hard-limited r(t) and VCO signals discussed in the previous paragraph. Function g is a 2π-periodic and odd function of its argument. Figure 2.13 illustrates two of the more commonly used, nonsinusoidal characteristic functions (when implemented by digital logic powered by a single positive supply voltage, these characteristic functions may be shifted in level so that they are nonnegative). Limiter phase detectors that implement these characteristic functions are discussed in Chapter 4.

For analysis purposes, the limiter phase detector output signal is split into a useful control component and an undesirable noise signal. The useful control component is described as a function of phase error; it is proportional to the *effective phase detector characteristic function*, a quantity defined in this section. When the input reference signal r(t) is noise free, this effective characteristic function is equal to the above-mentioned function g, and there is no noise component in the detector output. However, things are not this simple when signal r(t) contains additive noise as modeled by Equation 2.6-1. As discussed in this section, the effective phase detector characteristic function is degraded by hard limiting of the noisy input signal; in general, it is no longer equal to g. Instead, its shape and amplitude depend on input noise statistics and the input signal-to-noise ratio.

MODELING THE PHASE-LOCKED LOOP

2.6.1 Hard Limiting the Input and VCO Signals

To prepare for hard limiting, the input signal r(t) is rewritten in magnitude and phase form as

$$r(t) = \sqrt{2[A - N_s]^2 + 2N_c^2}\, \sin(\omega_0 t + \theta_1 + \theta_n), \tag{2.6-3}$$

where

$$N_c(t) = \eta_c(t)\cos\theta_1(t) + \eta_s(t)\sin\theta_1(t)$$
$$N_s(t) = \eta_s(t)\cos\theta_1(t) - \eta_c(t)\sin\theta_1(t) \tag{2.6-4}$$

$$\theta_n(t) = \tan^{-1}\left[\frac{N_c(t)}{A - N_s(t)}\right].$$

Notice that the phase noise $\theta_n(t)$ is a function of the input signal amplitude A, and the envelope of r(t) depends on the input noise components N_c and N_s.

Input signal r(t) is supplied as input to a hard limiter. The result of this limiting operation is sgn[r(t)], where

$$\operatorname{sgn}[r] = +1 \quad \text{if } r > 0$$
$$= -1 \quad \text{if } r < 0. \tag{2.6-5}$$

This produces a ±1 binary signal with the instantaneous phase $\omega_0 t + \theta_1 + \theta_n$. In a similar manner, the VCO output signal is hard limited to form a binary signal with the instantaneous phase $\omega_0 t + \theta_2$.

2.6.2 The Output of the Limiter Phase Detector

The limiter phase detector generates an output signal that is a function of the phase difference between the hard-limited input and hard-limited VCO signals. That is, the detector output is represented as

$$x = K_D\, g\big([\theta_1 + \theta_n] - \theta_2\big)$$
$$= K_D\, g(\phi + \theta_n), \tag{2.6-6}$$

where K_D is known as the *limiter phase detector gain*, and nonlinear, 2π-periodic g is an odd function of its argument. Figure 2.14 depicts a

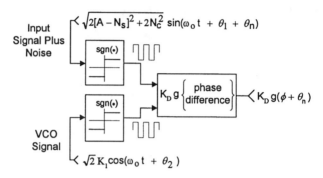

FIGURE 2.14
Limiter phase detector.

block diagram for the limiter phase detector. In a manner similar to that discussed in the first part of Section 2.5.3, the assumption is made that closed-loop phase error ϕ varies slowly compared to phase noise θ_n (i.e., the input noise bandwidth is large compared to the input signal and loop bandwidths). Hence, to the casual observer, output x might appear as a high-frequency, fast-varying noise signal (due to θ_n) that fluctuates around a low-frequency, slow-varying control signal (due to ϕ).

2.6.3 Splitting the Detector Output into Control and Noise Components

As given by Equation 2.6-6, output x depends on the random processes ϕ and θ_n. Unlike its counterpart Equation 2.5-7 for the sinusoidal phase detector without limiting, it does not split naturally into the sum of a control term (the desired loop-control function) and a noise term (a loop disturbance). However, to permit a relatively simple analysis of a PLL containing a limiter phase detector, Equation 2.6-6 must be split into a ϕ-dependent control component and a noise component that depends on both ϕ and θ_n (the goal here is to represent Equation 2.6-6 by an equation similar *in form* to Equation 2.5-7). As discussed at the end of this section, the desired ϕ-dependent control component is the best mean square approximation of $K_D g(\phi + \theta_n)$. But first, this control component is obtained by a simple intuitive argument based on the assumption that ϕ varies slowly relative to θ_n. This assumption allows the control component to be written by holding $\phi = \varphi_c$ constant and averaging out the phase noise to obtain

$$\text{Control Component} \equiv K_D \tilde{g}(\varphi_c) \equiv K_D E\left[g(\phi + \theta_n)\big|\phi = \varphi_c\right] \quad (2.6\text{-}7)$$

MODELING THE PHASE-LOCKED LOOP

(nonrandom φ_c is a realization of the random quantity ϕ). Function $\tilde{g}$ is known as the *effective phase detector characteristic function*. In terms of this effective function, the output of the limiter phase detector is a random process split according to

$$x(\phi,\theta_n) = K_D\left[\tilde{g}(\phi) + \eta(\phi,\theta_n)\right], \qquad (2.6\text{-}8)$$

where the noise component is given by

$$\eta(\phi,\theta_n) \equiv g(\phi+\theta_n) - \tilde{g}(\phi). \qquad (2.6\text{-}9)$$

Due to phase noise $\theta_n(t)$, noise term η is a rapidly varying function of time. Finally, for the remainder of this section, random ϕ and its realization are not distinguished by notation (when important, the difference between a random quantity and its realization is clear from context).

Equation 2.6-8 describes a general model of the limiter phase detector. Figure 2.15 depicts a block diagram of this model. Notice the similarity of the block diagram of this model with Figure 2.8, the model for the sinusoidal detector without hard limiting of the input and VCO signals. However, the similarity in Figures 2.8 and 2.15 is somewhat misleading; a cursory inspection of the figures does not reveal a fundamental difference in the detector models. For the sinusoidal detector without limiting (i.e., Figure 2.8), the statistics of the input noise and the input signal-to-noise ratio do not influence directly the shape and amplitude of $AK_1K_m\sin\phi$, the detector control component. However, when hard limiting of the input and VCO signals is incorporated (i.e., Figure 2.15), the level and statistical properties of the input noise can have a significant effect on the shape and amplitude of $K_D\tilde{g}(\phi)$, the control component in the detector output.

As an example of a significant and unsatisfactory effect, the input noise may cause the null point of the control component (i.e., the root of $\tilde{g}(\phi) = 0$) to shift from the point $\phi = 0$. However, such a shift in the

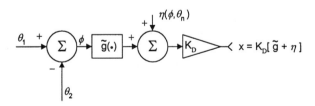

FIGURE 2.15
Equivalent model for limiter phase detector.

null point does not happen if the phase noise density is even (i.e., $p(\theta_n) = p(-\theta_n)$) and the noise-free phase characteristic is odd (i.e., $g(\phi_d) = -g(-\phi_d)$). For example, zero-mean Gaussian noise added to the input reference leads to an even phase noise density, and it does not cause a shift in the PLL equilibrium point.

By considering some elementary vector space and probability concepts, the dichotomy offered by Equations 2.6-7 through 2.6-9 can be given a descriptive geometric interpretation (see Sections 7.5 and 8.3 of Papoulis).[20] The control component $K_D \tilde{g}(\phi)$ is the random function that provides the best nonlinear mean square approximation of the limiter phase detector output $K_D g(\phi + \theta_n)$. That is, control Function 2.6-7 is chosen as the function that minimizes the mean square error

$$e_{ms} \equiv E\left[\left(g(\phi + \theta_n) - \tilde{g}(\phi)\right)^2\right]. \qquad (2.6\text{-}10)$$

Note that e_{ms} is equal to σ_η^2, the variance of η. Furthermore, from the theory of nonlinear mean square estimation, the error $g(\phi + \theta_n) - \tilde{g}(\phi)$ is orthogonal to ϕ in the sense that

$$E\left[\left(g(\phi + \theta_n) - \tilde{g}(\phi)\right)\phi\right] = 0. \qquad (2.6\text{-}11)$$

2.6.4 Practical Application of the Limiter Phase Detector Model

Equation 2.6-11 states that noise component η is orthogonal to ϕ. However, η is dependent on ϕ, at least to some extent as demonstrated by Springett and Simon.[51] For specific g and fixed ϕ, these authors have shown that noise variance $\sigma_\eta^2(\phi)$ depends only slightly on ϕ; even for ϕ as large as $\phi = \pi/4$, variance σ_η^2 does not change appreciably from its maximum value (that occurs at $\phi = 0$). That is, variance σ_η^2 is almost constant over the most important (from an applications standpoint) range of ϕ. Furthermore, with changes in ϕ, the deviation of $\sigma_\eta^2(\phi)$ from $\sigma_\eta^2(0)$ becomes less as the SNR decreases on the limiter phase detector input.

In many practical applications, the results developed by Springett and Simon[51] justify neglecting the dependence of noise η on the phase variable ϕ. This approximation results in a significant simplification, especially in many practical cases where $\tilde{g}(0) = 0$ and $E[\phi] = 0$. Under these conditions, the assumption that $\phi = 0$ can be used in η leads to

$$\left.\eta(\phi, \theta_n)\right|_{\phi=0} = g(\theta_n) \qquad (2.6\text{-}12)$$

as an approximation for the noise in the detector model depicted by Figure 2.15. Then, this simplified limiter phase detector model can be used to replace the sinusoidal detector model (enclosed by the dashed-line box) that appears on Figure 2.9. This substitution of phase detector models produces a relatively simple baseband model of the PLL containing a limiter phase detector.

In a manner similar to that discussed in Section 2.5.4, a linear analysis can be performed on the PLL model containing a limiter phase detector. Such an analysis requires the effective gain of the limiter phase detector. This effective gain is the slope, evaluated at $\phi = 0$, of the effective phase characteristic of the detector; once $\tilde{g}$ is known, this slope can be computed as

$$\text{Effective Phase Detector Gain} = K_D \left.\frac{d\tilde{g}}{d\phi}\right|_{\phi=0} \tag{2.6-13}$$

2.6.5 Example: Limiter Phase Detector with Sinusoidal g(ϕ)

This simple case utilizes the noise-free phase characteristic function

$$g(\varphi_c) = \sin\varphi_c. \tag{2.6-14}$$

Equation 2.6-7 can be used to write the effective phase detector characteristic function

$$\begin{aligned}\tilde{g}(\varphi_c) &= E\big[\sin(\phi+\theta_n)\big|\phi=\varphi_c\big] \\ &= E\big[\sin(\phi)\cos(\theta_n)+\cos(\phi)\sin(\theta_n)\big|\phi=\varphi_c\big] \quad (2.6\text{-}15) \\ &= E\big[\cos(\theta_n)\big]\sin(\varphi_c)\end{aligned}$$

for the usual case where phase noise θ_n has zero mean. In Appendix 2.6.1, the quantity $E[\cos(\theta_n)]$ is evaluated for the random phase noise process θ_n that results from applying to the limiter phase detector input a sinusoid embedded in zero-mean Gaussian noise. For this simple reference signal, the effective characteristic function is

$$\begin{aligned}\tilde{g}(\varphi_c) = \frac{1}{2}\sqrt{\pi(\text{SNR}_i)} &\exp[-\frac{\text{SNR}_i}{2}] \\ &\cdot\left[I_0\!\left(\frac{\text{SNR}_i}{2}\right)+I_1\!\left(\frac{\text{SNR}_i}{2}\right)\right]\sin(\varphi_c)\end{aligned} \tag{2.6-16}$$

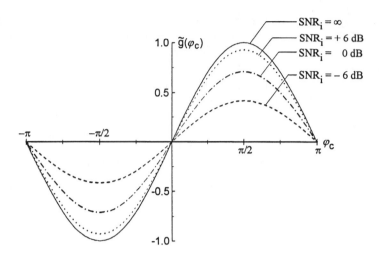

FIGURE 2.16
Effective phase characteristic for limiter phase detector with a sinusoidal noiseless phase characteristic. Plots show degradation due to zero-mean Gaussian noise on the input.

where

$$\text{SNR}_i \equiv A^2/\sigma_\eta^2 \tag{2.6-17}$$

is the signal-to-noise ratio on the input of the limiter phase detector (σ_η^2 is the variance of process $\eta_c \cos\omega_0 t + \eta_s \sin\omega_0 t$), and I_0 and I_1 are modified Bessel functions of order zero and one, respectively.

Figure 2.16 depicts plots of Equation 2.6-16 for several values of input SNR_i. As can be seen from inspection of these plots, hard limiting of the input and VCO processes does not cause a noise-induced degradation in the *shape* of a sinusoidal phase detector characteristic. However, as the plots on Figure 2.16 depict, it degrades the *amplitude* of the effective phase detector characteristic.

2.6.6 Example: Limiter Phase Detector with Triangular g(φ)

Consider the case where noise-free phase characteristic g is the triangular function depicted by Figure 2.13a. For use in Equation 2.6-7, a Fourier series expansion of $g(\phi + \theta_n)$ can be written as

MODELING THE PHASE-LOCKED LOOP

$$g(\phi+\theta_n) = \frac{4}{\pi} \sum_{\substack{k\geq 1 \\ k \text{ odd}}} \frac{(-1)^{(k-1)/2}}{k^2} \sin k(\phi+\theta_n)$$

$$= \frac{4}{\pi} \sum_{\substack{k\geq 1 \\ k \text{ odd}}} \frac{(-1)^{(k-1)/2}}{k^2} \{\sin k\phi \cos k\theta_n + \cos k\phi \sin k\theta_n\}.$$

(2.6-18)

Now, take the conditional expected value of Equation 2.6-18 to obtain

$$\tilde{g}(\varphi_c) = E\big[g(\phi+\theta_n)\,|\,\phi=\varphi_c\big]$$

$$= \frac{4}{\pi} \sum_{\substack{k\geq 1 \\ k \text{ odd}}} \frac{(-1)^{\frac{k-1}{2}}}{k^2} \{E[\cos(k\theta_n)]\sin(k\varphi_c) + E[\sin(k\theta_n)]\cos(k\varphi_c)\}$$

(2.6-19)

for the effective phase characteristic. The expectations $E[\cos(k\theta_n)]$ and $E[\sin(k\theta_n)]$ in Equation 2.6-19 are dependent on the statistics of the input noise process, and they must be determined.

In Appendix 2.6.1, the case of a sinusoidal reference embedded in additive zero-mean Gaussian noise is considered, and expressions are obtained for $E[\cos(k\theta_n)]$ and $E[\sin(k\theta_n)]$. These expressions can be used in Equation 2.6-19 to obtain

$$\tilde{g}(\varphi_c) = \frac{2}{\pi} \sum_{\substack{k\geq 1 \\ k \text{ odd}}} \frac{(-1)^{\frac{k-1}{2}}}{k^2} \mu_k(SNR_i) \sin(k\varphi_c)$$

$$\mu_k(SNR_i) \equiv \sqrt{\pi(SNR_i)} \exp\left(-\frac{SNR_i}{2}\right)$$

$$\cdot \left[I_{(k-1)/2}\left(\frac{SNR_i}{2}\right) + I_{(k+1)/2}\left(\frac{SNR_i}{2}\right)\right],$$

(2.6-20)

where SNR_i is given by Equation 2.6-17, and the I_k, $k \geq 0$ are modified Bessel functions of order k. The Series 2.6-20 converges very fast; in practical applications a few terms of this series can be summed to obtain excellent results.

Figure 2.17 depicts plots of $\tilde{g}$ for several values of input SNR_i (these were made by summing 20 terms of Equation 2.6-20). As can be seen from inspection of these plots, hard limiting of the input and VCO signals

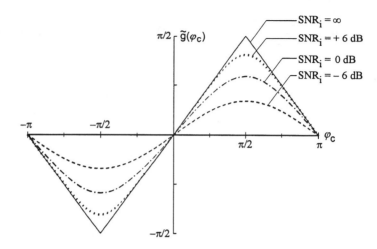

FIGURE 2.17
Effective phase characteristic for limiter phase detector with a triangular noiseless phase characteristic. Plots show degradation due to zero-mean Gaussian noise on the input.

causes a noise-induced degradation in the *shape* and *amplitude* of a triangular phase detector characteristic. The effective detector characteristic becomes more sinusoidal as the input SNR_i decreases. As it turns out, this observation applies to all practical nonsinusoidal phase detectors that are "memoryless" (as discussed in Section 4.1 in Chapter 4).

2.7 MODELING THE LONG LOOP

Often, phase-locked receivers are based on what is called a *long loop*. The distinguishing characteristic of a long loop is that it contains an IF signal path; an example of such a loop is illustrated by Figure 2.18. The received signal is heterodyned twice in order to form the baseband signal x(t) that drives the loop filter. First, the VCO output is used to heterodyne the received signal down to a bandpass IF signal $x_{bp}(t)$ at an IF frequency of ω_{if}; then, x_{bp} is passed through an IF filter/amplifier to produce $y_{bp}(t)$. Next, the output of a crystal oscillator at ω_{if} is used to heterodyne y_{bp} down to the baseband signal x(t) that drives the loop filter. On Figure 2.18, constant γ appears as an arbitrary phase angle in the crystal oscillator output. In what follows, this long loop is shown to be mathematically equivalent to a baseband, or short, loop (i.e., one that does not contain an IF signal path).

While the two loop architectures are mathematically equivalent, the long loop implementation is preferred in practical applications where the desired input reference signal is swamped in wideband noise (on Figure 2.18, η_{bp} represents wideband noise). The reason for this is

MODELING THE PHASE-LOCKED LOOP

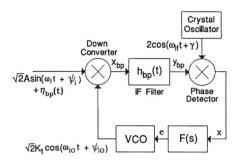

FIGURE 2.18
A long loop containing an IF signal path.

that a real phase comparator is imperfect; it has a finite dynamic range, and it exhibits a threshold phenomenon due to rectification of the input noise. Substantial wideband noise power applied on the phase comparator input could cause a DC component to appear at its output. This noise rectification-induced DC component could dominate the desired DC component due to normal operation of the loop. The occurrence of this phenomenon could degrade the signal acquisition and tracking abilities of the PLL.

This problem is minimized by using the long loop architecture depicted by Figure 2.18. Of course, there is still the potential for noise rectification in the first down converter shown on this figure. However, the resulting DC component in x_{bp} would be rejected by the IF filter. This IF filter serves to minimize the effects of noise rectification by limiting the total noise power that reaches an imperfect phase comparator.

2.7.1 Baseband Model of the Long Loop

In this subsection, a baseband model of the loop depicted by Figure 2.18 is developed. To simplify the analysis, it is assumed that the input reference is noise free (set $\eta_{bp} = 0$ on Figure 2.18). Of course, this assumption in no way influences the baseband architecture.

The output of the down converter on Figure 2.18 is given by

$$x_{bp} = A K_1 \sin(\omega_{if} t + \psi_{if}), \qquad (2.7\text{-}1)$$

where $\omega_{if} \equiv \omega_i - \omega_{lo}$, and $\psi_{if} \equiv \psi_i - \psi_{lo}$. Here it is assumed that low-side injection is used so that $\omega_i > \omega_{lo}$. Also, the sum frequency term in x_{bp} is omitted since it is rejected by the IF filter. Finally, note that x_{bp} can be written as

$$x_{bp}(t) = x_c(t)\cos\omega_{if}t - x_s(t)\sin\omega_{if}t$$

$$x_c(t) \equiv A\,K_1 \sin\psi_{if} \qquad (2.7\text{-}2)$$

$$x_s(t) \equiv -A\,K_1 \cos\psi_{if}.$$

In what follows, lowpass equivalent signals are used to analyze the IF path and find the baseband signal x(t) in the long loop depicted by Figure 2.18. As discussed in Appendix 2.5.1, IF signal x_{bp} can be represented as $x_{bp} = \text{Re}[x_{lp}\exp(j\omega_{if}t)]$, where

$$x_{lp} = x_c + jx_s \qquad (2.7\text{-}3)$$

is its lowpass equivalent. The bandpass IF filter $h_{bp} = \text{Re}[h_{lp}\exp j\omega_{if}t]$ in Figure 2.18 is assumed to be symmetrical as defined in Appendix 2.5.1. This means that it has a real-valued lowpass equivalent

$$h_{lp} = h_c. \qquad (2.7\text{-}4)$$

As discussed in Appendix 2.5.1, the lowpass equivalent y_{lp} of the IF filter bandpass output y_{bp} is

$$\begin{aligned}y_{lp} &= \frac{1}{2} x_{lp} * h_{lp} \\ &= \frac{1}{2}(x_c + jx_s) * h_c,\end{aligned} \qquad (2.7\text{-}5)$$

so that the bandpass output of the IF filter can be expressed as

$$y_{bp} = \frac{1}{2}\text{Re}\big[\{h_c * (x_c + jx_s)\}\exp(j\omega_{if}t)\big]. \qquad (2.7\text{-}6)$$

Now, the phase detector depicted on Figure 2.18 forms the product

$$\begin{aligned}&y_{bp}(t) \cdot 2\cos(\omega_{if}t + \gamma) \\ &= \frac{1}{2}\text{Re}\big[\,[(x_c + jx_s)*h_c]\exp(j\omega_{if}t)\,2\cos(\omega_{if}t + \gamma)\,\big]\end{aligned} \qquad (2.7\text{-}7)$$

of y_{bp} and the output of the crystal oscillator. To simplify this last result note that

MODELING THE PHASE-LOCKED LOOP

$$\exp(j\omega_{if}t)2\cos(\omega_{if}t+\gamma) = \exp[j(2\omega_{if}t+\gamma)] + \exp(-j\gamma), \quad (2.7\text{-}8)$$

and the baseband component in this product is the constant $e^{-j\gamma}$. Hence, on Figure 2.18, the baseband component in the output of the phase detector is

$$x = \frac{1}{2}\operatorname{Re}\left[h_c *(x_c + jx_s)e^{-j\gamma}\right], \quad (2.7\text{-}9)$$

a result obtained from inspection of Equation 2.7-7.

The IF signal path on Figure 2.18 can be removed by realizing that the IF filter/phase comparator combination can be replaced by a phase comparator followed by a scaled version of the IF filter lowpass equivalent. To see this, note that the product of x_{bp} and $2\cos(\omega_{if}t + \gamma)$ can be written as

$$x_{bp}(t)\cdot 2\cos(\omega_{if}t+\gamma) = \operatorname{Re}\left[(x_c+jx_s)\exp(j\omega_{if}t)2\cos(\omega_{if}t+\gamma)\right], \quad (2.7\text{-}10)$$

which has a baseband component given by

$$\operatorname{Re}\left[(x_c+jx_s)e^{-j\gamma}\right]. \quad (2.7\text{-}11)$$

As can be seen from Equation 2.7-9, the baseband signal x results if Equation 2.7-11 is passed through lowpass $\frac{1}{2}h_c$. Hence, parts (a) and (b) of Figure 2.19 depict mathematically equivalent methods of generating baseband x. However, the two successive down conversions shown on Figure 2.19b produce

$$\sqrt{2}A\sin(\omega_i t+\psi_i)\sqrt{2}K_1\cos(\omega_{lo}t+\psi_{lo})2\cos(\omega_{if}t+\gamma), \quad (2.7\text{-}12)$$

and the baseband component of this product is filtered by $\frac{1}{2}h_{lp}$. But, the baseband component in

$$\sqrt{2}A\sin(\omega_i t+\psi_i)\sqrt{2}K_1\cos(\omega_i t+\psi_{lo}+\gamma) \quad (2.7\text{-}13)$$

is identical to the one in Equation 2.7-12, so Figure 2.19c is mathematically equivalent to parts (a) and (b) of this diagram. Finally, this last observation implies that Figure 2.20 depicts a baseband PLL which is equivalent to the long loop illustrated by Figure 2.18. On Figure 2.20, the VCO has a center frequency that exceeds by ω_{if} the center frequency of the VCO depicted on Figure 2.18.

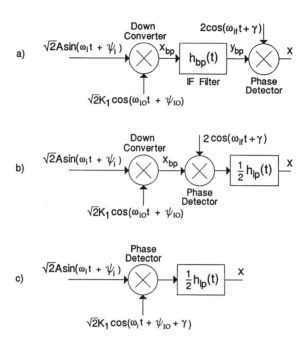

FIGURE 2.19
Three mathematically equivalent ways of producing baseband signal x.

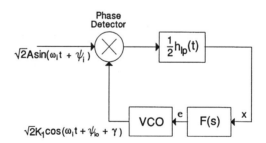

FIGURE 2.20
A baseband loop equivalent to the long loop discussed in Section 2.7.

As mentioned above, the long loop IF filter serves to limit the broadband noise power that reaches the phase detector. The goal of the IF filter is to limit the undesirable side effects that can occur when an imperfect phase detector is swamped with noise. To a satisfactory degree, this goal can be accomplished in many applications with an IF filter bandwidth that is large compared to the overall closed-loop bandwidth. In these cases, a common perception is that the IF filter only has a minor influence on the dynamics of the closed loop. However, this perception may not be true in practical applications. As shown in

MODELING THE PHASE-LOCKED LOOP

Section 5.4 in Chapter 5, the "small" effects of an IF filter can limit the pull-in range of the loop and cause other unacceptable behavior.

APPENDIX 2.5.1 NARROWBAND SIGNALS AND SYSTEMS

Narrowband signals and systems play an important role in phase-lock technology. Examples of this include the long loop and phase-locked receivers. This appendix contains some general modeling techniques for narrowband signals and systems.

2.5.1.1 Modeling Bandpass Signals and Systems

A real-valued bandpass signal $x_{bp}(t)$ can be represented as

$$x_{bp}(t) = \Gamma_{bp}(t)\cos(\omega_c t + \psi_{bp}(t)), \qquad (2.5.1\text{-}1)$$

where Γ_{bp} and ψ_{bp} are known as the signal *envelope* and *phase*, respectively. Also, constant ω_c is known as the *carrier frequency* of the signal. This equation can be written as

$$x_{bp}(t) = x_c(t)\cos\omega_c t - x_s(t)\sin\omega_c t$$

$$x_c(t) \equiv \Gamma_{bp}(t)\cos\psi_{bp}(t) \qquad (2.5.1\text{-}2)$$

$$x_s(t) \equiv \Gamma_{bp}(t)\sin\psi_{bp}(t).$$

The quantities x_c and x_s are the *quadrature components* of x_{bp}.

Signal x_{bp} is said to be *narrowband* if it has a bandwidth which is small compared to ω_c. The quadrature components of a narrowband signal vary slowly relative to $\cos\omega_c t$. Equivalently, these components are lowpass processes with bandwidths that are small compared to ω_c.

Linear, time-invariant bandpass systems can be treated in much the same manner as narrowband signals. Such a system has an impulse response that can be written as

$$h_{bp}(t) = h_c(t)\cos\omega_c t - h_s(t)\sin\omega_c t, \qquad (2.5.1\text{-}3)$$

where h_c and h_s are real-valued, lowpass functions. Equations 2.5.1-2 and 2.5.1-3 have similar forms; hence, many of the formal manipulations outlined below can be applied to both bandpass signals and systems.

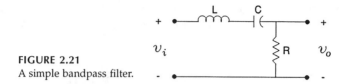

FIGURE 2.21
A simple bandpass filter.

A simple example of a narrowband system, that is used as a filter in many applications, is depicted by Figure 2.21. The transfer function of this simple network is given by

$$H_{bp}(s) = \frac{2\alpha_o s}{(s+\alpha_o)^2 + \omega_c^2}$$

$$= 2\alpha_o \left[\frac{s+\alpha_o}{(s+\alpha_o)^2 + \omega_c^2}\right] - \frac{2\alpha_o^2}{\omega_c}\left[\frac{\omega_c}{(s+\alpha_o)^2 + \omega_c^2}\right], \quad (2.5.1\text{-}4)$$

where $\alpha_o \equiv R/2L$, $\omega_c \equiv \sqrt{\omega_n^2 - \alpha_o^2}$, and $\omega_n \equiv \sqrt{1/LC}$. In most applications, the component values are chosen so that the narrowband condition $\omega_n^2 \gg \alpha_o^2$ applies and $\omega_c \approx \omega_n$. This filter has a narrowband impulse response with quadrature components given by

$$h_c(t) \equiv 2\alpha_o \exp(-\alpha_o t) U(t)$$

$$h_s(t) \equiv \frac{2\alpha_o^2}{\omega_c} \exp(-\alpha_o t) U(t), \quad (2.5.1\text{-}5)$$

where $U(t)$ denotes the unit step function.

2.5.1.2 Lowpass Equivalent Signals and Systems

Consider the bandpass signal in Equation 2.5.1-2. This bandpass function has a corresponding *lowpass equivalent* function defined by

$$x_{lp}(t) \equiv x_c(t) + jx_s(t). \quad (2.5.1\text{-}6)$$

The bandpass function x_{bp} can be written in terms of x_{lp} as

$$x_{bp}(t) = \text{Re}\left[x_{lp}(t)\exp(j\omega_c t)\right], \quad (2.5.1\text{-}7)$$

MODELING THE PHASE-LOCKED LOOP

where Re denotes that only the real part of the bracketed expression should be retained. As shown below, use of lowpass equivalents simplifies analysis involving bandpass signals and systems.

A simple relationship can be developed between the Fourier transforms of x_{lp} and x_{bp}. Equation 2.5.1-7 can be written as

$$x_{bp}(t) = \frac{1}{2}\left[x_{lp}(t)\exp(j\omega_c t) + x_{lp}^*(t)\exp(j\omega_c t)\right], \quad (2.5.1\text{-}8)$$

and the Fourier transform of this signal is

$$X_{bp}(j\omega) = \frac{1}{2}\left[X_{lp}(j\omega - j\omega_c) + X_{lp}^*(-j\omega - j\omega_c)\right], \quad (2.5.1\text{-}9)$$

where X_{lp} denotes the Fourier transform of X_{lp}. Now, all of the frequency components of lowpass $X_{lp}(j\omega)$ lie within a band whose upper frequency is small compared to ω_c. Hence, Equation 2.5.1-9 leads to the results

$$X_{bp}(j\omega)U(\omega) = \frac{1}{2}X_{lp}(j\omega - j\omega_c) \quad (2.5.1\text{-}10)$$

$$X_{lp}(j\omega) = 2X_{bp}(j\rho)U(\rho)\bigg|_{\rho = \omega + \omega_c} \quad (2.5.1\text{-}11)$$

where $U(\omega)$ denotes a step function in the frequency domain. Equation 2.5.1-11 serves as the basis of Figure 2.22, which illustrates the relationship between the magnitude and phase of X_{lp} and X_{bp}.

Equation 2.5.1-6 defines the lowpass function x_{lp} in terms of the quadrature components of bandpass x_{bp}. However, x_{lp} can be expressed directly in terms of this bandpass process. Take the inverse Fourier transform of Equation 2.5.1-11 and obtain

$$x_{lp}(t) = \frac{1}{2\pi}\left[x_{bp}(t) * 2\mathfrak{F}^{-1}[U(\omega)]\right]\exp(-j\omega_c t)$$

$$= \left[x_{bp}(t) * \left[\delta(t) + j\frac{1}{\pi t}\right]\right]\exp(-j\omega_c t) \quad (2.5.1\text{-}12)$$

$$= \left[x_{bp}(t) + j\hat{x}_{bp(t)}\right]\exp(-j\omega_c t),$$

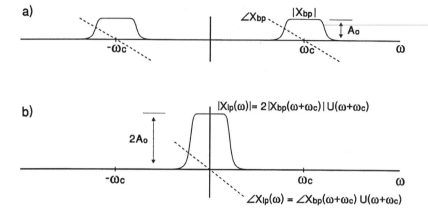

FIGURE 2.22
Magnitude and phase of (a) bandpass signal and (b) its lowpass equivalent.

where $\hat{x}_{bp}$ denotes the *Hilbert transform* of bandpass process x_{bp}. The quantity $x_{bp} + j\hat{x}_{bp}$ in Equation 2.5.1-12 is known as the *analytic signal* corresponding to x_{bp}; Equation 2.5.1-12 shows that x_{lp} is just this analytic signal translated down to baseband.

2.5.1.3 Symmetrical Bandpass Filter

The magnitude (phase) response of a *symmetrical bandpass filter* has ω_c as an axis of even (odd) symmetry. That is, the filter transfer function satisfies

$$\left|H_{bp}(j\omega_c + j\omega)\right| = \left|H_{bp}(j\omega_c - j\omega)\right|$$
$$\angle H_{bp}(j\omega_c + j\omega) = -\angle H_{bp}(j\omega_c - j\omega)$$
(2.5.1-13)

for $|\omega| < \omega_c$. Figure 2.22 illustrates magnitude and phase functions that satisfy Equation 2.5.1-13.

A symmetrical bandpass filter has a relatively simple impulse response. As can be seen from Equation 2.5.1-11, its lowpass equivalent $H_{lp}(j\omega)$ has an even magnitude and an odd phase response. This implies that lowpass equivalent $h_{lp} \equiv \mathcal{F}^{-1}[H_{lp}]$ is real valued, and the bandpass filter impulse response has the form

$$h_{bp}(t) = h_c(t)\cos\omega_c t.$$
(2.5.1-14)

MODELING THE PHASE-LOCKED LOOP

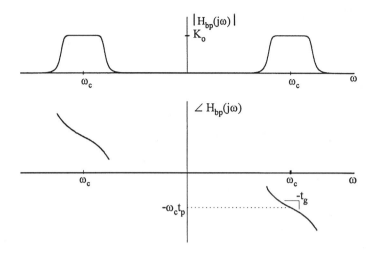

FIGURE 2.23
Phase delay t_p and Group delay t_g of a bandpass system.

The converse is also true; if $h_{bp}(t)$ has the form given by Equation 2.5.1-14, then the filter is symmetrical.

2.5.1.4 Phase and Group Delays of a Bandpass System

Figure 2.23 depicts the magnitude and phase response of a bandpass system with characteristics that are of interest in this subsection. Within the passband of this system, the magnitude response is almost flat and the phase response is almost linear. In addition to a bandwidth specification, times t_p and t_g can be used to characterize this system. These quantities are the subject of this subsection.

The quantities t_p and t_g are known as the *phase delay* and *group delay*, respectively, of the bandpass system (alternate terminology exists: also, t_p and t_g are known as the *carrier delay* and *envelope delay*, respectively, of the system). At its center frequency ω_c, the frequency-normalized phase of the system is $-t_p$, and $-t_g$ is the slope of the phase characteristic. In terms of the system transfer function, t_p and t_g are expressed as

$$t_p \equiv -\frac{\angle H_{bp}(j\omega_c)}{\omega_c}$$

$$t_g \equiv -\frac{d}{d\omega}\left[\angle H_{bp}(j\omega)\right]\bigg|_{\omega = \omega_c} \quad (2.5.1\text{-}15)$$

Both of these quantities have units of seconds.

A simple approximation of this bandpass system transfer function is useful in many applications. Since phase is nearly linear, it is possible to express

$$\angle H_{bp}(j\omega) \approx -\omega_c t_p - (\omega - \omega_c) t_g \qquad (2.5.1\text{-}16)$$

for ω within the passband around ω_c. Likewise, since the magnitude is almost constant, the approximation

$$|H_{bp}(j\omega)| \approx K_o \qquad (2.5.1\text{-}17)$$

holds for ω within the passband. Hence, within the passband centered at ω_c, the simple approximation

$$H_{bp}(j\omega) \approx K_o \exp\left[-j\omega_c t_p - j[\omega - \omega_c] t_g\right] \qquad (2.5.1\text{-}18)$$

follows from Equations 2.5.1-16 and 2.5.1-17. Finally, the lowpass equivalent of this system can be approximated as

$$H_{lp}(j\omega) \approx 2K_o \exp\left[-j\omega_c t_p - j\omega t_g\right]. \qquad (2.5.1\text{-}19)$$

Important properties of t_p and t_g can be obtained by considering the result of applying bandpass signal

$$x_{bp}(t) = x_d(t)\cos\omega_c t \qquad (2.5.1\text{-}20)$$

to the system. Assume that x_{bp} fits within the filter passband; that is, $X_{bp}(j\omega) \equiv \mathfrak{F}[x_{bp}]$ is approximately zero outside of the nearly flat passband of H_{bp}. The lowpass equivalent of the filter output can be approximated (see Equation 2.5.1-34) as

$$Y_{lp}(j\omega) = \frac{1}{2} H_{lp}(j\omega) X_{lp}(j\omega)$$
$$\approx \frac{1}{2}\left[2K_o \exp\left[-j(\omega_c t_p + \omega t_g)\right] X_d(j\omega)\right], \qquad (2.5.1\text{-}21)$$

where $X_{lp}(j\omega) = X_d(j\omega) \equiv \mathfrak{F}[x_d]$. In the time domain, the inverse of Equation 2.5.1-21 is

MODELING THE PHASE-LOCKED LOOP

$$y_{lp}(t) \approx K_o \exp(-j\omega_c t_p) x_d(t - t_g), \qquad (2.5.1\text{-}22)$$

and the bandpass output of the filter is approximated as

$$\begin{aligned} y_{bp}(t) &= \text{Re}[y_{lp}(t)\exp(j\omega_c t)] \\ &\approx K_o x_d(t - t_g)\cos\omega_c[t - t_p]. \end{aligned} \qquad (2.5.1\text{-}23)$$

Comparison of Equations 2.5.1-20 and 2.5.1-23 reveals why t_p and t_g are known as the system carrier delay and envelope delay, respectively.

2.5.1.5 Bandpass Input/Output

Apply signal x_{bp}, described by Equation 2.5.1-2, to the filter h_{bp}, described by Equation 2.5.1-3, to obtain the bandpass output y_{bp}. Clearly, y_{bp} can be obtained by convolving x_{bp} and h_{bp}. However, this is a messy and laborious operation involving the products of several trigonometric functions. As outlined below, a much easier approach to this problem utilizes lowpass equivalent functions.

Let y_{lp} denote the lowpass equivalent of the filter bandpass output. As shown in this subsection, y_{lp} can be computed as

$$y_{lp}(t) = \frac{1}{2} x_{lp} * h_{lp}, \qquad (2.5.1\text{-}24)$$

a computation that uses only lowpass functions. Of course, once Equation 2.5.1-24 is used to obtain y_{lp}, the filter bandpass output can be computed as

$$y_{bp} = \text{Re}[y_{lp}(t)e^{j\omega_c t}]. \qquad (2.5.1\text{-}25)$$

This simplified approach to bandpass input/output computations is summarized by Figure 2.24.

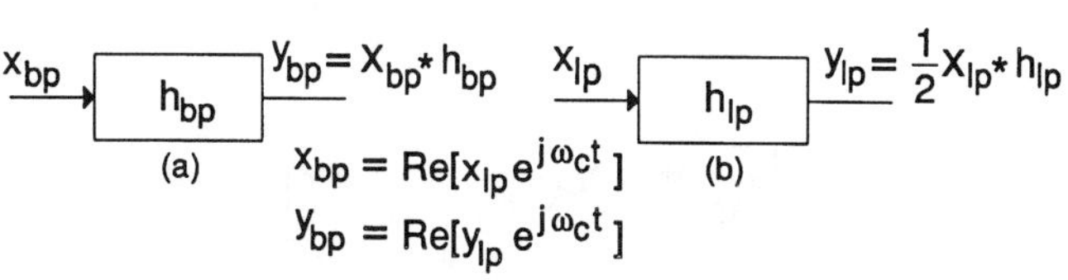

FIGURE 2.24
Input/output relationships for (a) bandpass functions and (b) lowpass functions.

The derivation of Equation 2.5.1-24 is straightforward. First, note that

$$y_{bp} = x_{bp} * h_{bp}$$
$$= [x_c(t)\cos\omega_c t - x_s(t)\sin\omega_c t] * [h_c(t)\cos\omega_c t - h_s(t)\sin\omega_c t], \quad (2.5.1\text{-}26)$$

a computation requiring the convolution of four bandpass functions. The first of these convolutions is

$$[x_c(t)\cos\omega_c t] * [h_c(t)\cos\omega_c t]$$

$$= \int_{-\infty}^{\infty} x_c(\tau)\cos\omega_c\tau [h_c(t-\tau)\cos\omega_c(t-\tau)]\, d\tau$$

$$= \int_{-\infty}^{\infty} x_c(\tau) h_c(t-\tau) \frac{1}{2}[\cos\omega_c t + \cos\omega_c(t-2\tau)]\, d\tau \quad (2.5.1\text{-}27)$$

$$= \frac{1}{2}\left[\int_{-\infty}^{\infty} x_c(\tau) h_c(t-\tau)\, d\tau\right]\cos\omega_c t$$

$$+ \frac{1}{2}\int_{-\infty}^{\infty} x_c(\tau) h_c(t-\tau)\cos\omega_c(t-2\tau)\, d\tau.$$

Now, the second integral on the right side of Equation 2.5.1-27 is very small compared to the first integral, and it can be neglected. To see this, use

$$\cos\omega_c(t-2\tau) = \cos\omega_c t \cos 2\omega_c\tau + \sin\omega_c t \sin 2\omega_c\tau \quad (2.5.1\text{-}28)$$

to write

$$\int_{-\infty}^{\infty} x_c(\tau) h_c(t-\tau)\cos\omega_c(t-2\tau)\, d\tau =$$

$$\left[\int_{-\infty}^{\infty}[x_c(\tau)\cos 2\omega_c\tau] h_c(t-\tau)\, d\tau\right]\cos\omega_c t \quad (2.5.1\text{-}29)$$

$$+ \left[\int_{-\infty}^{\infty}[x_c(\tau)\sin 2\omega_c\tau] h_c(t-\tau)\, d\tau\right]\sin\omega_c t.$$

Both convolutions on the right-hand side of Equation 2.5.1-29 are very small; they represent the response of lowpass $h_c(t)$ to a bandpass signal

MODELING THE PHASE-LOCKED LOOP

centered at $2\omega_c$. Hence, the second integral on the right-hand side of Equation 2.5.1-27 can be approximated as

$$\int_{-\infty}^{\infty} x_c(\tau) h_c(t-\tau) \cos\omega_c(t-2\tau) d\tau \approx 0, \qquad (2.5.1\text{-}30)$$

a result that leads to

$$[x_c(t)\cos\omega_c t] * [h_c(t)\cos\omega_c t] \approx \frac{1}{2}[x_c(t) * h_c(t)]\cos\omega_c t. \qquad (2.5.1\text{-}31)$$

The remaining three convolutions on the right-hand side of Equation 2.5.1-26 can be approximated in a similar manner. They are summarized here as

$$[x_s \sin\omega_c t] * [h_s \sin\omega_c t] \approx -\frac{1}{2}[x_s * h_s]\cos\omega_c t$$

$$[x_c \cos\omega_c t] * [h_s \sin\omega_c t] \approx \frac{1}{2}[x_c * h_s]\sin\omega_c t \qquad (2.5.1\text{-}32)$$

$$[x_s \sin\omega_c t] * [h_c \cos\omega_c t] \approx \frac{1}{2}[x_s * h_c]\sin\omega_c t.$$

Equation 2.5.1-26 can be simplified by using these approximations. With the use of Equations 2.5.1-31 and 2.5.1-32, Equation 2.5.1-26 can be written as

$$y_{bp} = \frac{1}{2}[x_c * h_c - x_s * h_s]\cos\omega_c t - \frac{1}{2}[x_s * h_c + x_c * h_s]\sin\omega_c t. \qquad (2.5.1\text{-}33)$$

Finally, this leads to the desired conclusion that

$$y_{lp} = \frac{1}{2}[x_c * h_c - x_s * h_s] + j\frac{1}{2}[x_s * h_c + x_c * h_s]$$

$$= \frac{1}{2}[(x_c + jx_s) * (h_c + jh_s)] \qquad (2.5.1\text{-}34)$$

$$y_{lp} = \frac{1}{2} x_{lp} * h_{lp},$$

a result that is summarized by Figure 2.24.

The response of a bandpass filter can be calculated simply by using Equations 2.5.1-34 and 2.5.1-25. As an example, consider the bandpass filter depicted by Figure 2.21. Assume that this filter has an input given by

$$x_{bp}(t) = x_c(t)\cos\omega_c t, \qquad (2.5.1\text{-}35)$$

where

$$x_c(t) = \begin{cases} 1, & 0 \le t \le t_o \\ 0, & \text{elsewhere} \end{cases}. \qquad (2.5.1\text{-}36)$$

In many applications, this filter is designed so that $1/LC \gg (R/2L)^2$. Under this condition, the frequency ω_c is approximately equal to $1/\sqrt{LC}$, and the filter is narrowband. This narrowband condition is assumed to hold here. As can be seen from Equation 2.5.1-5, the narrowband assumption implies that $h_c(t) \gg h_s(t)$ for all time so that

$$h_{lp}(t) \approx 2\alpha_o \exp(-\alpha_o t) U(t), \qquad (2.5.1\text{-}37)$$

and a symmetrical filter can be assumed. Apply Equation 2.5.1-34 to the last two equations and obtain

$$y_{lp}(t) = \begin{cases} 1-\exp(-\alpha_o t), & 0 \le t \le t_o \\ [\exp(\alpha_o t_o)-1]\exp(-\alpha_o t), & t_o < t \end{cases}. \qquad (2.5.1\text{-}38)$$

for the lowpass equivalent of the output. Finally, the bandpass output is obtained by using Equation 2.5.1-25 to write

$$y_{bp}(t) = \begin{cases} [1-\exp(-\alpha_o t)]\cos\omega_c t, & 0 \le t \le t_o \\ [\exp(\alpha_o t_o)-1]\exp(-\alpha_o t)\cos\omega_c t, & t_o < t \end{cases}. \qquad (2.5.1\text{-}39)$$

Figure 2.25 depicts an example plot of the response described by Equation 2.5.1-39 when $\alpha_o = 31.831$, $\omega_c = 1000$, and $t_0 = 0.079$.

APPENDIX 2.5.2   NARROWBAND NOISE

Let $\eta(t)$ denote a zero-mean, stationary Gaussian noise process with autocorrelation function $R_\eta(\tau)$. Process $\eta(t)$ is said to be *narrowband* if

MODELING THE PHASE-LOCKED LOOP

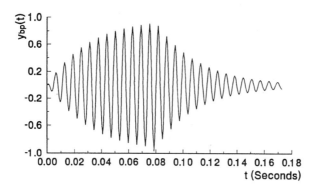

FIGURE 2.25
Response of LC bandpass filter to a gated sinusoid.

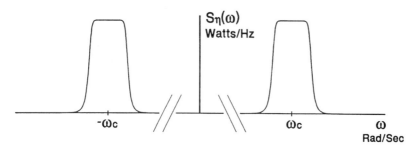

FIGURE 2.26
Example spectrum of narrowband noise.

its spectral density $S_\eta(\omega)$ is zero except for a narrow band around the frequencies $\pm\omega_c$. Spectral density $S_\eta(\omega)$ satisfies the properties

$$S_\eta(\omega) \geq 0$$
$$S_\eta(\omega) = S_\eta(-\omega) \qquad (2.5.2\text{-}1)$$

for all ω. Figure 2.26 depicts an example spectrum of a narrowband process.

Sample functions of η can be written as

$$\eta(t) = \eta_c(t)\cos\omega_c t - \eta_s(t)\sin\omega_c t, \qquad (2.5.2\text{-}2)$$

where η_c and η_s are real-valued, lowpass Gaussian processes with autocorrelation functions $R_{\eta_c}(\tau)$ and $R_{\eta_s}(\tau)$, respectively. Often, lowpass processes η_c and η_s are referred to as the quadrature components of the noise η. That η is narrowband implies that its quadrature components have very small spectra for ω larger than some ω_1, where $0 < \omega_1 \ll \omega_c$.

To avoid confusion when reviewing the engineering literature on narrowband noise, the reader should remember that different authors use slightly different definitions for the cross correlation of jointly stationary, real-valued random processes x(t) and y(t). As used in this book (and most of the recent PLL literature), the cross correlation of x and y is defined as $R_{xy}(\tau) \equiv E[x(t + \tau)y(t)]$. However, when defining R_{xy}, some authors shift (by τ) the time variable of the function y instead of the function x. Fortunately, this possible discrepancy is accounted for easily when comparing the work of different authors.

2.5.2.1 Relationships Between Correlations R_η, R_{η_c}, R_{η_s}, and $R_{\eta_c \eta_s}$

Let the Hilbert transform of a time–domain function $\eta(t)$ be denoted in the usual way by the use of a circumflex; that is, $\hat{\eta}(t)$ denotes the Hilbert transform of $\eta(t)$. Then, it is easy to show that (see Papoulis)[20]

$$R_{\eta\hat{\eta}}(\tau) \equiv E[\eta(t+\tau)\hat{\eta}(t)] = -\hat{R}_\eta(\tau)$$

$$R_{\hat{\eta}\eta}(\tau) \equiv E[\hat{\eta}(t+\tau)\eta(t)] = \hat{R}_\eta(\tau)$$

$$R_{\hat{\eta}\eta}(0) = R_{\eta\hat{\eta}}(0) = 0 \qquad (2.5.2\text{-}3)$$

$$R_{\hat{\eta}}(\tau) = R_\eta(\tau).$$

Also, the Hilbert transform of the noise signal η can be expressed as

$$\hat{\eta}(t) = \overline{\eta_c(t)\cos\omega_c t - \eta_s(t)\sin\omega_c t}$$

$$= \eta_c(t)\overline{\cos\omega_c t} - \eta_s(t)\overline{\sin\omega_c t} \qquad (2.5.2\text{-}4)$$

$$= \eta_c(t)\sin\omega_c t + \eta_s(t)\cos\omega_c t$$

This result follows from the fact that ω_c is much higher than any frequency component in η_c or η_s so that the Hilbert transform is only applied to the high-frequency sinusoidal functions.

MODELING THE PHASE-LOCKED LOOP

The quadrature components can be expressed in terms of η and $\hat{\eta}$. This can be done by solving Equations 2.5.2-2 and 2.5.2-4 for

$$\eta_c(t) = \eta(t)\cos\omega_c t + \hat{\eta}(t)\sin\omega_c t$$
$$\eta_s(t) = \hat{\eta}(t)\cos\omega_c t - \eta(t)\sin\omega_c t. \quad (2.5.2\text{-}5)$$

It is easy to compute the autocorrelation of the quadrature components. Use Equation 2.5.2-5 and compute the autocorrelation

$$\begin{aligned}R_{\eta_c}(\tau) &= E[\eta_c(t)\eta_c(t+\tau)] \\ &= E[\eta(t)\eta(t+\tau)]\cos\omega_c t \cos\omega_c(t+\tau) \\ &\quad + E[\hat{\eta}(t)\eta(t+\tau)]\sin\omega_c t \cos\omega_c(t+\tau) \quad (2.5.2\text{-}6)\\ &\quad + E[\eta(t)\hat{\eta}(t+\tau)]\cos\omega_c t \sin\omega_c(t+\tau) \\ &\quad + E[\hat{\eta}(t)\hat{\eta}(t+\tau)]\sin\omega_c t \sin\omega_c(t+\tau).\end{aligned}$$

This last result can be simplified by using Equation 2.5.2-3 to obtain

$$\begin{aligned}R_{\eta_c}(\tau) &= R_\eta(\tau)[\cos\omega_c t \cos\omega_c(t+\tau) + \sin\omega_c t \sin\omega_c(t+\tau)] \\ &\quad + \hat{R}_\eta(\tau)[\cos\omega_c t \sin\omega_c(t+\tau) - \sin\omega_c t \cos\omega_c(t+\tau)]\end{aligned}$$

or

$$R_{\eta_c}(\tau) = R_\eta(\tau)\cos\omega_c\tau + \hat{R}_\eta(\tau)\sin\omega_c\tau. \quad (2.5.2\text{-}7)$$

R_{η_s} can be computed in a manner identical to that used to obtain Equation 2.5.2-7; the result

$$R_{\eta_c}(\tau) = R_{\eta_s}(\tau) \quad (2.5.2\text{-}8)$$

follows in this manner. Hence, the autocorrelation functions of the quadrature components are equal.

Some properties of the quadrature components η_c and η_s can be obtained from Equations 2.5.2-5 through 2.5.2-8. First, the quadrature components are stationary. Next, they have equal spectral density functions. Finally, the average power in each quadrature component is equal to the average power in η. This result follows by setting $\tau = 0$ in Equation 2.5.2-7, and realizing that $R_{\eta_c}(0) = R_{\eta_s}(0) = R_\eta(0)$ represents average power.

It is easy to compute the cross correlation of the quadrature components. From Equation 2.5.2-5 it follows that

$$R_{\eta_c \eta_s}(\tau) = E[\eta_c(t+\tau)\eta_s(t)]$$

$$= E[\eta(t+\tau)\hat{\eta}(t)]\cos\omega_c(t+\tau)\cos\omega_c t$$

$$- E[\eta(t+\tau)\eta(t)]\cos\omega_c(t+\tau)\sin\omega_c t \qquad (2.5.2\text{-}9)$$

$$+ E[\hat{\eta}(t+\tau)\hat{\eta}(t)]\sin\omega_c(t+\tau)\cos\omega_c t$$

$$- E[\hat{\eta}(t+\tau)\eta(t)]\sin\omega_c(t+\tau)\sin\omega_c t.$$

This result can be simplified by using Equation 2.5.2-3 to obtain

$$R_{\eta_c \eta_s}(\tau) = R_\eta(\tau)\left[-\sin\omega_c t \cos\omega_c(t+\tau) + \cos\omega_c t \sin\omega_c(t+\tau)\right]$$

$$- \hat{R}_\eta(\tau)\left[\cos\omega_c t \cos\omega_c(t+\tau) + \sin\omega_c t \sin\omega_c(t+\tau)\right],$$

or

$$R_{\eta_c \eta_s}(\tau) = R_\eta(\tau)\sin\omega_c\tau - \hat{R}_\eta(\tau)\cos\omega_c\tau. \qquad (2.5.2\text{-}10)$$

In a manner identical to that used to obtain Equation 2.5.2-10, it can be shown that $-R_{\eta_s \eta_c}$ is given by the right-hand side of Equation 2.5.2-10. Hence, the cross-correlation functions have the property that

$$R_{\eta_c \eta_s}(\tau) = -R_{\eta_s \eta_c}(\tau). \qquad (2.5.2\text{-}11)$$

The cross correlation of the quadrature components is an odd function of τ. To see this, first note that

$$R_{\eta_c \eta_s}(\tau) = E[\eta_c(t+\tau)\eta_s(t)]$$

$$= E[\eta_c(t)\eta_s(t-\tau)] \qquad (2.5.2\text{-}12)$$

$$= R_{\eta_s \eta_c}(-\tau),$$

a result that follows from the fact that the quadrature components are real valued. Next, compare Equations 2.5.2-11 and 2.5.2-12 and obtain

$$R_{\eta_s \eta_c}(\tau) = -R_{\eta_s \eta_c}(-\tau), \qquad (2.5.2\text{-}13)$$

which implies that the cross correlation $R_{\eta_s \eta_c}$ is an odd function of τ. Finally, the fact that this cross correlation is odd implies that $R_{\eta_s \eta_c}(0) = 0$; taken at the same time, the samples of η_c and η_s are uncorrelated and independent.

The autocorrelation R_η of the narrowband noise can be expressed in terms of the autocorrelation and cross correlation of the quadrature components η_c and η_s. This important result follows from using Equations 2.5.2-7 and 2.5.2-10 in

$$R_{\eta_c}(\tau)\cos\omega_c\tau + R_{\eta_c\eta_s}(\tau)\sin\omega_c\tau$$

$$= \left[R_\eta(\tau)\cos\omega_c\tau + \hat{R}_\eta(\tau)\sin\omega_c\tau\right]\cos\omega_c\tau \qquad (2.5.2\text{-}14)$$

$$+ \left[R_\eta(\tau)\sin\omega_c\tau - \hat{R}_\eta(\tau)\cos\omega_c\tau\right]\sin\omega_c\tau.$$

However, R_η results from simplification of the right-hand side of Equation 2.5.2-14, and the desired relationship

$$R_\eta(\tau) = R_{\eta_c}(\tau)\cos\omega_c\tau + R_{\eta_c\eta_s}(\tau)\sin\omega_c\tau \qquad (2.5.2\text{-}15)$$

follows.

2.5.2.2 Symmetrical Bandpass Processes

Narrowband process $\eta(t)$ is said to be a *symmetrical bandpass process* if

$$S_\eta(\omega + \omega_c) = S_\eta(-\omega + \omega_c) \qquad (2.5.2\text{-}16)$$

for $0 < \omega < \omega_c$. Such a bandpass process has its center frequency ω_c as an axis of local symmetry. In what follows, it is shown that this local symmetry attribute is equivalent to the condition $R_{\eta_c\eta_s}(\tau) = 0$ for all τ.

The desired result follows from inspecting the Fourier transform of Equation 2.5.2-10; this transform is the cross spectrum of the quadrature components, and it vanishes when the narrowband process has spectral symmetry as defined by Equation 2.5.2-16. To compute this cross spectrum, first note the Fourier transform pairs

$$R_\eta(\tau) \leftrightarrow S_\eta(\omega)$$
$$\hat{R}_\eta(\tau) \leftrightarrow -j\mathrm{Sgn}(\omega)S_\eta(\omega), \qquad (2.5.2\text{-}17)$$

where

$$\mathrm{Sgn}(\omega) \equiv \begin{cases} +1 & \text{for } \omega > 0 \\ -1 & \text{for } \omega < 0 \end{cases}. \qquad (2.5.2\text{-}18)$$

Now, use Equation 2.5.2-17 to obtain the Fourier transform pairs

$$R_\eta(\tau)\sin\omega_c t \leftrightarrow \frac{1}{2j}\left[S_\eta(\omega-\omega_c) - S_\eta(\omega+\omega_c)\right]$$

$$\hat{R}_\eta(\tau)\cos\omega_c t \leftrightarrow \begin{array}{c} -j\dfrac{1}{2}\mathrm{Sgn}(\omega-\omega_c)S_\eta(\omega-\omega_c) \\ -j\dfrac{1}{2}\mathrm{Sgn}(\omega+\omega_c)S_\eta(\omega+\omega_c) \end{array} \qquad (2.5.2\text{-}19)$$

Finally, use this last equation and Equation 2.5.2-10 to compute the cross spectrum

$$S_{\eta_c\eta_s}(\omega) = \mathfrak{F}[R_{\eta_c\eta_s}(\tau)]$$

$$= \frac{1}{2j}S_\eta(\omega-\omega_c)[1-\mathrm{Sgn}(\omega-\omega_c)] \qquad (2.5.2\text{-}20)$$

$$-\frac{1}{2j}S_\eta(\omega+\omega_c)[1+\mathrm{Sgn}(\omega+\omega_c)].$$

Figure 2.27 depicts example plots useful for visualizing important properties of Equation 2.5.2-20. From parts (b) and (c) of this plot, note that the products on the right-hand side of Equation 2.5.2-20 are low-pass processes. Then it is easily seen that

$$S_{\eta_c\eta_s}(\omega) = \begin{cases} 0 & , \quad \omega > \omega_c \\ -j[S_\eta(\omega-\omega_c) - S_\eta(\omega+\omega_c)] & , \quad -\omega_c < \omega < \omega_c \\ 0 & , \quad \omega < -\omega_c \end{cases} \qquad (2.5.2\text{-}21)$$

Finally, note that $S_{\eta_c\eta_s}(\omega) = 0$ is equivalent to the narrowband process η satisfying the symmetry Condition 2.5.2-16. Since the cross spectrum

MODELING THE PHASE-LOCKED LOOP

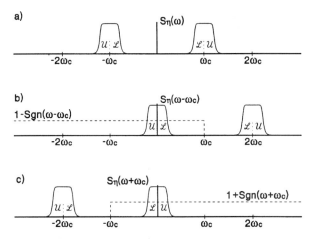

FIGURE 2.27
Symmetrical bandpass processes have $\eta_c(t_1)$ and $\eta_s(t_2)$ uncorrelated for all t_1 and t_2.

is the Fourier transform of the cross correlation, this last statement implies that, for *all* t_1 and t_2 (not just $t_1 = t_2$), $\eta_c(t_1)$ and $\eta_s(t_2)$ are uncorrelated if and only if Equation 2.5.2-16 holds. On Figure 2.27, symmetry implies that the spectral components labeled with $\mathcal{U}$ can be obtained from those labeled with $\mathcal{L}$ by a simple folding operation.

System analysis is simplified greatly if the noise encountered has a symmetrical spectrum. Under these conditions, the quadrature components are uncorrelated, and Equation 2.5.2-15 simplifies to

$$R_\eta(\tau) = R_{\eta_c}(\tau)\cos\omega_c\tau. \quad (2.5.2\text{-}22)$$

Also, the spectrum S_η of the noise is obtained easily by scaling and translating $S_{\eta_c} \equiv \mathfrak{F}[R_{\eta_c}]$ as shown by

$$S_\eta(\omega) = \frac{1}{2}\left[S_{\eta_c}(\omega - \omega_c) + S_{\eta_c}(\omega + \omega_c)\right]. \quad (2.5.2\text{-}23)$$

This result follows directly by taking the Fourier transform of Equation 2.5.2-22.

APPENDIX 2.61 EVALUATION OF $E[\cos\theta_n]$ AND $E[\sin\theta_n]$ FOR THE GAUSSIAN NOISE CASE

Consider signal $r(t)$ given by Equation 2.6-1, a sinusoid embedded in additive, zero-mean Gaussian noise. This signal can be expressed as

Equation 2.6-3, a signal in magnitude and phase form. The envelope and phase of this bandpass signal are random processes that have been analyzed by many authors (for example, see Section 10.6 of Peebles).[21] The density function of phase noise θ_n is used to obtain expressions for $E[\cos\theta_n]$ and $E[\sin\theta_n]$, the main results in this appendix.

As given by Peebles and others, phase noise θ_n, $-\pi < \theta_n \leq \pi$, is described by the density function

$$f(\theta_n) = \frac{1}{2\pi} \exp[-SNR_i]$$

$$+ \frac{\cos\theta_n}{\sqrt{\pi/SNR_i}} \left(\frac{1 + \text{erf}[\sqrt{SNR_i} \cos\theta_n]}{2} \right) \exp[-(SNR_i)\sin^2\theta_n] \quad (2.6.1\text{-}1)$$

where SNR_i is the input signal-to-noise ratio given by Equation 2.6-16, and

$$\text{erf}(x) = \frac{2}{\sqrt{\pi}} \int_0^x \exp(-u^2)\, du \quad (2.6.1\text{-}2)$$

is the error function (the reader should note that alternate definitions for erf(x) appear in the literature). Density 2.6.1-1 is an even function of θ_n; hence, the phase noise has zero mean.

The zero-mean nature of θ_n simplifies calculation of $E[\cos\theta_n]$ and $E[\sin\theta_n]$. First, the odd nature of $\sin(k\theta_n)$ implies

$$E[\sin(k\theta_n)] = 0, \quad (2.6.1\text{-}3)$$

and the even symmetry of $\cos(k\theta_n)$ implies

$$E[\cos(k\theta_n)] = \int_{-\pi}^{\pi} \cos(k\theta_n) f(\theta_n)\, d\theta_n$$

$$= 2\int_0^{\pi} \cos(k\theta_n) f(\theta_n)\, d\theta_n. \quad (2.6.1\text{-}4)$$

Now, substitute Equation 2.6.1-1 into this last result to obtain

MODELING THE PHASE-LOCKED LOOP

$E[\cos(k\theta_n)]$

$$= 2\int_0^\pi \frac{1}{2\pi}\exp(-SNR_i)\cos(k\theta_n)\,d\theta_n$$

$$+ 2\sqrt{\frac{SNR_i}{\pi}} \int_0^\pi \exp[-(SNR_i)\sin^2\theta_n]\cos(\theta_n) \qquad (2.6.1\text{-}5)$$

$$\cdot \cos(k\theta_n)\left[\frac{1+\text{erf}[\sqrt{SNR_i}\,\cos(\theta_n)]}{2}\right]d\theta_n.$$

On the right-hand side of Equation 2.6.1-5, the first integral is zero. The second integral can be simplified by using

$$\cos(\theta_n)\cos(k\theta_n) = \frac{1}{2}[\cos(k+1)\theta_n + \cos(k-1)\theta_n] \qquad (2.6.1\text{-}6)$$

$$\frac{1+\text{erf}[\sqrt{SNR_i}\,\cos\theta_n]}{2}$$

$$= \frac{1}{\sqrt{\pi}}\int_{-\sqrt{SNR_i}\cos\theta_n}^{\infty} \exp(-u^2)\,du$$

$$= \frac{1}{\sqrt{\pi}}\int_0^\infty \exp[-(u-\sqrt{SNR_i}\,\cos\theta_n)^2]\,du \qquad (2.6.1\text{-}7)$$

$$= \frac{1}{\sqrt{\pi}}\exp[-SNR_i(1-\sin^2\theta_n)]$$

$$\cdot \int_0^\infty \exp[-u^2 + (2\sqrt{SNR_i}\,\cos\theta_n)u]\,du$$

to obtain

$E[\cos(k\theta_n)]$

$$= \sqrt{SNR_i}\,\exp(-SNR_i)$$

$$\cdot \int_0^\infty \exp(-u^2)\frac{1}{\pi}\int_0^\pi [\cos(k+1)\theta_n + \cos(k-1)\theta_n] \qquad (2.6.1\text{-}8)$$

$$\cdot \exp[2(\sqrt{SNR_i}\,\cos\theta_n)u]\,d\theta_n\,du.$$

But, the inner integral can be expressed in terms of modified Bessel functions; this observation leads to

$$E[\cos(k\theta_n)] = \sqrt{SNR_i}\, e^{-SNR_i}$$
$$\times \int_0^\infty e^{-u^2}\left[I_{k+1}\left(2\sqrt{SNR_i}\, u\right) + I_{k-1}\left(2\sqrt{SNR_i}\, u\right)\right] du. \tag{2.6.1-9}$$

Finally, the integrals in Equation 2.6.1-9 are tabulated (see 11.4.31, page 487 of Abramowitz and Stegun);[22] these tabulated integrals can be used to write

$$E[\cos(k\theta_n)] = \sqrt{\frac{\pi}{4} SNR_i}\, \exp\left(-\frac{SNR_i}{2}\right)$$
$$\cdot \left[I_{(k+1)/2}\left(\frac{SNR_i}{2}\right) + I_{(k-1)/2}\left(\frac{SNR_i}{2}\right)\right]. \tag{2.6.1-10}$$

This formula is used in Equation 2.6-18, a series representation of the equivalent phase detector characteristic.

Chapter 3

LINEAR ANALYSIS OF COMMON FIRST- AND SECOND-ORDER PLLs

Phase-lock technology has reached a high level of maturity. In addition to military and space systems, the technology has appeared in a wide range of consumer electronics, where it has produced improvements in performance, reliability, and cost. Undoubtedly, this trend will continue as phase-lock technology becomes more ubiquitous over time.

Many of the PLLs in current applications are modeled by their designers as either first- or second-order loops. There are several reasons for this. It is commonly perceived that many applications are well served by a PLL no higher than second order. Also, the global nonlinear behavior of third- and higher-order PLL is not well understood. Finally, a wealth of design information is available for first- and second-order loops, and the operation of these loops is documented extensively in the literature.

This chapter serves the design philosophy that first- and second-order loops are popular enough that separate coverage is warranted of their elementary linear theory. It contains material which is consulted most often by practicing design engineers during the application of phase-lock technology. At the end of each section, a summary is provided of the most important formulas developed in the section. Design engineers should remember that the extensive body of linear control theory can be applied to analyze the PLL model once it is linearized.

In Section 3.1, an analysis is given of the linear model for the first-order PLL. The linear model for the second-order PLL containing a perfect integrator loop filter is analyzed in Section 3.2, and Section 3.3 accomplishes a similar task for the second-order PLL containing a loop filter based on an imperfect integrator. The elementary linear theory of PLLs is covered extensively in several of the texts listed in the bibliography; to gain a wider perspective, the reader should consult this reference material.

3.1 THE FIRST-ORDER PLL

The first-order PLL uses no loop filter. Instead, there is a direct connection between the phase comparator output and the VCO input. Chapter 2 presented the development of a general PLL model; this model becomes first order when $F(s) \equiv 1$ is used. The closed-loop transfer function is developed in this section for the linearized first-order PLL. This transfer function is used to analyze the transient and steady-state responses of the loop to various inputs. Finally, it is used to develop a simple expression for the noise-equivalent bandwidth B_L.

3.1.1 Closed-Loop Transfer Function

Figure 2.5 with $F \equiv 1$ illustrates the Laplace domain model of the first-order PLL. From Equation 2.3-4, the closed-loop transfer function of this PLL is

$$H(s) = \frac{G}{s+G}, \qquad (3.1\text{-}1)$$

where G denotes a closed-loop gain factor. This result can be used with Equation 2.3-6 to determine the VCO phase θ_2 for an arbitrary input θ_1. Coupled with Equation 2.3-7, it can be used to determine the closed-loop phase error for an arbitrary input.

For $G > 0$, the pole of $H(s)$ lies in the left half of the s-plane, and the first-order PLL is unconditionally stable. An inspection of Equation 3.1-1 might lead to the naive and incorrect conclusion that phase lock is not possible for $G < 0$, since the pole of H would be in the right-half s-plane. As discussed in Section 2.2.1, the PLL with negative G has stable lock points that are displaced by π radians from the stable lock points that exist for positive G. Hence, without loss of generality, $G > 0$ can be assumed.

3.1.2 Transient and Steady-State Tracking Errors

The first-order PLL response is calculated easily for common inputs. Sample results are given below for the cases when the input reference is subjected to a phase step, frequency step, frequency ramp, and sinusoidal angle modulation. Of course, superposition applies for the linear model under consideration, so the results given below can be combined to produce more complicated input/output pairs.

As a first example, consider a reference subjected to a phase step. In this case, the reference is a sinusoid at frequency ω_0; the phase of this signal jumps by the constant θ_Δ so that

$$\theta_1(t) = \theta_\Delta U(t)$$
$$\Theta_1(s) = \frac{\theta_\Delta}{s}. \tag{3.1-2}$$

Along with Equation 2.3-7, use this last result to obtain

$$\Phi(s) = (1 - H(s))\Theta_1(s)$$
$$= \frac{s}{s+G}\left[\frac{\theta_\Delta}{s}\right] \tag{3.1-3}$$

as the transform of the closed-loop phase error. Hence, the response of the first-order PLL to a phase step is simply

$$\phi(t) = \mathcal{L}^{-1}[\Phi(s)]$$
$$= \theta_\Delta e^{-Gt} U(t). \tag{3.1-4}$$

Finally, note that the steady-state response of the loop to a phase step is

$$\lim_{t \to \infty} \phi(t) = 0. \tag{3.1-5}$$

Consider applying a frequency step to the first-order PLL. In this case, at $t = 0$, the reference sinusoid jumps in frequency from ω_0 to ω_i. This implies an input phase given by

$$\theta_1(t) = \omega_\Delta t U(t), \tag{3.1-6}$$

where $\omega_\Delta \equiv \omega_i - \omega_0$ denotes the input frequency jump. In a manner similar to that used to obtain Equation 3.1-4, the response of the PLL to this input is

$$\phi(t) = \frac{\omega_\Delta}{G}(1 - e^{-Gt}). \tag{3.1-7}$$

Hence, the steady-state error to a frequency step is

$$\lim_{t \to \infty} \phi(t) = \frac{\omega_\Delta}{G}. \tag{3.1-8}$$

The first-order PLL incurs a constant steady-state phase error as the result of a frequency-stepped reference.

A first-order loop cannot track a frequency ramp. In this case, the relative phase of the signal is given in the time and Laplace domains as

$$\theta_1(t) = \frac{1}{2} R t^2 U(t) \tag{3.1-9}$$

and

$$\Theta_1(s) = \frac{R}{s^3}, \tag{3.1-10}$$

respectively, where R is a constant with units of radians per second². Due to the effects of Doppler, such a signal could be received from a constant frequency transmitter aboard a vehicle moving with a constant radial acceleration of Rc/ω_i meters/sec², where ω_i denotes the frequency of the transmitted signal, and the constant c denotes the speed of light in meters per second. The closed-loop phase error for this input is

$$\phi(t) = \frac{R}{G}\left(Gt + e^{-Gt} - 1\right) U(t), \tag{3.1-11}$$

which was obtained by using techniques similar to those given by Equations 3.1-3 and 3.1-4. Finally, note that this result is unbounded as t approaches infinity, so the first-order loop cannot track a frequency-ramped reference.

As a last example of steady-state phase error calculation in a first-order PLL, consider the case involving an angle-modulated reference. Assume that a sinusoidal modulating signal is employed so that θ_1 has the form given by Equation 2.4-8, where β is the modulation index, and ω_m is the frequency of the modulating signal. Now, use the transfer function given by Equation 3.1-1 with the phase angle expression given by Equation 2.4-15; the result of this combination is

$$\phi(t) = \frac{\omega_\Delta}{G} + \beta \left[\sin \omega_m t - \frac{G}{\sqrt{G^2 + \omega_m^2}} \sin\left(\omega_m t - \operatorname{Tan}^{-1}(\omega_m/G)\right) \right]. \tag{3.1-12}$$

LINEAR ANALYSIS OF COMMON FIRST- AND SECOND-ORDER PLLS

This equation represents the steady-state tracking error in a first-order PLL with an angle-modulated reference.

Consider the steady-state phase error given by Equation 3.1-12 as a function of modulating frequency ω_m. Note that the sinusoidal portion of ϕ is small when $\omega_m \ll G$. In this case, the frequency of the external sinusoidal modulation is inside of the closed-loop bandwidth G, and the PLL tracks it with only a small error. However, the amplitude of the sinusoidal portion of ϕ is approximately β for $\omega_m \gg G$; in this case, the frequency of the external sinusoidal modulation is outside of the closed-loop bandwidth, and the PLL ignores it. Under these conditions, the PLL is locked to the reference carrier component, and it ignores the data sidebands.

3.1.3 Noise-Equivalent Bandwidth

The first-order PLL noise-equivalent bandwidth can be calculated easily with the use of Equation 2.5-27. First, form the product

$$H(s)H(-s) = \frac{G^2}{(s+G)(-s+G)}. \qquad (3.1\text{-}13)$$

Now, note that the residue of this product at the left-half plane pole $s = -G$ is simply $G/2$. From Equation 2.5-27, the first-order PLL noise-equivalent bandwidth is

$$B_L = \frac{G}{4}. \qquad (3.1\text{-}14)$$

3.1.4 Summary of the First-Order PLL Linear Model

The first-order PLL model describes a relatively simple feedback loop containing nothing but a voltage-controlled oscillator (VCO) and a phase comparator. While it is not often used in applications, the first-order model serves as a good starting point for a study of PLLs. Often, the general nonlinear PLL theory can be applied to the first-order PLL model to produce relatively simple closed-form results. For this reason, the first-order PLL model must be understood, and its linear version is the natural place to start.

Table 3.1 contains a summary of the main results developed in this section. Tabulated for easy access when using the first-order linear model, results are given on steady-state tracking errors and noise-equivalent bandwidth. These formulas should be compared with their counterparts for the second-order PLLs discussed in Sections 3.2 and

TABLE 3.1

Summary for First-Order PLL Model

Input $\theta_1(t) \Rightarrow \phi_{ss} = \lim_{t \to \infty} \phi(t)$	
$\theta_1 = \theta_\Delta U(t) \Rightarrow \phi_{ss} = 0$	$H(s) = \dfrac{G}{s+G}$
$\theta_1 = \omega_\Delta t\, U(t) \Rightarrow \phi_{ss} = \omega_\Delta/G$	$B_L = \dfrac{G}{4}$
$\theta_1 = \tfrac{1}{2} Rt^2 U(t) \Rightarrow \phi_{ss} = \infty$	

3.3. Such a comparison reveals performance improvements that can be realized by using second-order loops.

3.2 THE SECOND-ORDER PLL WITH A PERFECT INTEGRATOR

Consider the second-order PLL with the loop filter

$$F(s) = 1 + \frac{\alpha}{s}. \qquad (3.2\text{-}1)$$

As discussed in Chapter 4, this loop filter is implemented easily with modern operational amplifier technology. Note that the loop filter contains a perfect integrator. As discussed in this section and in Chapter 5, the PLL based on Equation 3.2-1 has a number of important properties that make it a very attractive choice for many applications.

3.2.1 Transfer Functions

The linear, Laplace domain model of the PLL under consideration is depicted by Figure 2.5 when loop filter F(s) is given by Equation 3.2-1. For this model, the open-loop transfer function is obtained easily; simply substitute Equation 3.2-1 into Equation 2.3-2 and write the open-loop transfer function

$$G_o(s) = G\left[\frac{s+\alpha}{s^2}\right]. \qquad (3.2\text{-}2)$$

Applying terminology from classical control theory, the PLL with open-loop transfer function Equation 3.2-2 is referred to as a *Type II* loop since $G_o(s)$ has two poles at the origin of the s-plane.

LINEAR ANALYSIS OF COMMON FIRST- AND SECOND-ORDER PLLS

From Equation 2.3-4, the closed-loop transfer function of this PLL is calculated easily as

$$H(s) = G\left[\frac{s+\alpha}{s^2 + Gs + \alpha G}\right], \qquad (3.2\text{-}3)$$

which can be written in the commonly used form

$$H(s) = \frac{2\omega_n \zeta s + \omega_n^2}{s^2 + 2\zeta \omega_n s + \omega_n^2}$$

$$\zeta = \sqrt{G/4\alpha} \qquad (3.2\text{-}4)$$

$$\omega_n = \sqrt{\alpha G}.$$

The quantities ζ and ω_n are the damping factor and natural frequency, respectively. Equation 3.2-3 can be used with Equations 2.3-6 and 2.3-7 to determine the output phase and phase error for an arbitrary input.

Classical feedback control theory provides guidance in the selection of ζ and ω_n. In general, practical designs of this PLL strive to maintain $\zeta \approx 1/\sqrt{2}$ in order to achieve a fast step response (without ringing and excessive overshoot) for a given value of ω_n. Then, by using the results given in Section 3.2.3 below, select ω_n large enough to meet specified settling time requirements.

3.2.2 Loop Stability

The poles of Equation 3.2-3 remain in the left-half s-plane for all $G > 0$; hence, this loop is unconditionally stable. This is best illustrated by the root locus plot depicted by Figure 3.1. This diagram shows the locus of the closed-loop poles as gain G goes from zero to infinity. At $G = 0$, the poles break away from the real axis at $s = 0$; at $G = 4\alpha$, they return at the value $s = -2\alpha$. For $G > 4\alpha$, the poles remain real valued; as $G \to \infty$, one pole approaches $s = -\alpha$, and the other tends to minus infinity. This type of plot can be constructed easily using graphical techniques and the known location of the open-loop poles and zeros.

3.2.3 Transient and Steady-State Tracking Errors

The transient phase error response of the second-order Type II loop under consideration is calculated easily for common inputs. First, the phase error is calculated in the Laplace domain by using

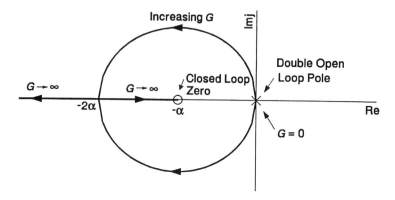

FIGURE 3.1
Root locus for a second-order PLL with a perfect integrator.

$$\Phi(s) = (1 - H(s))\Theta_1(s)$$

$$= \frac{s^2}{s^2 + 2\zeta\omega_n s + \omega_n^2}\Theta_1(s). \tag{3.2-5}$$

Then the desired results are obtained by taking the inverse Laplace transform of Equation 3.2-5.

This method was used to produce the results provided in Table 3.2. The tabulated error responses were obtained for inputs consisting of the phase step described by Equation 3.1-2, the frequency step described by Equation 3.1-6, and the frequency ramp described by Equation 3.1-9. As expected of the data in Table 3.2, the entries in the frequency step column are obtained by replacing θ_Δ by ω_Δ and integrating over [0, t] the entries in the phase step column. A similar operation yields the frequency ramp column data from the frequency step column data.

Plots of these phase error functions are depicted by Figures 3.2 through 3.4. Each figure contains a plot for the heavily overdamped case ($\zeta = 2$) and the severely underdamped case ($\zeta = 0.3$). Also, on each figure is a plot for $\zeta = 1/\sqrt{2}$, a highly desirable value of damping factor. For a fixed value of ζ, a plot similar to the ones depicted here can be used to determine a value for ω_n once the desired settling time is known.

The steady-state error response can be obtained from Table 3.2. Let $t \to \infty$ in these results to obtain

$$\lim_{t \to \infty} \phi(t) = \begin{cases} 0 & \text{for a phase step} \\ 0 & \text{for a frequency step} \\ R/\omega_n^2 & \text{for a frequency ramp} \end{cases} \tag{3.2-6}$$

TABLE 3.2
Transient Response of a Second-Order, Type II PLL Containing a Perfect Integrator

	Phase step (θ_Δ radians)	Frequency step (ω_Δ rad/sec)	Frequency ramp (R rad/sec^2)
$\zeta<1$	$\theta_\Delta \left(\cos\sqrt{1-\zeta^2}\,\omega_n t - \dfrac{\zeta}{\sqrt{1-\zeta^2}} \sin\sqrt{1-\zeta^2}\,\omega_n t \right) e^{-\zeta\omega_n t}$	$\dfrac{\omega_\Delta}{\omega_n}\left(\dfrac{1}{\sqrt{1-\zeta^2}}\sin\sqrt{1-\zeta^2}\,\omega_n t\right) e^{-\zeta\omega_n t}$	$\dfrac{R}{\omega_n^2} - \dfrac{R}{\omega_n^2}\left(\cos\sqrt{1-\zeta^2}\,\omega_n t + \dfrac{\zeta}{\sqrt{1-\zeta^2}}\sin\sqrt{1-\zeta^2}\,\omega_n t\right) e^{-\zeta\omega_n t}$
$\zeta=1$	$\theta_\Delta(1-\omega_n t)e^{-\omega_n t}$	$\dfrac{\omega_\Delta}{\omega_n}(\omega_n t)e^{-\omega_n t}$	$\dfrac{R}{\omega_n^2} - \dfrac{R}{\omega_n^2}(1+\omega_n t)e^{-\omega_n t}$
$\zeta>1$	$\theta_\Delta \left(\cosh\sqrt{\zeta^2-1}\,\omega_n t - \dfrac{\zeta}{\sqrt{\zeta^2-1}} \sinh\sqrt{\zeta^2-1}\,\omega_n t \right) e^{-\zeta\omega_n t}$	$\dfrac{\omega_\Delta}{\omega_n}\left(\dfrac{1}{\sqrt{\zeta^2-1}}\sinh\sqrt{\zeta^2-1}\,\omega_n t\right) e^{-\zeta\omega_n t}$	$\dfrac{R}{\omega_n^2} - \dfrac{R}{\omega_n^2}\left(\cosh\sqrt{\zeta^2-1}\,\omega_n t + \dfrac{\zeta}{\sqrt{\zeta^2-1}}\sinh\sqrt{\zeta^2-1}\,\omega_n t\right) e^{-\zeta\omega_n t}$

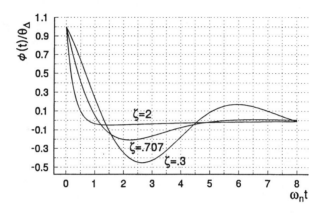

FIGURE 3.2
Normalized phase error due to a phase step input.

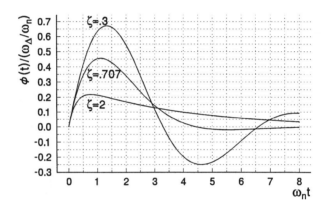

FIGURE 3.3
Normalized phase error due to a frequency step input.

for a second-order PLL with loop filter based on a perfect integrator.

It is interesting to compare Equation 3.2-6 with similar results for the first-order PLL. This comparison points out some of the performance improvements which can be achieved by using a loop filter based on a perfect integrator. For example, the PLL that contains a perfect integrator in its loop filter outperforms the first-order PLL when the reference is subjected to either a frequency step or a frequency ramp.

As a last example of steady-state tracking error in this second-order Type II PLL, consider the application of an angle-modulated reference. Assume that angle θ_1 has the form given by Equation 2.4-8, where β is the modulation index, and ω_m is the frequency of the modulating signal. In this case, the VCO steady-state relative phase and the steady-state phase error are

LINEAR ANALYSIS OF COMMON FIRST- AND SECOND-ORDER PLLS

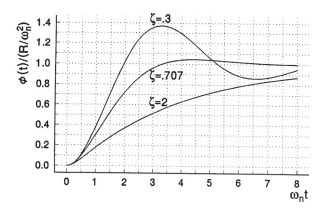

FIGURE 3.4
Normalized phase error due to a frequency ramp input.

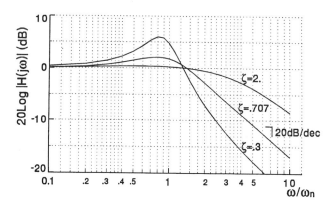

FIGURE 3.5
Frequency response of a second-order Type II PLL.

$$\theta_2(t) = \omega_\Delta t + \beta |H(j\omega_m)| \sin(\omega_m t + \angle H(j\omega_m)) \quad (3.2\text{-}7)$$

$$\phi(t) = \beta \left[\sin \omega_m t - |H(j\omega_m)| \sin(\omega_m t + \angle H(j\omega_m)) \right], \quad (3.2\text{-}8)$$

respectively, where H is given by Equation 3.2-4. Note the absence of a constant component in the phase error. This follows from the fact that this loop can track a frequency step with zero phase error.

Figure 3.5 depicts frequency response plots of $|H(j\omega)|$ for several values of damping factor ζ. This type of plot is useful for determining the magnitude of the sinusoidal component in θ_2. A similar type of plot for the relative error $|\Phi(j\omega)/\Theta_1(j\omega)|$ is given by Figure 3.6. Note that the

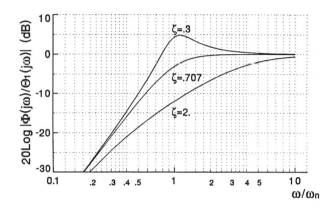

FIGURE 3.6
Error response of a second-order Type II PLL.

PLL tracks well the sinusoidal modulation of the reference as long as the frequency of this modulation is significantly less than the natural frequency ω_n. However, the tracking error increases greatly as the modulation frequency approaches ω_n.

3.2.4 Noise-Equivalent Bandwidth

Equation 2.5-27 can serve as a basis for computing the noise-equivalent bandwidth of the second-order PLL containing a perfect integrator in its loop filter. Use Equation 3.2-3 to compute

$$H(s)H(-s) = \frac{G^2(s+\alpha)(-s+\alpha)}{(s^2+Gs+\alpha G)(s^2-Gs+\alpha G)}$$

$$= \frac{G^2(-s^2+\alpha^2)}{(s-p_1)(s-p_1^*)(s+p_1)(s+p_1^*)},$$

(3.2-9)

where

$$p_1 \equiv -\frac{G}{2} + j\sqrt{\alpha G - G^2/4}$$

(3.2-10)

and p_1^* are the left-half s-plane poles of Equation 3.2-9. Now, the residue at p_1 is

$$r_1 = \left.\frac{G^2(-s^2+\alpha^2)}{(s-p_1^*)(s+p_1)(s+p_1^*)}\right|_{s=p_1} \quad (3.2\text{-}11)$$

$$= \frac{G^2(-p_1^2+\alpha^2)p_1^*}{2j\,\text{Im}[p_1](2|p_1|^2)2\,\text{Re}[p_1]}.$$

Finally, this last result yields the noise-equivalent bandwidth

$$B_L = \frac{1}{2}[r_1+r_1^*] = \text{Re}[r_1]$$

$$= \frac{G^2}{8|p_1|^2\,\text{Re}[p_1]\,\text{Im}[p_1]}\,\text{Im}\!\left[-|p_1|^2 p_1 + \alpha^2 p_1^*\right] \quad (3.2\text{-}12)$$

$$= \frac{G+\alpha}{4}$$

for a second-order PLL with a loop filter based on a perfect integrator. This result can be written as

$$B_L = \frac{\omega_n \zeta}{2}\left[1+\frac{1}{4\zeta^2}\right], \quad (3.2\text{-}13)$$

where damping factor ζ and natural frequency ω_n are given by Equation 3.2-4. Figure 3.7 depicts a plot of B_L/ω_n as a function of ζ. Note that for a fixed natural frequency ω_n, the noise-equivalent bandwidth B_L is minimized for $\zeta = 1/2$.

3.2.5 Summary for the Second-Order PLL Containing a Perfect Integrator Loop Filter

In this section, a PLL model is considered that contains a perfect integrator in its loop filter. In practical system design, this model is used widely, perhaps more than any other. The reasons for this boil down to the fact that the model offers an attractive blend of good performance and relative simplicity. In Chapters 5 and 7, nonlinear methods are used to analyze the general model of this PLL; as a prerequisite to this effort, an understanding of the material in this section is crucial.

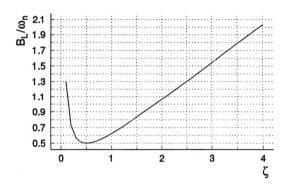

FIGURE 3.7
Noise bandwidth as a function of ζ.

To aid the system engineer in the use of this PLL model, Table 3.3 contains a summary of the important formulas developed in this section. The often-needed results on damping factor and natural frequency are included along with formulas for noise-equivalent bandwidth and steady-state tracking errors. Due to the complexity of the formula, the table does not list the steady-state tracking error when the reference is angle-modulated with a tone. However, the tracking error for this reference is given by the generic formula Formula 3.2-8, a result that applies to all linear PLL models.

3.3 THE SECOND-ORDER PLL WITH IMPERFECT INTEGRATOR

This PLL has the loop filter

$$F(s) = \frac{s+a}{s+b}, \qquad (3.3\text{-}1)$$

TABLE 3.3

Summary for Second-Order PLL Model with $F(s) = 1 + \alpha/s$.

Input $\theta_1(t) \Rightarrow \phi_{ss} = \lim_{t \to \infty} \phi(t)$		
$\theta_1 = \theta_\Delta U(t) \Rightarrow \phi_{ss} = 0$	$H(s) = \dfrac{2\zeta\omega_n s + \omega_n^2}{s^2 + 2\zeta\omega_n s + \omega_n^2}$	
$\theta_1 = \omega_\Delta t\, U(t) \Rightarrow \phi_{ss} = 0$	$B_L = \dfrac{G+\alpha}{4}$	$\omega_n = \sqrt{\alpha G}$
$\theta_1 = \tfrac{1}{2} R t^2 U(t) \Rightarrow \phi_{ss} = R/\omega_n^2$		$\zeta = \sqrt{G/4\alpha}$

LINEAR ANALYSIS OF COMMON FIRST- AND SECOND-ORDER PLLS

where a and b are constants with a > b > 0 in application. As discussed in Chapter 4, the advantage of this loop filter is its simplicity: up to a multiplicative scalar gain factor, it can be implemented with passive components. Often, these components are selected to make b small (relative to G) and a/b large in an effort to approximate the perfect integrator case discussed in Section 3.2. For this reason, in the PLL literature, it is often stated that this PLL contains an *imperfect integrator*.

3.3.1 Transfer Functions

The linear model of this PLL is depicted by Figure 2.5 with F(s) given by Equation 3.3-1. This model has the open-loop transfer function

$$G_0(s) = G\frac{(s+a)}{s(s+b)}, \qquad (3.3\text{-}2)$$

a result obtained by using Equation 2.3-2. Borrowing from the terminology of classical control theory, a PLL with $G_0(s)$ given by Equation 3.3-2 is referred to as a *Type I* loop since its open-loop transfer function has one pole at the origin of the s-plane.

The closed-loop transfer function for this case is

$$H(s) = \frac{G(s+a)}{s^2 + (b+G)s + aG}, \qquad (3.3\text{-}3)$$

a result that follows from substituting Equation 3.3-1 into Equation 2.3-4. Using standard control system terminology, this transfer function can be written as

$$H(s) = \frac{Gs + \omega_n^2}{s^2 + 2\zeta\omega_n s + \omega_n^2}, \qquad (3.3\text{-}4)$$

where

$$\omega_n = \sqrt{aG} \qquad (3.3\text{-}5)$$

and

$$\zeta = \frac{b+G}{2\sqrt{aG}} \qquad (3.3\text{-}6)$$

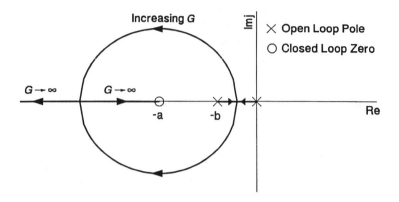

FIGURE 3.8
Root locus for a second-order PLL with an imperfect integrator.

are the natural frequency and damping factor of the loop, respectively. As is evident from Equation 3.3-4, gain parameter G remains in transfer function H(s) since the three parameters G, a, and b cannot be expressed as functions of ω_n and ζ alone.

3.3.2 Loop Stability

As given by Equation 3.3-4, transfer function H(s) has poles which remain in the left half of the s-plane for all $G > 0$. Hence, this PLL is unconditionally stable. This fact is seen best by considering Figure 3.8, a root locus plot. At $G = 0$, the closed-loop poles start at $s = 0$ and $s = -b$; these values are the poles of the open-loop transfer function $G_0(s)$. Also note that the locus contains complex-valued segments for G in a positive interval of finite length. Finally, as $G \to \infty$, one pole of H(s) approaches $s = -\infty$, and the other pole approaches $s = -a$.

3.3.3 Transient and Steady-State Tracking Errors

The response of this Type I loop to a phase step, frequency step, and frequency ramp can be computed easily. Simply apply to this loop the standard frequency domain techniques that are utilized in Sections 3.1 and 3.2.

A comparison of Equations 3.2-3 and 3.3-3 reveals that the transfer functions for the Type I and II loops will be close to one another if constants a and b (used in the imperfect integrator loop filter) satisfy

$$b \ll G \qquad (3.3\text{-}7)$$

LINEAR ANALYSIS OF COMMON FIRST- AND SECOND-ORDER PLLS 87

and

$$a = \alpha. \tag{3.3-8}$$

Under these conditions, and with a small value of $\omega_\Delta/(Ga/b)$, the PLL under consideration can be used (in many practical applications) as a substitute for a PLL containing a perfect integrator in its loop filter.

The steady-state error response of the Type I loop can be obtained from

$$\lim_{t \to \infty} \phi(t) = \lim_{s \to 0} \left[s(1 - H(s))\Theta_1(s) \right] \tag{3.3-9}$$

once input Θ_1 is known. Use this formula with Equation 3.3-3 to obtain

$$\lim_{t \to \infty} \phi(t) = \begin{cases} 0 & \text{for a phase step input} \\ \omega_\Delta/(Ga/b) & \text{for a frequency step input} \\ \infty & \text{for a frequency ramp input.} \end{cases} \tag{3.3-10}$$

By comparing Equation 3.3-10 with the results obtained in Section 3.1, note that the second-order, Type I PLL under consideration and the first-order PLL have the same steady-state responses, once the DC gain (a/b) of the loop filter is taken into consideration.

As a final example of steady-state tracking error when the PLL loop filter contains an imperfect integrator, consider the application of an angle-modulated reference. In this example, θ_1 has the form given by Equation 2.4-8. Under these conditions, the VCO relative phase and the closed-loop phase error are

$$\theta_2(t) = \omega_\Delta t - \frac{\omega_\Delta}{G} + \beta |H(j\omega_m)| \sin(\omega_m t + \angle H(j\omega_m)) \tag{3.3-11}$$

and

$$\phi(t) = \frac{\omega_\Delta}{G} + \beta \left[\sin \omega_m t - |H(j\omega_m)| \sin(\omega_m t + \angle H(j\omega_m)) \right], \tag{3.3-12}$$

respectively, where H is given by Equation 3.3-4. These results follow from application of superposition in the linear model; the constant and sinusoidal terms in Equation 3.3-11 are the steady-state responses to the frequency step and sinusoidal components, respectively, in θ_1.

Under the conditions Equations 3.3-7 and 3.3-8, the frequency responses of the Type I and II loops are similar. In this case, the PLL with imperfect integrator has frequency and error responses that are close to those depicted by Figures 3.5 and 3.6, respectively.

3.3.4 Noise-Equivalent Bandwidth

The second-order PLL containing an imperfect integrator in its loop filter has a noise-equivalent bandwidth that can be calculated with the aid of Equation 2.5-24. Using an approach similar to that shown in Section 3.2, B_L for the loop containing an imperfect integrator can be computed as

$$B_L = G \frac{(G+a)}{4(G+b)}. \qquad (3.3\text{-}13)$$

Under the conditions given by Equation 3.3-7, this B_L is closely approximated by Equation 3.3-13 which is the noise-equivalent bandwidth for the perfect integrator case.

3.3.5 Summary for the Second-Order PLL Based on the Imperfect Integrator Loop Filter

In this section, a PLL model is considered that contains an imperfect integrator in its loop filter. While inferior in performance to the perfect integrator case discussed in the previous section, this type of PLL is used where simplicity and low cost are paramount. As shown in Chapter 4, the imperfect integrator case can be realized by using a loop filter synthesized from passive components, and this is its most attractive attribute. In Chapter 5, the model discussed in this section is generalized to the nonlinear case, and a global phase plane analysis is given.

To aid in the utilization of this PLL model, Table 3.4 contains a summary of the important formulas relevant to the imperfect integrator case. In the table, formulas are given for the often-used steady-state tracking errors, noise-equivalent bandwidth, damping factor, and natural frequency. A comparison of steady-state tracking errors in Tables 3.3 and 3.4 reveals some performance costs that are incurred by using a loop filter based on an imperfect integrator. Phase tracking ability is important in many applications characterized by nonzero, unknown values of loop detuning. In these applications, an imperfect integrator loop filter may not produce adequate performance.

TABLE 3.4

Summary for Second-Order PLL Model with $F(s) = (s + a)/(s + b)$

Input $\theta_1(t) \Rightarrow \phi_{ss} = \lim\limits_{t \to \infty} \phi(t)$		
$\theta_1 = \theta_\Delta U(t) \Rightarrow \phi_{ss} = 0$	$H(s) = \dfrac{Gs + \omega_n^2}{s^2 + 2\zeta\omega_n s + \omega_n^2}$	
$\theta_1 = \omega_\Delta t\, U(t) \Rightarrow \phi_{ss} = \omega_\Delta/(Ga/b)$	$B_L = \dfrac{G(G+a)}{4(G+b)}$	$\omega_n = \sqrt{aG}$
$\theta_1 = \tfrac{1}{2} R t^2 U(t) \Rightarrow \phi_{ss} = \infty$		$\zeta = \dfrac{G+b}{2\omega_n}$

Chapter 4

PHASE-LOCKED LOOP COMPONENTS AND TECHNOLOGIES

Phase detectors, loop filters, and VCOs are components incorporated in PLL applications. These devices take several different forms, and they have been implemented by using several different technologies. In most cases, the technical requirements of an application and the economics that control the design process tend to limit the selection of PLL component technologies that can be employed.

Some of the more common methods of implementing PLL components are discussed in this chapter. However, PLL technology is changing rapidly, and the subject is only introduced here. Section 4.1 is devoted to a discussion of phase detectors. The designer must be aware of the many different phase detector implementations that are available. Each of these implementations has strengths and weaknesses that must be considered during the design process. Section 4.2 contains a short discussion of commonly used loop filters. Most modern applications employ active loop filters based on an operational amplifier connected in a feedback configuration. Commonly used VCO technologies are discussed in Section 4.3. VCO design and construction can be very challenging; the outcome of these efforts greatly influences PLL performance. Finally, Section 4.4 is devoted to discussion of a quadrature detector capable of detecting the occurrence of phase lock.

Large-scale integrated (LSI) circuit technology has an overwhelming influence on the design and utilization of PLLs. Many integrated circuits exist that incorporate all of the functions necessary to implement a PLL. In particular, PLL-based frequency synthesizer technology has been revolutionized by LSI circuits. Obviously, the rapidly changing subject of PLLs based on LSI circuit technology cannot be covered in one chapter of this book (that is archival in nature). It is best covered in the technical literature published by the major semiconductor manufacturers in the PLL applications market.

4.1 PHASE DETECTORS — ANALOG AND DIGITAL

Many varieties of phase detectors have been used in PLLs. Some accept sinusoidal inputs and operate like the analog multiplier phase detector discussed in Chapter 2. Others are based on switching mechanisms, and they either hard limit their inputs or accept digital signals directly. In most cases, a practical phase detector has an output that is either a sinusoidal, sawtooth, or triangular function of the phase error. Finally, phase detectors may be based on simple passive circuits that require no external power, or they may be complex integrated circuits.

As used in PLLs, phase detectors generally fall into two broad classes. The first class contains those detectors that form their output at time t by using their inputs at time t. In these detectors, no memory of previous inputs is used to form their output. The second class consists of those detectors that utilize only the locations of input signal zero crossings (or negative-going input transitions in the case of digital signals) to determine an output signal. At time t, members of this class base their output on negative edges that occurred prior to time t; that is, they utilize memory of previous inputs to determine their output.

In general, members of the first class are much simpler in construction than members of the second class. Usually, detectors that do not utilize memory are based on multiplier circuits; such phase detectors form their output from a simple product of their inputs. Common examples of memoryless, multiplier-type phase detectors are the exclusive OR gate and the diode ring mixer. Phase detectors in the second class, those with memory, are sequential circuits that can be constructed from flip-flops and logic gates. There are several commercially available integrated circuit sequential phase detectors.

Each class of phase detectors has its advantages and disadvantages. Multiplier-type phase detectors have the advantage when it comes to frequency of operation; some phase detectors of the multiplier type can be made to work well into the microwave region. Also, they are the only ones which provide adequate performance when the applied signals are buried in noise; for example, they are used almost exclusively in phase-locked receivers. Generally, sequential logic phase detectors have the advantage when it comes to functionality (i.e., some serve as phase and frequency detectors). Also, their outputs are a linear function of phase error ϕ over a wider range than what is found in the typical multiplier-type phase detector. Generally, sequential logic phase detectors are used in applications where the input signals are free of significant noise and have well-defined level transitions. For example, they are used almost exclusively in PLL-based frequency synthesizers.

While not universally accepted, special terminology is used in some cases to classify phase detectors and PLLs based on the nature

PHASE-LOCKED LOOP COMPONENTS AND TECHNOLOGIES   93

of their input signals. Sometimes in the PLL literature, the name *digital phase detector* is used to describe phase detectors that employ digital inputs (however, this application of the terminology is controversial). The phase detectors described in Sections 4.1.3, 4.1.4, and 4.1.5 belong to this class. Also, the name *digital phase-locked loop* (DPLL) is used sometimes to describe a PLL based on a digital phase detector (for example, see Chapter 3 of Best).[10] Sometimes, this name is applied even if the VCO and/or loop filter operate on analog principles (again, this is a source of controversy in the literature).

4.1.1 Integrated Circuit Four-Quadrant Analog Multipliers

The phase detector discussed in Chapter 2 is referred to as a *four-quadrant analog multiplier* since it accepts signals of both polarities on both inputs. It is an example of a *double-balanced analog mixer* since no components from either input show up as additive terms in the detector output (there is no direct feedthrough of either input to the output). Additional sources of information on the four-quadrant multiplier include Gilbert[53] and the data sheets which are published by manufactures of four-quadrant multiplier integrated circuits.

True analog multiplier phase detectors of this type usually incorporate Gilbert multiplier circuitry; Figure 4.1 illustrates a simplified circuit for such an analog multiplier. The quantities v_i and v_o denote inputs, while v_d denotes the multiplier output. On the figure, the quantities I and K are circuit-dependent constants which serve to limit the range of voltages and currents in the multiplier circuitry.

To understand how this multiplier functions, assume that constant current I is split into

$$i_1 = \frac{I}{2K}(v_o + K)$$
$$i_2 = I - \frac{I}{2K}(v_o + K),$$
(4.1-1)

$-K \leq v_o \leq K$, by the differential transistor pair controlled by v_o. Furthermore, these currents are split into

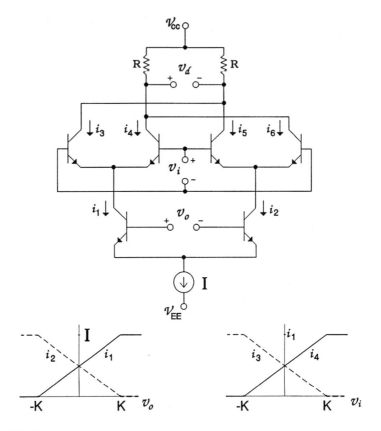

FIGURE 4.1
Four-quadrant, double-balanced analog multiplier.

$$i_3 = i_1 - \frac{i_1}{2K}(v_i + K)$$

$$i_4 = \frac{i_1}{2K}(v_i + K)$$

$$i_5 = \frac{i_2}{2K}(v_i + K)$$

$$i_6 = i_2 - \frac{i_2}{2K}(v_i + K),$$

(4.1-2)

$-K \leq v_i \leq K$, by the two differential pairs controlled by v_i. Now, the output voltage can be expressed as

$$v_d = (i_4 + i_6 - i_3 - i_5)R. \qquad (4.1\text{-}3)$$

Substitute Equation 4.1-2 into Equation 4.1-3 and obtain

$$v_d = (i_1 - i_2)\left[\frac{v_i + K}{K} - 1\right]R \qquad (4.1\text{-}4)$$

after combining terms that depend on $i_1 - i_2$. Finally, substitute Equation 4.1-1 into Equation 4.1-4 and obtain

$$v_d = \frac{I}{K^2}(v_o v_i) \qquad (4.1\text{-}5)$$

after some simple algebra. Hence, the output of the Gilbert multiplier is proportional to the product of the applied voltages.

A *singly balanced* version of this multiplier results if the transistors carrying i_2, i_5, and i_6 are removed. Also, replace the current source I and the transistor carrying i_1 by a voltage-controlled current source generating i_1 as approximated by Equation 4.1-1. The resulting circuit would have an output of

$$v_d = \frac{RI}{2K^2}[v_o v_i + K v_i]. \qquad (4.1\text{-}6)$$

This multiplier is singly balanced since input signal v_i does not balance out of v_d.

In circuits incorporating this type of multiplier, a significant problem is imbalance in the differential transistor pairs and other components. Also, thermal drift can be a significant problem. For these reasons, use of this circuit would not be successful if it were built by using only discrete components. However, it is possible to overcome these problems and obtain good performance by using modern integrated circuit technology where electrical and thermal characteristics can be matched closely.

Several integrated circuit four-quadrant multiplier chips are available commercially. Typically, they require one or more external resistors which must be adjusted during circuit alignment. Excellent performance can be obtained at the expense of a relatively complicated alignment procedure that might have to be repeated on a periodic schedule to reduce the problem of drift. Typically, these integrated circuits are limited to applications where the input signals are relatively low in frequency (a few megahertz at most). Finally, integrated circuit

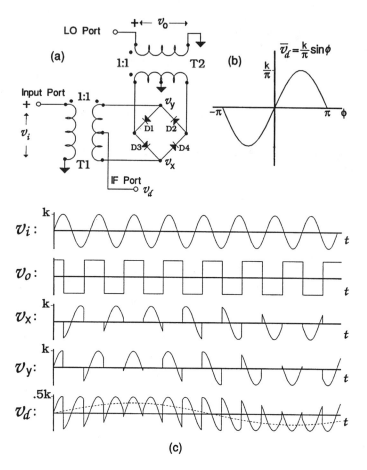

FIGURE 4.2
Diode ring double-balanced mixer.

four-quadrant multiplier chips are expensive relative to most other phase comparator technologies.

4.1.2 Diode Ring Mixer

Figure 4.2 depicts a popular form of analog multiplier that is commonly called a *diode ring double-balanced mixer*. It contains no active components, and versions of it can operate at frequencies well into the microwave region. High-performance multipliers of this type are available in many different package configurations as commercial products, and they are relatively inexpensive. See Krauss et al.,[23] Wolaver,[9] and the handbooks in References 24, 25 for additional sources of information on diode ring mixers.

In typical applications, v_i and v_d are low-level signals at the input and intermediate frequency (IF) ports, respectively. Good performance requires that the IF port termination be resistive (usually 50 Ω) at the sum and difference frequencies that appear there. Signal v_o is applied at the local oscillator (LO) port, and it represents a high-level (approximately 7 dBm) signal responsible for switching the diodes. Like the IF port, proper termination at the LO port is required in most practical applications. Finally, conversion gains on the order of –8 to –6 dB are typical.

This type of mixer can operate with a wide variety of applied waveforms, but a basic understanding of circuit operation is easier to obtain if v_o is a square wave as depicted on Figure 4.2. Also note that v_i and v_o are at slightly different frequencies, and the phase difference ϕ between them is increasing slowly with time. Note that the waveforms are defined so that, at t = 0, phase difference ϕ starts at $\pi/2$.

Square wave v_o is a switching waveform; when negative, it turns on diodes D1 and D2, and it turns on diodes D3 and D4 when it is positive. When D3 and D4 are on, v_x is forced to ground through the center tap on the secondary of T2, and v_d is one half of v_i. In a similar manner, when D1 and D2 are conducting, v_y is forced to ground through the center tap, and v_d is minus one half of v_i. This produces the waveform v_d illustrated on the figure.

As a dashed-line curve, a beat note component is superimposed on the graph of v_d. It varies as a sinusoidal function of ϕ, the phase difference between v_i and v_o. The magnitude of this difference frequency component is

$$V_p = \left(\frac{1}{2}k\right)\left(\frac{4}{\pi}\right)\left(\frac{1}{2}\right)$$
$$= \frac{k}{\pi},$$
(4.1-7)

where $4/\pi$ represents the peak value of the fundamental component of v_o.

Part (b) of Figure 4.2 depicts the sinusoidal phase detector characteristic for this multiplier. On the figure, the quantity $\bar{v}_d$ denotes the DC component that is produced when v_i and v_o are out of phase (but of the same frequency) by ϕ radians. The phase detector gain constant is

$$K_m = \frac{k}{\pi},$$
(4.1-8)

a result that follows from Equation 4.1-7.

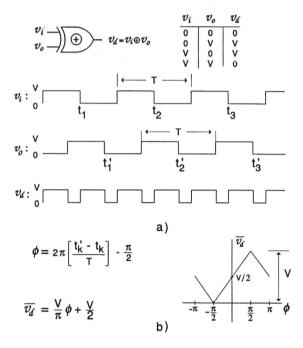

FIGURE 4.3
An exclusive OR gate phase detector with (a) waveforms and (b) phase characteristics.

4.1.3 Exclusive OR Gate

The phase detectors considered so far have sinusoidal phase characteristics. However, this is more a function of the applied waveforms than it is of phase comparator circuitry. In most cases, multiplier-type phase detectors have triangular phase characteristics when both of their inputs are supplied with square waves.

As shown on Figure 4.3, the *exclusive OR gate* is a simple multiplier-type phase detector that uses digital-input waveforms. The figure depicts a truth table that defines the gate operation; the output goes high when only one input goes high, and the gate output remains at a logic low for all other inputs. Also, note that the figure depicts inputs v_i and v_o as having the same frequency. Note that the output waveform v_d is a 50% duty cycle logic signal with a period of $T/2$ when the input waveforms are in phase quadrature. Under these conditions, the average value of v_d (denoted as $\bar{v}_d$) is zero. For this reason, phase difference ϕ is defined to be zero when the input signals are in phase quadrature.

On Figure 4.3b, the plot of $\bar{v}_d$ displays a triangular function of the phase difference ϕ; this plot is the characteristic function of the phase detector. It is linear over the range $-\pi/2 \le \phi \le \pi/2$ and centered around

V/2 instead of zero. Depending on the application, the nonzero offset in the phase detector characteristic can be a problem, or it can be helpful. In some applications, it is possible to exploit this nonzero offset when designing the circuitry that follows the phase detector.

The phase detector gain follows from inspection of the phase detector characteristic function. From Figure 4.3b, note that

$$K_m = \frac{V}{\pi} \qquad (4.1\text{-}9)$$

since the average output voltage $\bar{v}_d$ changes by V volts as input phase changes over a range of π radians.

As should be expected, the operation of the exclusive OR gate is sensitive to the duty cycle of the input logic waveforms since it is a multiplier-type phase detector. In most applications, the gate is operated with logic waveforms that have a 50% duty cycle. It is easily shown that this maximizes the range of phase ϕ over which the detector has a linear phase characteristic. See Wolaver[9] for a discussion of the effects that input duty cycle has on phase detector operation.

As can be seen from Figure 4.3, the output of this phase detector contains a strong frequency component at twice the input frequency. This is similar to the other multiplier-type phase detectors considered so far. Undesirable frequency components in the phase comparator output can lead to corruption of the VCO output spectrum. There are applications where this is a potential problem which must be dealt with.

4.1.4 RS Flip-Flop

The *RS flip-flop* is a simple example of a sequential phase detector; its operation is summarized by Figure 4.4. As shown on the figure, the flip-flop changes state on the negative-going transitions of the input signals. In contrast to the exclusive OR phase detector, the duty cycle of the input waveforms is not critical. Also different is the definition of phase ϕ; as defined for the RS flip-flop, phase ϕ is zero when one input is the complement of the other. The flip-flop Q output, defined as v_d on Figure 4.4, is of interest. This output has a duty cycle which ranges from zero to unity. As shown on the figure, this duty cycle maps onto the ranges $-\pi \leq \phi \leq \pi$ and $0 \leq \bar{v}_d \leq V$.

The RS flip-flop is a two-state device as the state diagram on Figure 4.4 illustrates. The occurrence of a negative-going transition on an input signal is denoted on the state diagram by a downward-pointing arrow. The state $v_d = 0$ remains unchanged under a negative-going transition on v_o, but under a like transition on v_i, the state changes to $v_d = V$. An analogous statement can be made about the effects of negative-going transitions on the state $v_d = V$.

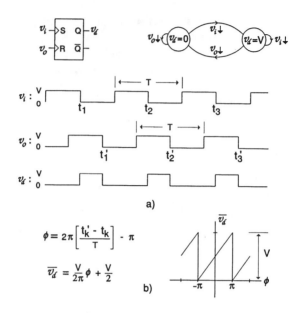

FIGURE 4.4
An RS flip-flop phase detector with (a) waveforms and (b) phase characteristic.

On Figure 4.4b, the plot of $\bar{v}_d$ is the phase detector characteristic function for an RS flip-flop. Note that it is a linear function of ϕ over a 2π range. Also different from previously discussed phase detector characteristics is the fact that this one has discontinuities. Like the exclusive OR phase detector, $\bar{v}_d$ is centered around $V/2$. Finally, inspection of Figure 4.4b reveals that an RS flip-flop phase detector has a gain of

$$K_m = \frac{V}{2\pi}. \qquad (4.1\text{-}10)$$

As might be expected, the operation of this sequential detector is relatively sensitive to missing or poorly defined transitions on the input signals. Hence, it is more sensitive to noise than multiplier-type phase detectors. Essentially, this detector remembers an error due to noise, while the multiplier-type phase detector does not. For this reason, sequential-type phase detectors are useful in applications where the input has a high signal-to-noise ratio.

As can be seen from the plot of v_d on Figure 4.4a, the output of this circuit contains a strong undesirable spectral component at the frequency of the input signals. On this point, the circuit is at a disadvantage relative to the exclusive OR detector (which has its undesirable output component at twice the input frequency). These undesirable output components can degrade the VCO output spectrum. In general,

4.1.5 Sequential Phase/Frequency Detectors

A sequential *phase/frequency detector* is a digital logic circuit that acts as an extended-range phase detector; also, it generates an output that is indicative of a frequency error between its input signals. For this reason, circuits of this type have found widespread use in frequency synthesizer applications where signal-to-noise ratios are high, and signal level transitions are well defined and predictable. Because of this popular application, many vendors offer integrated circuits that incorporate sequential phase/frequency detectors. See the texts by Gardner,[6] Wolaver,[9] Best,[10] and the references on frequency synthesizers (15 through 19) for additional information on sequential phase/frequency detectors.

This type of circuit has general characteristics that are of interest here. The circuit has two inputs, denoted here as v_i and v_o. Both of these inputs accept digital waveforms, and circuit changes occur on the negative-going edges of these input waveforms (as discussed in the introduction to this section, the circuit employs memory). In most implementations, the circuit has two outputs; the first (second) output is labeled as *up (down)*, and it generates the digital waveform v_u (v_d).

Consider this type of circuit as a phase detector with input waveforms that are slightly out of phase. As discussed in this section, most sequential phase/frequency detectors produce a pulse-train output for v_u if input v_i leads input v_o in phase; the nonactive output v_d is held low. On the other hand, the pulse-train output appears on v_d (and output v_u is held low) if input v_o leads input v_i. The duty cycle (and DC average) of this pulse-train output is proportional to the magnitude of the phase error. The output pulse train is processed by the loop filter to generate a control voltage that drives the VCO to a zero-phase error condition.

As a frequency detector, consider the case where input v_i has a slightly higher frequency that input v_o. As discussed in this section, output v_u is a periodic waveform with a significant DC component, and output v_d is held low. By interchanging the output variables v_u and v_d, a similar statement can be made that describes the case when the input v_o has a slightly higher frequency than the input v_i. The DC component in the active output is integrated by the loop filter to generate a voltage that slews the VCO towards a frequency-locked state.

Sequential phase/frequency detectors tend to be complicated sequential circuits constructed from flip-flops and gates. One of the simplest examples of this class of detectors is depicted by Figure 4.5.

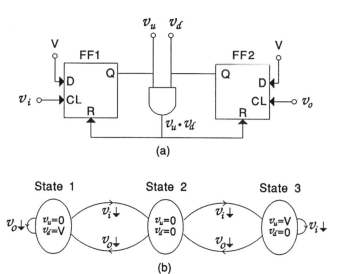

FIGURE 4.5
Phase/frequency detector with (a) block and (b) state diagrams.

The operation of this circuit is based on two D-type flip-flops and a simple AND gate. First, note that each flip-flop has its D input wired high. Under this condition, the flip-flop with a low Q output will transition high (Q goes high) on the next negative-going edge of its clock input. Also, if such an input transition occurs when Q is high there will be no change in the flip-flop state. A high signal on a reset input (denoted by R on the figure) will force Q low as soon as the reset signal is applied. Finally, a logical high on both of the Q outputs causes the resetting of both flip-flops.

Based on this description of the circuit, it is easy to see that Figure 4.5 depicts a three logic state circuit with digital inputs v_i and v_o and digital outputs v_d and v_u. Note that state changes are possible on negative-going transitions of the input data. This is indicated on the state diagram Figure 4.5b by a downward-directed arrow immediately after an input. The state where both v_d and v_u are high is not stable (and it is not included on Figure 4.5b) since it generates a signal that resets both flip-flops.

On Figure 4.5, outputs v_u and v_d are the up and down outputs, respectively. The reason for this nomenclature comes from how these outputs are used in most applications; often, v_u and v_d are used to drive a circuit similar to that depicted by Figure 4.6. On this figure, each field effect transistor (FET) acts as a simple switch that closes when its input goes high. Hence, the terminal common to both FETs goes high when v_u goes high, and it is grounded when v_d goes high. In most applications,

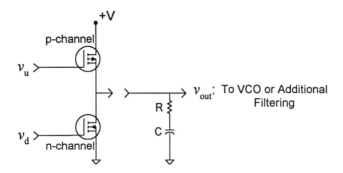

FIGURE 4.6
Output circuitry for use with phase/frequency detector.

a high v_u (v_d) causes the loop filter to integrate some current I_p ($-I_p$). This generates a VCO control voltage that slews the oscillator in the proper direction. Because of this operation, the circuit depicted by Figure 4.6 is part of what is called a *charge pump*.

Figure 4.7 describes use of the sequential phase/frequency detector circuit as a phase detector. On the figure, all depictions of v_i and v_0 have the common period T; however, these waveforms may differ in phase. Also, times t_k (t_k'), $1 \leq k \leq 6$, correspond to the falling edges of v_i (v_o). Consider first the waveforms v_i and v_o shown on Figure 4.7a, where $t_k = t_k'$, $1 \leq k \leq 6$. There is zero phase difference between these input waveforms, and both v_u and v_d are inactive. Next, consider Figure 4.7b, the middle set of waveforms with $t_k < t_k'$, $1 \leq k \leq 6$. For this case, waveform v_i leads v_o in phase by

$$\phi = 2\pi(t_k' - t_k)/T \qquad (4.1\text{-}11)$$

radians (i.e., ϕ is positive). Falling edges of v_i force output v_u high, and falling edges of v_o resets v_u; output v_d is inactive when v_u toggles in this manner. Figure 4.7b corresponds to the circuit toggling between states 2 and 3 (see Figure 4.5b). Finally, consider Figure 4.7c, the bottom set of waveforms with $t_k > t_k'$, $1 \leq k \leq 6$. For this case, ϕ is negative, and the circuit toggles between states 1 and 2. Hence, when used as a phase detector, the circuit must toggle between a pair of states, and *the exact pair of states changes as phase ϕ passes through zero*. As ϕ changes sign, this "swapping" of the pair of "toggling states" is essential for the circuit to act as a phase detector (as discussed below, "swapping" does not occur when the circuit acts as a frequency detector).

On Figure 4.7, note that the duty cycle of $v_u - v_d$ is proportional to the phase difference between the inputs. Furthermore, the average value of this difference can be expressed as

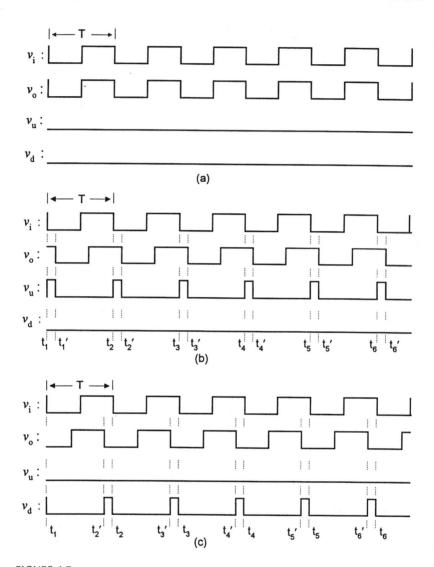

FIGURE 4.7
Timing diagram for phase/frequency detector.

$$\overline{v_u - v_d} = \frac{V(t_k' - t_k)}{T}$$
$$= \frac{V}{2\pi}\phi ,$$
(4.1-12)

for $-2\pi \leq \phi \equiv 2\pi(t_k' - t_k)/T \leq 2\pi$. This relationship is depicted by Figure 4.8a.

PHASE-LOCKED LOOP COMPONENTS AND TECHNOLOGIES

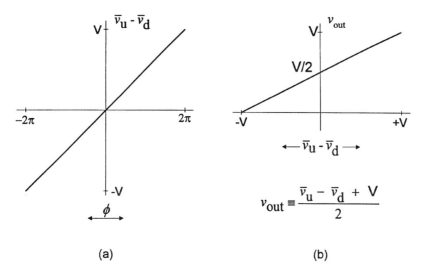

FIGURE 4.8
(a) Mapping from ϕ to $\overline{v_u - v_d}$. (b) Mapping from $\overline{v_u - v_d}$ to phase detector output v_{out}.

Average value $\overline{v_u - v_d}$ is used to form the phase detector output. Most practical implementations of the phase detector are in the form of an integrated circuit powered by a single supply voltage. Because of this single supply voltage constraint, the phase comparator output v_{out} must be nonnegative, and the average value $\overline{v_u - v_d}$ must be mapped onto a range $0 \leq v_{out} \leq V$, for some value V. Figure 4.8b illustrates the mapping

$$v_{out} = \frac{V}{2} + \frac{V}{4\pi}\phi, \qquad (4.1\text{-}13)$$

$-2\pi \leq \phi \leq 2\pi$, used to form the useful output.

As shown by Figure 4.8b, this phase detector characteristic function is linear over a 4π range of phase ϕ. The slope of this characteristic function is

$$K_m = \frac{V}{4\pi} \qquad (4.1\text{-}14)$$

volts/radians. This value of phase detector gain can be used in a linear analysis of a PLL that employs the sequential logic phase/frequency detector.

A more complete investigation of the sequential phase/frequency detector circuit leads to the conclusion that its output characteristic function is multiple valued. For the inputs v_i and v_o shown on

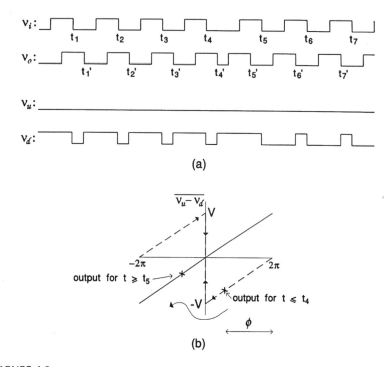

FIGURE 4.9
Phase characteristic is a multivalued function of ϕ.

Figure 4.7b, it is possible for the fall of v_i at t_k to force v_d low and the fall of v_o at t_k' to force v_d high. That is, for the inputs v_i and v_o shown on Figure 4.7b, the circuit could be toggling between states 1 and 2 (instead of 2 and 3 as shown of Figure 4.7b).

This scenario is depicted by Figure 4.9a. For $t < t_4'$, inputs v_i and v_o on Figure 4.9a are identical in phase relationship to the inputs depicted on Figure 4.7b. However, Figure 4.9a shows the circuit toggling between states 1 and 2 instead of states 2 and 3, as shown by Figure 4.7b.

This multiple-valued characteristic of the circuit allows an output signal that is both phase and frequency-error dependent. The solid-line graph on Figure 4.9b shows the previously discussed (see Figure 4.8a) phase response that is linear over the range $-2\pi < \phi \le 2\pi$. However, other responses of the circuit are given by the two dashed-line plots (one for each polarity of ϕ). Arrows on Figure 4.9b indicate that both dashed-line response curves lead to the solid-line curve as ϕ passes through the origin.

As shown on Figure 4.9a, signal v_i leads v_o in phase for time up to t_4'. Then a decrease in $\phi \equiv 2\pi(t_k' - t_k)/T$ occurs, and v_i lags v_o for $t \ge t_5$. For $t \le t_4$, $\phi > 0$, the circuit is toggling between states 1 and 2, and output $\overline{v_u - v_d}$ lies on the dashed-line segment in the fourth quadrant of

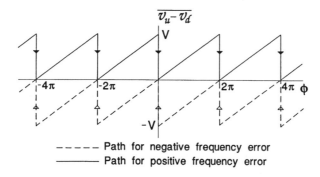

FIGURE 4.10
Frequency detector characteristic of phase/frequency detector.

Figure 4.9b. Then a decrease in ϕ occurs; for $t \geq t_5$, $\phi < 0$, the circuit is *still* toggling between states 1 and 2, and output $\overline{v_u - v_d}$ moves (following the arrows) from the dashed-line curve in the fourth quadrant to the segment of the solid-line curve that lies in the third quadrant. It is important to note that ϕ decreased through zero *without a change in the pair of states that toggle*. For a second example, the output will go from the dashed-line segment in the second quadrant of Figure 4.9b to the solid-line segment in the first quadrant as ϕ increases through zero (this example is not annotated on Figure 4.9b). In this second example (which corresponds to the circuit toggling between states 2 and 3), phase ϕ increased through zero without a change in the pair of states that toggle. In both of these examples, movement from the origin (0,0) cannot occur if it is against the arrows on the vertical axis.

The circuit can generate a frequency error-sensitive output. The main reason for this follows from behavior described in the previous paragraph. First, consider the circuit operation with a small positive constant frequency error; in this case, phase ϕ is an increasing ramp. The circuit toggles between states 2 and 3; it never enters state 1. That is, as phase $\phi(t)$ increases through multiples of 2π, no "swap" occurs in the pair of states which the circuit toggles between. The solid-line graph on Figure 4.10 depicts the output $\overline{v_u - v_d}$ as a function of increasing ϕ. Since this output is never negative, it contains a positive DC component, and this indicates a positive frequency error. On the other hand, if a small negative constant frequency error exists, then the output contains a negative DC component since it follows the dashed-line path on Figure 4.10 as phase ϕ ramps downward.

Figure 4.11 shows the case where v_i has a slightly higher frequency than v_o. To simplify this figure, v_i and v_o are not drawn completely. Instead, for integers $k \geq 0$, markers are drawn that indicate the times t_k and t_k' associated with the falling edges of v_i and v_o, respectively. For every $k \geq 0$, output v_u is high for the interval (t_k, t_j') if t_j' satisfies $t_k <$

$t_j' < t_{k+1}$. A second case of practical interest occurs when integers k and j exist such $(t_k, t_{k+1}) \subset (t_j', t_{j+1}')$; on Figure 4.11, examples of this case are $(t_7, t_8) \subset (t_6', t_7')$ and $(t_{16}, t_{17}) \subset (t_{14}', t_{15}')$. For this case, output v_u is high for the interval (t_k, t_{j+1}'). The duty cycle of output $v_u - v_d$ changes from one period of v_i to the next, and it is approximated by the positive-going sawtooth graph on the figure. This output has a positive DC component that indicates a positive frequency error. An output consisting of a negative-going sawtooth waveform, with a negative DC component, results if v_i has a lower frequency than v_o. Hence, this detector can sense the sign of frequency error.

A potential problem must be mentioned before leaving the topic of sequential phase/frequency detectors. This problem deals with the device behavior at $\phi = 0$. At this point, the device output should be zero. However, noise and other anomalies (delay-related race problems) cause the outputs v_u and v_d to toggle, usually at a low rate. That is, noise, rather than input signals, controls the phase–voltage characteristic in the zero phase difference region (in effect, a form of "noise amplification" occurs). In some applications, this low-rate, noise-induced toggling can cause a noticeable degradation in the VCO spectrum. Hence, sometimes, it is beneficial to insert a small bias in the loop in order to avoid an operating point of $\phi = 0$ (this occurs naturally in most practical designs).

4.2 LOOP FILTERS

Both active and passive loop filters can be used in the PLL. Most modern applications utilize an active filter based on high-performance operational amplifier technology. The reasons for this are based on PLL performance: as discussed in Chapter 3, active loop filters provide the PLL designer with superior performance. Modern integrated circuit operational amplifiers have come a long way in reducing the thermal drift, noise, and cost problems which once plagued DC amplifiers (Graeme et al.).[26] The problems of DC offset, thermal drift, and noise in modern operational amplifiers are usually less severe than the corresponding problems which can plague phase comparators and VCOs.

Figure 4.12 illustrates a popular active loop filter which can be used to build the second-order Type II PLL discussed in Section 3.2. An integrated circuit operational amplifier serves as the active device of the filter, and it is used in an inverting configuration as is indicated by the minus sign which appears with the transfer function on Figure 4.12. The fact that Equation 3.2-1 does not contain this minus sign is of little practical significance. The minus sign can be accounted for in the PLL discussed in Section 3.2 by introducing a π-radian shift in the phase error variable.

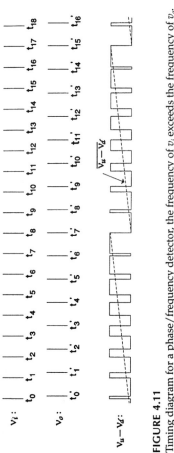

FIGURE 4.11
Timing diagram for a phase/frequency detector, the frequency of v_i exceeds the frequency of v_o.

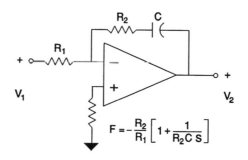

FIGURE 4.12
A simple active loop filter.

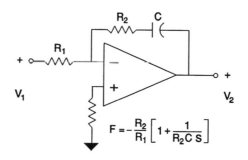

FIGURE 4.13
A simple passive loop filter.

A common passive loop filter used in PLLs is depicted by Figure 4.13. In most applications resistor R_1 is large compared to R_2 so that $\tau_1 \gg \tau_2$. Use of this filter results in a second-order Type I PLL. Section 3.3 contains a description of the linear model of this PLL.

4.3 VOLTAGE-CONTROLLED OSCILLATORS

A large number of VCO circuit configurations can be realized. Also, a wide range of technologies can be utilized in the construction of VCOs. Different circuit configurations and/or technologies offer different capabilities and performances. As might be expected, no single VCO design or technology can serve all applications. Instead, the selection of circuit configuration and technology is driven by the requirements of an application, and the selection process usually involves tradeoffs and compromises.

An application can impose many different requirements that influence VCO design. Some of the more important requirements are the following:

1. Center frequency,
2. High spectral purity and low phase noise,

PHASE-LOCKED LOOP COMPONENTS AND TECHNOLOGIES 111

3. Electrical tuning range and linearity of frequency vs. tuning voltage,
4. Response time or ability to accept wideband modulation, and
5. Low system cost.

VCO center frequencies above a few tens of megahertz usually require circuits employing some form of resonance condition to tune out device reactances and establish the frequency of oscillation. In general, oscillators employing high-Q resonant circuits are required in applications where it is necessary to achieve an output with high spectral purity (low phase noise). Often, large linear tuning ranges and the ability to accept wideband modulation are obtained at the expense of high spectral purity. Finally, low system cost usually requires that the VCO be based on a common integrated circuit.

Some of the more common technologies and/or devices used in VCO construction include the following:

1. Quartz crystals in high-Q circuits,
2. Other resonators (LC, coaxial, and cavity circuits), and
3. RC multivibrators and other oscillators based on nonresonant circuits.

These stalwarts have been joined by newer technologies such as the dielectric resonant oscillator (DRO). Also, new resonator technologies have appeared in oscillators; these technologies include the surface acoustic wave (SAW) device and resonators which use the piezoelectric properties of some ceramic materials.

4.3.1 Voltage-Controlled Crystal Oscillators

A VCO that relies on pulling the frequency of a quartz crystal is known as a *voltage-controlled crystal oscillator* (*VCXO*). These oscillators have excellent spectral purity and stability, and they can be made physically small. Their main limitation is tuning range. In most applications that use fundamental mode crystals, VCXOs have a tuning range of not more than 0.1% of their operating frequency. Often, VCXOs are used in clock recovery and synchronization applications where the frequency of the clock is known to within a small tolerance. See Parzen[27] and the handbook in Reference 24 for a discussion of crystal oscillator technology.

Figure 4.14 depicts a simplified circuit model for a quartz crystal. The quantity R_x is used to model the crystal losses. The quantities L_x and C_x are called the *motional arm inductance* and *motional arm capacitance*, respectively. Their values are determined by a number of factors including crystal cut (orientation: AT, BT, etc.), type of vibration (shear,

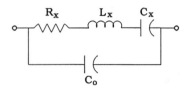

FIGURE 4.14
Simplified equivalent circuit of a quartz crystal.

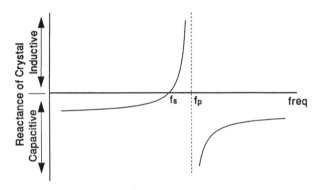

FIGURE 4.15
Reactance as a function of frequency for a quartz crystal.

extension, etc.), and size of the quartz slab. The quantity C_o denotes the total static shunt capacitance. It results from the capacitance between the electrodes attached to the quartz and the capacitance due to the crystal case and leads.

Figure 4.15 depicts a plot of device reactance vs. frequency for a hypothetical quartz crystal. Frequencies f_s and f_p denote where series resonance and parallel resonance, respectively, occur. The separation between these frequencies is small; typically, it is less than 0.1% of f_s for commonly available, fundamental-mode, AT-cut crystals.

Depending on circuit design, an oscillator can use its crystal in either the parallel- or series-resonant mode. To place the oscillator frequency under voltage control, the applied tuning voltage must be used to shift the frequency of the chosen resonant mode. Usually, this is accomplished with the aid of a varactor diode in a network that applies a voltage-controlled reactance to the crystal.

Figure 4.16 shows a simplified schematic of a VCXO that uses its crystal in the series-resonant mode. For proper operation, the oscillator LC tank circuit should resonate at or near f_s. Loosely coupled link L_1 provides a positive-feedback path to facilitate circuit oscillation. At some nominal value of tuning voltage, the capacitance of the tuning diodes should series resonate with circuit inductance. Then L_1 and the diodes would be capacitive (inductive) if the control voltage were above (below) this nominal value. This voltage-variable reactance

PHASE-LOCKED LOOP COMPONENTS AND TECHNOLOGIES 113

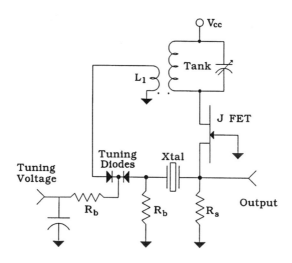

FIGURE 4.16
VCXO using series-resonant mode of crystal.

would combine with the crystal reactance and make the circuit oscillate at a frequency which could be tuned by a small amount to both sides of f_s.

It is relatively easy to design and build VCXO circuits that oscillate and tune over a narrow range of frequencies. It is much more difficult to optimize one or more parameters of the circuit operation. Optimizing VCXO performance requires significant experience and skill in RF circuit design.

4.3.2 RC Multivibrators

A VCO based on an *RC multivibrator* should be considered when a large tuning range and low cost are required. Unfortunately, such oscillators often suffer from poor stability and high output noise. The upper operating frequency of practical multivibrators is limited to a few tens of megahertz. Oscillators of this type are usually constructed from a single low-cost integrated circuit augmented with a few passive components. For additional information on multivibrator technology see Best[10] and the handbook in Reference 24.

Figure 4.17 depicts a functional block diagram of a simple multivibrator. This multivibrator contains a current source which switches between $+kv_{in}$ and $-kv_{in}$, where k is a positive constant, and v_{in} is a positive control voltage. The switching function is controlled by circuitry which senses capacitor voltage v_c; the source is switched when v_c reaches either of the programmed limits V_1 or V_2. The oscillator period is determined by how long it takes this current source to ramp

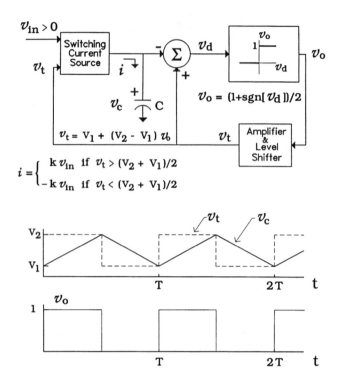

FIGURE 4.17
RC multivibrator block diagram and timing diagram. Operation requires k > 0 and $v_{in} > 0$.

the capacitor voltage between V_1 and V_2. Voltage v_o is the oscillator output.

The multivibrator operates in a simple manner. First, assume that the circuit is in the $v_o = 1$ state where $v_t = V_2$, $i = kv_{in} > 0$, and $v_c < V_2$. Under this condition, the capacitor integrates i, and v_c increases linearly with time until it reaches V_2. Then the circuit switches to the $v_o = 0$ state with $v_t = V_1$ and $i = -kv_{in} < 0$; when this happens, v_c decreases in value. Finally, the circuit switches back to the $v_o = 1$ state when the linearly decreasing v_c reaches V_1.

Noise impacts this circuit by causing jitter in the times associated with state changes. A significant amount of this timing jitter can be related to noise-induced changes which cause time displacements in the above-mentioned threshold operations. Noise-induced time displacements are inversely proportional to the magnitude of the slope of the capacitor voltage as it integrates toward a threshold value. The output timing jitter decreases when the capacitor is changed to increase the magnitude of this slope; this increase in the slope magnitude decreases the integration times associated with one period of the output and increases the frequency of oscillation.

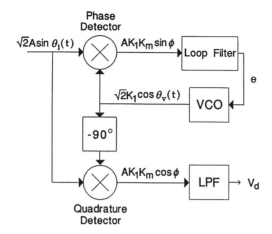

FIGURE 4.18
PLL and quadrature lock detector.

In some applications this observation can be exploited to reduce the timing jitter problem discussed above. To reduce jitter in the times associated with state changes, it is useful to select the multivibrator capacitor so that oscillation occurs at Nf_o, where f_o is the desired output frequency. The signal produced in this manner can be used to drive a divide-by-N counter to produce an output at the desired frequency f_o. This method will produce an output signal with less jitter than if the multivibrator was operated at f_o directly by using a larger capacitor.

4.4 LOCK DETECTION

An electronic method of detecting phase lock is needed in most applications. Often, the signal indicating lock is used to activate a panel light or meter monitored by the equipment operator. In more complex applications, it is used to change some aspect of the loop or system configuration. For example, the lock-detection signal may be used to indicate that a change should be made in loop filter parameters, it may be used as a source of coherent automatic gain control (AGC) voltage, or it may be used to inhibit or allow selected aspects of system operation.

Figure 4.18 illustrates use of a quadrature phase detector as a lock detector. The received (reference) signal is supplied to one input of the detector. The other input of the detector is a 90° phase-shifted version of the VCO output. Hence, the quadrature detector output is proportional to $\cos\phi$. Under phase-locked conditions, ϕ is small so that $\cos\phi \approx 1$, and the quadrature detector output is proportional to the amplitude of the received signal. When the loop is out of lock, a beat note appears at the quadrature detector output. It has a DC component which is, at most, only a small fraction of the input signal amplitude. A lowpass filter is used to reduce noise and brief fluctuations in the quadrature detector output.

The lowpass filter output (V_d on Figure 4.18) can be subjected to a threshold or used as an analog signal. A threshold operation on V_d produces a binary signal which indicates that the loop is either locked or unlocked. Analog voltage V_d is indicative of received signal strength, and it can be used as an AGC voltage to ensure that a constant signal (reference) level is applied to the PLL input. One or both of these uses of the quadrature lock detector serve in most PLL applications.

Part II
Nonlinear PLL Analysis

Chapter 5

NONLINEAR PLL BEHAVIOR IN THE ABSENCE OF NOISE

This chapter describes the basic nonlinear theory of the first- and second-order PLLs that are introduced in Chapter 3. The approach taken utilizes the phase plane method to illustrate known qualitative properties of the loops. Also, bifurcation theory is used to explain structural changes that occur in the phase plane of the second-order loop containing an imperfect integrator. The phenomena discussed in this chapter cannot be described by the linear methods covered in Chapter 3.

The nonlinear PLL model discussed in Chapter 2 is the starting point for the coverage given in this chapter. Figure 5.1 serves to summarize this model. The quantities A and K_1 on the figure correspond to the amplitudes (in RMS volts) of the input reference and VCO output, respectively, while K_m corresponds to the phase detector gain. As discussed in Section 2.1.3, the constant ω_0 represents the VCO center frequency. The angles θ_i and θ_v represent the instantaneous phases of the input reference and VCO, respectively, and $\phi \equiv \theta_1 - \theta_2$ denotes the closed-loop phase error. In this chapter, the reference signal is assumed to have a constant frequency of ω_i so that $\theta_1(t) \equiv \omega_\Delta t$, where $\omega_\Delta \equiv (\omega_i - \omega_0)$. By using Equations 2.1-5 and 2.2-3, the PLL can be modeled by the differential equation

$$\left[b_n \frac{d^n}{dt^n} + b_{n-1}\frac{d^{n-1}}{dt^{n-1}} + \ldots + b_0\right]\left[\frac{d\phi}{dt} - \omega_\Delta\right]$$
$$= -G\left[a_m \frac{d^m}{dt^m} + a_{m-1}\frac{d^{m-1}}{dt^{m-1}} + \ldots + a_0\right]\sin\phi, \quad (5.0\text{-}1)$$

where $G = AK_1K_mK_v$ denotes a closed-loop gain term. Without loss of generality, gain G is assumed to be positive. A negative value of G can be accounted for by increasing ϕ by π radians (see Section 2.2.1).

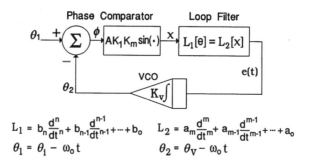

FIGURE 5.1
Nonlinear time domain PLL model.

The first three sections of this chapter cover the basic nonlinear behavior of first- and second-order PLLs. Section 5.1 contains a discussion of the nonlinear behavior of the first-order PLL. For the first-order case, phase lock is shown to occur if $|\omega_\Delta| < G$, and a simple result is developed for the phase acquisition time. A second-order PLL containing a perfect integrator loop filter is discussed in Section 5.2. It is argued that this loop has an infinite pull-in range. However, the time required to pull in increases at a rate that is proportional to the square of the initial frequency error if this value is large. Section 5.3 contains a discussion of the second-order PLL containing an imperfect integrator in its loop filter. In this case, bifurcations of periodic limit cycles cause the structure of the phase plane to depend on the parameter ω_Δ. Finally, Section 5.4 shows how pull-in range can be decreased by the inclusion of intermediate frequency (IF) filtering within the PLL. This undesirable side effect of IF filtering is of importance in the design of phase-locked receivers.

As is apparent from a quick inspection of this chapter, nonlinear PLL theory is more subtle and intricate than the linear theory covered in Part I. The main results in this chapter are qualitative in nature, and they are illustrated by phase plane plots and bifurcation diagrams. To understand this material, the reader should have a good working knowledge of control systems and differential equations. At times, specialized ideas and techniques are required. Here, these are described in an elementary manner by using words and pictures, but they are described elsewhere in a more comprehensive manner. The mathematical terminology and ideas utilized here are standard, and they are described fully in the works of Perko[30] and Andronov et al.[31] Perko's book is widely available, and it is written at an undergraduate level. It is highly recommended reading for those who desire a comprehensive and in-depth understanding of the first- and second-order PLL models discussed in this book.

5.1 FIRST-ORDER PLL WITH CONSTANT FREQUENCY REFERENCE

This PLL contains no loop filter (F(s) = 1), and its reference consists of a sinusoid at frequency ω_i. Set n = m = 0, $a_o = b_o = 1$, and $\theta_1 = \omega_\Delta t$ in Equation 5.0-1 to obtain

$$\frac{d\phi}{dt} = \omega_\Delta - G\sin\phi \qquad (5.1\text{-}1)$$

as the equation that describes this first-order PLL. Loop detuning $\omega_\Delta \equiv \omega_i - \omega_0$ is assumed to be positive in what follows (if $\omega_\Delta < 0$, then replace ϕ by $-\phi$ to obtain the case $\omega_\Delta > 0$). The books by Viterbi,[3] Lindsey,[4] and Meyr and Ascheid[8] can be consulted as additional sources for the information supplied in this section.

The number of explicit constants in Equation 5.1-1 can be reduced by utilizing $\tau \equiv Gt$ to obtain

$$\frac{d\phi}{d\tau} = \omega_\Delta' - \sin\phi, \qquad (5.1\text{-}2)$$

where $\omega_\Delta' \equiv \omega_\Delta/G > 0$. Since G is significantly larger than unity in practical applications, τ is referred to as the *slow-time variable*. The equilibrium points of Equation 5.1-2 are examined in what follows. Also discussed is the time required for this PLL to achieve phase lock.

5.1.1 Phase Plane Analysis of a First-Order PLL

Figure 5.2 depicts typical plots of $d\phi/d\tau$ vs. ϕ for the first-order PLL; it represents graphically the differential equation given by Equation 5.1-2. Plots of this type are referred to as *phase planes*, and they are useful in analyzing first- and second-order nonlinear differential equations that have constant (i.e., time-invariant) coefficients. They are used extensively in this chapter.

The solid-line plot on Figure 5.2 was drawn to represent the case $0 \leq \omega_\Delta' < 1$. For ω_Δ' in this range, values of phase ϕ exist such that $d\phi/d\tau = 0$. These values of phase are known as *equilibrium points* of the equation. For Equation 5.1-2, these points can be divided into two classes described by

$$\phi_{1k} = 2\pi k + \sin^{-1}\omega_\Delta' \qquad (5.1\text{-}3)$$

$$\phi_{2k} = (2k-1)\pi - \sin^{-1}\omega_\Delta', \qquad (5.1\text{-}4)$$

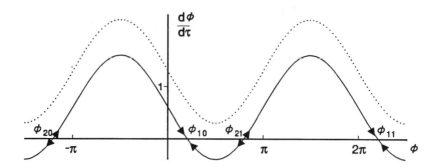

FIGURE 5.2
Phase plane for a first-order PLL. Solid (dotted) line graph depicts typical results for $0 < \omega_\Delta' < 1$ ($\omega_\Delta' > 1$).

where k is an integer. At the points described by Equation 5.1-3, the phase plane plot crosses the ϕ axis with a negative slope. The plot crosses the ϕ axis with a positive slope at the points described by Equation 5.1-4.

The set of equilibrium points given by Equation 5.1-3 represent stable phase-lock points of the PLL under consideration. This follows from a simple argument that can be applied to point ϕ_{10} displayed on Figure 5.2. First, assume that the phase error ϕ has a value which is slightly smaller than ϕ_{10} so that $d\phi/d\tau$ is positive. At this point, the phase error ϕ is increasing towards ϕ_{10}. Next, assume that ϕ has a value slightly larger than ϕ_{10} so that $d\phi/d\tau$ is negative. At this point, ϕ is decreasing towards ϕ_{10}. Hence, the point ϕ_{10} is a stable phase-lock point of the PLL, and it follows that Equation 5.1-3 describes the set of stable equilibrium points. In a similar manner, it is easy to argue that the equilibrium points described by Equation 5.1-4 are unstable.

In general, the equilibrium points may vanish as ω_Δ increases through some positive value Ω_h. Now, a loop that is phase locked becomes unlocked when its equilibrium points vanish. For this reason, Ω_h is known as the PLL *hold-in range*. On Figure 5.2, it is seen easily that the equilibrium points vanish for $\omega_\Delta' > 1$. For this reason, the hold-in range of the first-order PLL is $\Omega_h = G$.

Denoted here as Ω_p, the *pull-in range* of a first- or second-order PLL with a constant frequency reference is the largest value of ω_Δ for which there are no limit cycles on the phase plane. It is possible to argue that, regardless of initial conditions, phase lock is achieved if ω_Δ has a magnitude that is smaller than Ω_p. However, if $|\omega_\Delta|$ is larger than Ω_p, there exists initial conditions from which the PLL can start and never pull in successfully. From Figure 5.2 and the discussion provided above, it is seen easily that the pull-in range of a first-order PLL is $\Omega_p = G$. Hence, a first-order PLL has identical pull-in and hold-in ranges. As will be discussed in Section 5.3, this is not true in a second-order PLL that contains an imperfect integrator loop filter.

5.1.2 Phase Acquisition in a First-Order PLL

As described in the previous subsection, the first-order PLL always phase locks if $|\omega_\Delta'| < 1$. The amount of time required for the PLL to achieve phase lock can be found easily. First, solve Equation 5.1-2 for

$$d\tau = \frac{d\phi}{\omega_\Delta' - \sin\phi}. \tag{5.1-5}$$

Given that it starts at $\phi(0)$, the time τ required for the first-order PLL to reach phase $\phi(\tau)$ can be found by integrating Equation 5.1-5 to obtain

$$\tau = \int_{\phi(0)}^{\phi(\tau)} \frac{1}{-\sin x} dx$$

$$= \ln\left[\frac{\tan[\phi(0)/2]}{\tan[\phi(\tau)/2]}\right] \tag{5.1-6}$$

for the case $\omega_\Delta' = 0$ and

$$\tau = \int_{\phi(0)}^{\phi(\tau)} \frac{1}{\omega_\Delta' - \sin x} dx$$

$$= \frac{1}{\sqrt{1-(\omega_\Delta')^2}} \ln\left[\frac{\omega_\Delta' \tan\left[\frac{\phi(\tau)}{2}\right] - 1 - \sqrt{1-(\omega_\Delta')^2}}{\omega_\Delta' \tan\left[\frac{\phi(\tau)}{2}\right] - 1 + \sqrt{1-(\omega_\Delta')^2}}\right] \tag{5.1-7}$$

$$- \frac{1}{\sqrt{1-(\omega_\Delta')^2}} \ln\left[\frac{\omega_\Delta' \tan\left[\frac{\phi(0)}{2}\right] - 1 - \sqrt{1-(\omega_\Delta')^2}}{\omega_\Delta' \tan\left[\frac{\phi(0)}{2}\right] - 1 + \sqrt{1-(\omega_\Delta')^2}}\right]$$

for the case $0 < \omega_\Delta' < 1$. Given that phase starts between two equilibrium points, these results can be used to compute the amount of gain-normalized time that is required for the phase to reach a specified neighborhood of a stable equilibrium point.

Note that $d\phi/d\tau$ becomes small near an equilibrium point. This implies that τ, given by Equations 5.1-6 or 5.1-7, becomes large for either $\phi(\tau)$ or $\phi(0)$ near an equilibrium point. A general problem, known as *hang-up*, occurs when a PLL state dwells for a relatively long time near an unstable equilibrium point. Hang-up can be very troublesome in applications where rapid phase acquisition is essential.

Consider the dashed-line plot on Figure 5.2; this plot is typical of the case $|\omega_\Delta'| > 1$. Note that the value of $(\phi, d\phi/d\tau)$ never leaves this path. Also, the path does not intersect the ϕ axis; the system never makes it to an equilibrium point, so phase lock is not possible. In terms of quantities that can be observed in a first-order PLL, this path corresponds to a periodic beat note in the phase comparator output.

The period of this beat note can be computed. For the case $|\omega_\Delta'| > 1$, integrate Equation 5.1-5 from $\phi(0)$ to $\phi(0) + 2\pi$ to obtain

$$T_p \equiv \frac{2\pi}{\sqrt{(\omega_\Delta')^2 - 1}} \qquad (5.1\text{-}8)$$

for the period of the beat note. Note that T_p approaches infinity as ω_Δ' approaches unity from the right ($T_p \to \infty$ as $\omega_\Delta' \to 1^+$).

For the case $|\omega_\Delta'| > 1$, a simple numerical computation yields the beat note in the phase comparator output. After starting from any initial condition, Equation 5.1-2 can be integrated numerically over one period, and this result can be used to calculate the beat note. Figures 5.3a and 5.3b depict typical plots of $\sin\phi(\tau/T_p)$ for $\omega_\Delta' = 1.1$ and $\omega_\Delta' = 5.0$, respectively. Both plots show two complete periods of the beat note. The nearly sinusoidal results depicted by Figure 5.3b is characteristic of the large ω_Δ' case. However, as ω_Δ' decreases towards unity, the beat note becomes more unsymmetrical (so that it has a larger DC component) as is shown by Figure 5.3a.

5.2 A SECOND-ORDER PLL USING A PERFECT INTEGRATOR

The PLL considered in this section has a loop filter given by

$$F(s) = 1 + \alpha/s, \qquad (5.2\text{-}1)$$

where $\alpha > 0$ is the integrator gain constant. Also, the PLL under consideration has a reference source that supplies a sinusoid at the constant frequency ω_i. This section contains a phase-plane analysis of this loop. It is shown that this loop can achieve phase lock for an arbitrary value

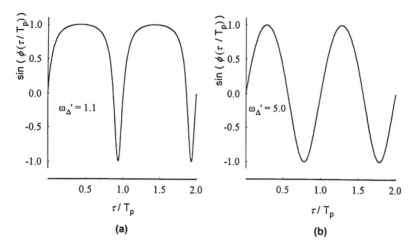

FIGURE 5.3
Normalized phase comparator output for a first-order PLL.

of ω_i and for arbitrary initial conditions. Finally, an approximation is developed for the time required for the loop to pull in to a frequency-locked state. The texts by Viterbi,[3] Lindsey,[4] and Meyr and Ascheid[8] can be consulted as additional sources of information on the nonlinear behavior of second-order PLL.

The equation that describes this loop can be obtained from Equation 5.0-1 by setting $n = m = 1$, $\theta_1 = (\omega_i - \omega_0)t = \omega_\Delta t$, $a_1 = b_1 = 1$, and $a_0 = \alpha$. In this equation, all remaining coefficients are set to zero. This procedure yields

$$\frac{d}{dt}\left[\frac{d\phi}{dt} - \omega_\Delta\right] = -G\left[\frac{d}{dt} + \alpha\right]\sin\phi, \quad (5.2\text{-}2)$$

or

$$\frac{d^2\phi}{dt^2} + G\cos\phi \frac{d\phi}{dt} + \alpha G \sin\phi = 0, \quad (5.2\text{-}3)$$

as the equation that describes the closed-loop phase error. Equation 5.2-3 is nonlinear and second order; it is said to be autonomous, or time invariant, since its coefficients α and G do not vary with time.

The gain constant G in Equation 5.2-3 can be eliminated by normalizing the time variable. This task is accomplished by using $\tau \equiv Gt$ so that $d\phi/dt = G d\phi/d\tau$ and $d^2\phi/dt^2 = G^2 d^2\phi/d\tau^2$. When τ is used as the independent variable, Equation 5.2-3 becomes

$$\frac{d^2\phi}{d\tau^2} + \cos\phi \frac{d\phi}{d\tau} + \alpha' \sin\phi = 0, \qquad (5.2\text{-}4)$$

where $\alpha' \equiv \alpha/G > 0$.

5.2.1 Stable and Unstable Equilibrium Points

Equation 5.2-4 can be represented by the first-order system

$$\frac{d\phi}{d\tau} = \dot{\phi}$$

$$\frac{d\dot{\phi}}{d\tau} = -(\cos\phi)\dot{\phi} - \alpha' \sin\phi \qquad (5.2\text{-}5)$$

in the variables ϕ and $\dot{\phi} \equiv d\phi/d\tau$.

This system can be written in a compact form by using state variable notation. First, define the 2 × 1 vector

$$\vec{X}(\tau) = \begin{bmatrix} x_1(\tau) \\ x_2(\tau) \end{bmatrix} \equiv \begin{bmatrix} \phi(\tau) \\ \dot{\phi}(\tau) \end{bmatrix}. \qquad (5.2\text{-}6)$$

Next, use $\vec{X}$ to write Equation 5.2-5 as

$$\frac{d\vec{X}}{d\tau} = \vec{F}(\vec{X})$$

$$\vec{F}(\vec{X}) = \begin{bmatrix} f_1(\vec{X}) \\ f_2(\vec{X}) \end{bmatrix} \equiv \begin{bmatrix} x_2 \\ -(\cos x_1)x_2 - \alpha' \sin x_1 \end{bmatrix}. \qquad (5.2\text{-}7)$$

Use of vector notation in this manner simplifies the symbolic representation and formal algebraic manipulation of Equation 5.2-5.

Another simplification in notation is realized in what follows by employing the standard *Euclidean norm* of the vector $\vec{X} = [x_1 \; x_2]^T$. This scalar quantity is denoted as $\|\vec{X}\|$, and it is defined as

$$\|\vec{X}\| \equiv \sqrt{x_1^2 + x_2^2}. \qquad (5.2\text{-}8)$$

Clearly, Equation 5.2-8 generalizes the notion of length to vectors, and it serves to simplify the developments that follow.

The *equilibrium points* of Equation 5.2-7 are defined as those values of $\vec{X}$ that cause $\vec{F}(\vec{X})$ to vanish. Values of $\vec{X}$ that are not equilibrium points are referred to as *ordinary points* of the equation. The equilibrium points of Equation 5.2-7 are found by inspection to be

$$\vec{X}_k \equiv \begin{bmatrix} k\pi \\ 0 \end{bmatrix}, \qquad (5.2\text{-}9)$$

where k is an integer. These points may be stable or unstable; if a point is stable, then it represents a possible lock point for the PLL.

The question of stability of an equilibrium point can be answered easily. This can be accomplished by considering solutions to Equation 5.2-7 of the form

$$\vec{X}(\tau) = \vec{X}_k + \vec{X}_\Delta(\tau), \qquad (5.2\text{-}10)$$

where $\vec{X}_\Delta(\tau)$ can be thought of as a "small" perturbation of the equilibrium point $\vec{X}_k$. Then $\vec{X}_k$ is *stable* if, for each $\varepsilon > 0$, there exists an $\delta > 0$ such that $\|\vec{X}_\Delta(0)\| < \delta$ implies that $\|\vec{X}_\Delta(\tau)\| < \varepsilon$ for all $\tau > 0$; that is, solutions that start "close" to $\vec{X}_k$ remain "close" to $\vec{X}_k$ for all time. A stable equilibrium point $\vec{X}_k$ is *asymptotically stable* if there exists an $\varepsilon > 0$ such that $\|\vec{X}_\Delta(0)\| < \varepsilon$ implies that $\|\vec{X}_\Delta(\tau)\|$ approaches zero as τ approaches infinity.

Fortunately, a simple test for asymptotic stability can be applied to the equilibrium points described by Equation 5.2-9. First, substitute Equation 5.2-10 into Equation 5.2-7, and use the Taylor expansion to write

$$\begin{aligned}\frac{d}{d\tau}(\vec{X}_k + \vec{X}_\Delta) &= \vec{F}(\vec{X}_k + \vec{X}_\Delta) \\ &= \vec{F}(\vec{X}_k) + F_x(\vec{X}_k)\vec{X}_\Delta \qquad (5.2\text{-}11) \\ &\quad + [\text{2nd and Higher - Order Terms}],\end{aligned}$$

where

$$\begin{aligned}F_x &= \begin{bmatrix} \partial f_1/\partial x_1 & \partial f_1/\partial x_2 \\ \partial f_2/\partial x_1 & \partial f_2/\partial x_2 \end{bmatrix} \\ &= \begin{bmatrix} 0 & 1 \\ x_2 \sin(x_1) - \alpha' \cos(x_1) & -\cos(x_1) \end{bmatrix}\end{aligned} \qquad (5.2\text{-}12)$$

denotes the 2 × 2 Jacobian matrix of $\vec{F} = [f_1 \ f_2]^T$. In Equation 5.2-11, the Jacobian is evaluated at the equilibrium point $\vec{X}_k$. Now, the first term cancels on both sides of Equation 5.2-11; the question of the stability of $\vec{X}_k$ can be answered by considering only the first-order terms

$$\frac{d\vec{X}_\Delta}{d\tau} = F_x(\vec{X}_k)\vec{X}_\Delta \qquad (5.2\text{-}13)$$

that remain after second- and higher-order terms are truncated. Equation 5.2-13 has constant coefficients (since $\vec{X}_k$ is constant), it is linear, and it is called the *first-variation equation* with respect to $\vec{X}_k$.

For every integer k, the equilibrium point $\vec{X}_k$ has the same stability properties as the equilibrium point $\vec{X}_\Delta = \vec{0}$ of Equation 5.2-13. Hence, the equilibrium point $\vec{X}_k$ is asymptotically stable if the two eigenvalues of $F_x(\vec{X}_k)$ lie in the left half of the complex plane. It is unstable if one or both eigenvalues lie in the right half of the plane.

The matrix $F_x(\vec{X}_k)$ can be computed by substituting Equation 5.2-9 into Equation 5.2-12 to obtain

$$F_x(\vec{X}_k) = \begin{bmatrix} 0 & 1 \\ -\alpha'(-1)^k & -(-1)^k \end{bmatrix}. \qquad (5.2\text{-}14)$$

This matrix has eigenvalues given by

$$\lambda_\pm = -(-1)^k/2 \pm \sqrt{1/4 - \alpha'(-1)^k}. \qquad (5.2\text{-}15)$$

For k even, both eigenvalues are in the left half of the plane, and $\vec{X}_k$ is asymptotically stable. In this case, $\vec{X}_k$ is a stable *node (focus)* if the eigenvalues are real (complex). For k odd, there is one eigenvalue in each of the right- and left-half planes, and $\vec{X}_k$ is unstable. Under these conditions, equilibrium point $\vec{X}_k$ is known as a *saddle point*.

5.2.2 A Phase Plane Analysis of the Perfect Integrator Case

The second-order PLL containing a perfect integrator in its loop filter has a very simple phase plane. On the phase plane, the equilibrium point locations are given by Equation 5.2-9, and they are independent of positive α'. As discussed in this section, the phase plane contains no limit cycles; hence, there is nothing to inhibit the natural pull-in process. In fact, the phase plane structure is invariant to changes in positive α', the only parameter in Equation 5.2-5 (with changes in a system parameter, phase plane *structural changes* are said to have occurred if equilibrium points and/or limit cycles appear or disappear).

The phase plane trajectories obey a first-order, ordinary differential equation. By simple division, it is possible to eliminate τ from Equation 5.2-5 and obtain

$$\frac{d\dot{\phi}}{d\phi} = -\cos\phi - \alpha' \left[\frac{\sin\phi}{\dot{\phi}}\right]. \qquad (5.2\text{-}16)$$

This equation describes $\dot{\phi}$ as a function of ϕ. Note that the right-hand side of this equation is 2π-periodic in ϕ. This implies that ϕ can be taken modulo-2π; solutions of Equation 5.2-16 may be restricted to the interval $-\pi \le \phi < \pi$. Note that the right-hand side of Equation 5.2-16 is indeterminate at the equilibrium points $(\phi, \dot{\phi}) = (k\pi, 0)$.

For the second-order PLL under consideration, plots of $\dot{\phi}$ vs. ϕ are known as *phase plane plots*. Such plots can be made by numerically integrating Equation 5.2-16 or 5.2-5. It can be shown that a unique solution of Equation 5.2-16 passes through every ordinary point (i.e., where the right-hand side of Equation 5.2-5 is not zero) on the phase plane. Such a solution curve is referred to as a *trajectory* of the equation; the uniqueness property implies that two trajectories cannot cross each other.

Other elementary properties of these trajectories can be determined by inspection of Equation 5.2-16. For very large $|\dot{\phi}|$, the first term on the right-hand side dominates the second term, and the trajectories are nearly sinusoidal. Also, note that trajectories that pass through ordinary points on the $\dot{\phi}$ axis (ϕ axis) do so with a slope of -1 (infinity).

As τ approaches either plus or minus infinity, an equilibrium point can be approached by trajectories; however, a trajectory cannot pass through an equilibrium point. As τ approaches infinity, there are two trajectories that approach each saddle point $(\phi, \dot{\phi}) = (k\pi, 0)$, where k is an odd integer. Likewise, there are two other trajectories that approach each saddle point as $\tau \to -\infty$. These four trajectories are known as *separatrices*. In a small neighborhood of each saddle point, separatrices appear to approach the saddle point along straight lines that form well-defined angles with the ϕ axis of the phase plane (Perko[30] gives simple formulas for these angles).

Figure 5.4 depicts an example of four separatrices in a small neighborhood of the saddle point φ_s. The separatrices appear on the figure as dashed lines. The separatrices with a negative (positive) slope approach the saddle point as $\tau \to \infty$ ($\tau \to -\infty$). Also drawn are a few trajectories that lie close to the separatrices. As is suggested by this figure, separatrices divide the phase plane into regions with distinctly

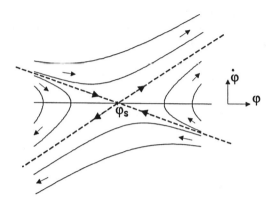

FIGURE 5.4
Trajectories in the neighborhood of a saddle point.

different properties; because of this, they play an important role in PLL analysis.

Figure 5.5a depicts a phase plane plot for $\alpha' = 1$; for this value of integrator gain, the loop has $\zeta = 1/\sqrt{2}$, a desirable value of damping factor. Trajectories with $\dot{\phi} > 0$ ($\dot{\phi} < 0$) move in the direction of increasing (decreasing) ϕ on the phase plane. Note the symmetrical nature of Figure 5.5a (and Equation 5.2-16); a trajectory remains a trajectory if both ϕ and $\dot{\phi}$ are negated. Also, when an upper half plane trajectory reaches $\phi = \pi$, it is restarted at $\phi = -\pi$ with the same value of $\dot{\phi}$; likewise, lower-half plane trajectories that reach $\phi = -\pi$ are restarted at $\phi = \pi$. For every 2π increase in ϕ, $\dot{\phi}$ decays by an amount that is small for large $\dot{\phi} > 0$, but the amount of decay increases with decreasing $\dot{\phi}$. In a neighborhood of $(\phi, \dot{\phi}) = (0, 0)$ on Figures 5.5a and 5.5b, trajectories appear to spiral towards the origin. This behavior is characteristic of the underdamped case when Equation 5.2-15, with k = 0, produces complex eigenvalues which lie in the left-half plane. Under these conditions on the eigenvalues, the phase plane origin is said to be a stable *focus*.

Figure 5.6a depicts a phase plane plot for $\alpha' = 1/8$; to a significant extent, the PLL is overdamped for this value of α'. For every 2π change in ϕ, the amount of change in $\dot{\phi}$ is smaller than was observed on Figure 5.5a (pull-in towards a phase-locked state occurs more slowly). In a neighborhood of $(\phi, \dot{\phi}) = (0, 0)$ on Figures 5.6a and 5.6b, trajectories appear to approach the origin along a tangent line. This behavior is characteristic of the overdamped case when Equation 5.2-15, with k = 0, produces real and unequal eigenvalues which lie in the left-half plane. Under these conditions on the eigenvalues, the phase plane origin is said to be a stable *node*.

Figures 5.5b and 5.6b depict plots of separatrices for the cases under investigation. The separatrices were computed by using an algorithm

NONLINEAR PLL BEHAVIOR IN THE ABSENCE OF NOISE

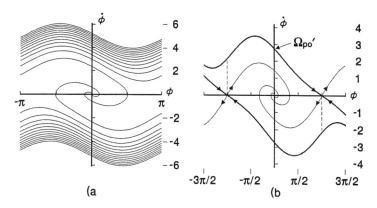

FIGURE 5.5
(a) Phase plane plot and (b) separatrices for a Type II PLL with $\alpha' = 1$.

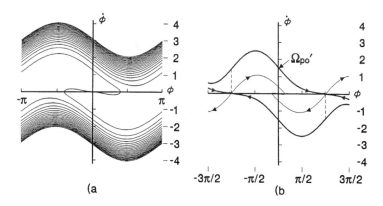

FIGURE 5.6
(a) Phase plane plot and (b) separatrices for a Type II PLL with $\alpha' = 1/8$.

similar to that given in Appendix 5.3.2. On plots in this chapter, separatrices are depicted by thick (thin) lines if they approach a saddle point as $\tau \to \infty$ ($\tau \to -\infty$). Also, thin-line separatrices serve to connect saddle and stable equilibrium points (on the figures, they *appear* to leave a saddle point and terminate on a stable equilibrium point). Between the thick-line separatrices on Figures 5.5b and 5.6b are trajectories that lead to a stable phase-lock point, and this is done without ϕ advancing or retarding multiples of 2π (trajectories are said to *slip cycles* when such 2π changes in ϕ occur). Trajectories that lie outside of these thick-lined separatrices slip one or more cycles before arriving at a stable equilibrium point.

The PLL is said to be *frequency locked* once its trajectory crosses one of the vertical dashed lines on Figures 5.5b or 5.6b. On each of these figures, the vertical dashed lines start on saddle points, and they serve as left and right boundaries on the *lock-in range*. The thick-line separatrices serve as upper and lower boundaries on the lock-in range. Once a trajectory enters the lock-in range, the PLL is frequency locked, and the trajectory will approach the stable equilibrium point as $\tau \to \infty$.

A general description can be given for the lock-in range associated with p_s, a stable equilibrium point. On the phase plane, draw vertical straight lines through the two saddle points that bracket p_s. Also, consider the lengths of these lines to be infinite. Then, the lock-in range for p_s consists of the set S of phase plane points that satisfy two conditions. First, p_s is in S, and it is the only equilibrium point contained in S. Secondly, through every ordinary point of S passes a trajectory that approaches p_s in the limit as time approaches infinity. And this is done without the trajectory crossing a vertical line drawn through a saddle point.

For the second-order PLL discussed in this section, the lock-in range for an equilibrium point decreases in size with decreasing α. This is illustrated by comparing Figures 5.5b and 5.6b which use similar vertical scales on gain-normalized $\dot{\phi}$. Also, Figure 5.7 supports this claim. For several values of α', this figure depicts pairs of separatrices plotted on a common $\dot{\phi} = G^{-1}d\phi/dt$ scale (G is fixed and α varies from one plot to the next). Each separatrix pair serves as upper and lower boundaries on a lock-in range.

The instantaneous phase and frequency errors are zero when the PLL under consideration is phase locked. Under these conditions, there is a maximum frequency step that can be applied to the input without the PLL slipping cycles. This value is known as the *pull-out frequency* Ω_{po}, and it is an important parameter in practical PLL design. The gain-normalized pull-out frequency Ω_{po}' is depicted on Figures 5.5b and 5.6b as the point where the thick-line separatrix crosses the $\dot{\phi}$ axis.

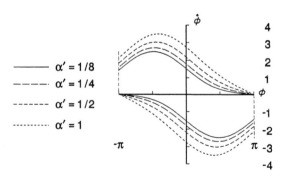

FIGURE 5.7
Lock-in range as a function of α' for a Type II PLL.

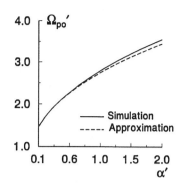

FIGURE 5.8
Normalized pull-out frequency as a function of α'.

The solid-line plot on Figure 5.8 depicts numerically computed values of Ω_{po}' for the range $0.1 \leq \alpha' \leq 2$ (this corresponds to values of damping factor that satisfy $0.353 \leq \zeta \leq 1.58$). These data were obtained by recording where computed separatrices cross the $\dot{\phi}$ axis. Also, Figure 5.8 displays a dashed-line plot of

$$\Omega_{po}' \approx 1.8\left(0.5 + \sqrt{\alpha'}\right). \quad (5.2\text{-}17)$$

This formula was obtained by gain-normalizing the commonly used approximation

$$\Omega_{po} \approx 1.8\omega_n(\zeta + 1), \quad (5.2\text{-}18)$$

where ζ and ω_n are the PLL damping factor and natural frequency, respectively (see p. 57 of Gardner).[6] As can be seen from Figure 5.8, the simple Approximation 5.2-17 is accurate for values of α' most often used in applications.

Figures 5.5 and 5.6 are generic phase plane examples for the second-order Type II PLL that contains a perfect integrator in its loop filter. They illustrate a very simple phase plane structure that is based on fixed equilibrium points and no limit cycles. As stated earlier, this structure is invariant to changes in positive α', the only parameter. That is, equilibrium points and limit cycles do not appear and/or disappear with changes in α', $\alpha' > 0$.

In the field of nonlinear PLL dynamics, this level of simplicity is exceptional. For example, it is in stark contrast to the relatively complex phase plane structure for the second-order Type I PLL that contains an imperfect integrator in its loop filter. As described in Section 5.3, both equilibrium points and limit cycles are on the phase plane of this Type I PLL. Furthermore, these quantities appear and disappear with changes in the PLL parameters.

5.2.3 Pull-In Properties of a Second-Order Type II PLL

A process known as *pull in* is illustrated by Figures 5.5a and 5.6a. It is the phenomenon by which a PLL achieves phase lock naturally and without assistance. The PLL discussed in this section has an infinite pull-in range. As shown in this subsection, phase lock occurs regardless of the value of ω_Δ and the initial conditions in effect when the loop is closed. Of course, these theoretical properties cannot be realized in applications; all practical PLL have an upper limit on their pull-in range. In fact, the pull-in phenomenon is unreliable and slow in many practical applications. In these applications, additional circuits, that aid the acquisition process, are added to the PLL (Gardner).[6]

A simple argument leads to the fact that the PLL considered here has an infinite pull-in range. First, multiply both sides of Equation 5.2-16 by $\dot{\phi}$ and integrate over $(-\pi, \pi)$ to obtain

$$\frac{1}{2}\left[\dot{\phi}^2(\pi) - \dot{\phi}^2(-\pi)\right] = -\int_{-\pi}^{\pi} \dot{\phi} \cos\phi \, d\phi - \alpha' \int_{-\pi}^{\pi} \sin\phi \, d\phi. \qquad (5.2\text{-}19)$$

The second integral on the right-hand side evaluates to zero, and the first can be evaluated by using integration by parts; these observations yield

$$\frac{1}{2}\left[\dot{\phi}^2(\pi) - \dot{\phi}^2(-\pi)\right] = \int_{-\pi}^{\pi} \sin\phi \, d\dot{\phi}. \qquad (5.2\text{-}20)$$

Now, Equation 5.2-16 can be used to produce

$$d\dot{\phi} = -(\cos\phi)d\phi - \alpha'\left[\frac{\sin(\phi)}{\dot{\phi}}\right]d\phi, \qquad (5.2\text{-}21)$$

and this result can be substituted into Equation 5.2-20 to obtain

$$\frac{1}{2}\left[\dot{\phi}^2(\pi) - \dot{\phi}^2(-\pi)\right] = -\alpha' \int_{-\pi}^{\pi} \frac{\sin^2\phi}{\dot{\phi}} d\phi. \qquad (5.2\text{-}22)$$

Consider what Equation 5.2-22 implies about trajectories which lie in the upper half of the phase plane. For these trajectories, the integrand $(\sin^2\phi)/\dot{\phi}$ is positive, and the right-hand side of Equation 5.2-22 is negative. Hence, $\dot{\phi}$ decreases each time the PLL slips a cycle and the phase increases by 2π. In a similar manner, it is seen easily that trajectories in the lower-half plane experience an increase in $\dot{\phi}$ each time the PLL slips a cycle and the phase decreases by 2π. Therefore, regardless of initial conditions, $|\dot{\phi}|$ decreases each time the PLL slips a cycle; this fact implies an infinite pull-in range for the PLL under consideration.

The *pull-in time* is a function of the PLL initial condition, and it is defined as the time required for the PLL to go from this initial condition to a frequency-locked state. A frequency-locked state is reached when the PLL phase plane trajectory enters the lock-in region and cycle slippage no longer occurs. In terms of Figure 5.7, the lock-in region is bounded on its sides by the vertical dashed lines, and it is bounded above and below by separatrices. The nonlinear nature of the pull-in phenomenon prevents the determination of an exact formula for the pull-in time.

A simple approximation is developed in Appendix 5.2.1 for the pull-in time of the second-order PLL considered in this section. It employs a number of related assumptions. The initial frequency error is set equal to the quantity ω_Δ, and the assumption is made that ω_Δ is large compared to the frequency error in effect when the PLL trajectory enters the lock-in region and cycle slipping no longer occurs. This means that the PLL will slip a large number of cycles before locking up. Also, over most of the pull-in time, the rate of frequency error decrease is small. By using these assumptions, the pull-in time can be approximated as

$$T_d \approx \frac{\omega_\Delta^2}{\alpha G^2}. \qquad (5.2\text{-}23)$$

Hence, the fact that the PLL discussed in this section has an infinite pull-in range is tempered by the observation that the time required to pull in grows at a rate that is proportional to the square of the initial frequency error when this value is large. In many applications, the pull-in phenomenon is too slow and unreliable. In these cases, the PLL contains additional circuitry specifically designed to aid the process of obtaining phase lock.

5.3 A SECOND-ORDER PLL CONTAINING AN IMPERFECT INTEGRATOR LOOP FILTER

The PLL considered in this section has a loop filter given by Equation 3.3-1 that is repeated here as

$$F(s) = \frac{s+a}{s+b}, \qquad (5.3\text{-}1)$$

where it is assumed that a > b > 0. Also, the PLL has a reference source that supplies a sinusoid at the constant frequency ω_i. As discussed in the sections that follow, this PLL has a more complex phase plane structure than the Type II loop considered in Section 5.2. The texts by Viterbi,[3] Lindsey,[4] and Meyr and Ascheid[8] provide additional sources of information on the nonlinear behavior of the second-order PLL with loop filter given by Equation 5.3-1. Also, the works of Perko[30] and Andronov et al.[31,32] should be consulted for bifurcation theory that can be applied directly to the PLL under consideration. Finally, the papers by Endo and Tada[54] and Stensby[55-57,61] should be consulted as alternate sources of information on bifurcations in the second-order PLL containing an imperfect integrator.

From Equation 5.0-1, the equation that describes the PLL with loop Filter 5.3-1 is

$$\left[\frac{d}{dt}+b\right]\left[\frac{d\phi}{dt}-\frac{d\theta_1}{dt}\right]=-G\left[\frac{d}{dt}+a\right]\sin\phi. \tag{5.3-2}$$

However, the relative input phase is $\theta_1 \equiv \omega_\Delta t$; this restriction and simple algebra lead to

$$\frac{d^2\phi}{dt^2}+(b+G\cos\phi)\frac{d\phi}{dt}+aG\sin\phi=b\omega_\Delta \tag{5.3-3}$$

as the second-order, autonomous (i.e., time invariant) equation that describes the PLL under consideration. As discussed in Section 2.2.1, it is assumed that G and $\omega_\Delta \equiv \omega_i - \omega_0$ are positive.

The explicit dependence of Equation 5.3-3 on G can be eliminated by normalizing time so that $\tau = Gt$. This substitution leads to

$$\frac{d^2\phi}{d\tau^2}+(b'+\cos\phi)\frac{d\phi}{d\tau}+a'\sin\phi=b'\omega_\Delta', \tag{5.3-4}$$

where

$$a' \equiv a/G$$
$$b' \equiv b/G \tag{5.3-5}$$
$$\omega_\Delta' \equiv \omega_\Delta/G.$$

A similar time-normalization procedure was carried out for the PLL studied in Sections 5.1 and 5.2.

The second-order Equation 5.3-4 can be written as the system

$$\frac{d\phi}{d\tau} = \dot{\phi}$$

$$\frac{d\dot{\phi}}{d\tau} = -(b' + \cos\phi)\dot{\phi} - a'\sin\phi + b'\omega'_\Delta .$$

(5.3-6)

This system was numerically integrated to produce several of the phase plane plots (by plotting $\dot{\phi}$ against ϕ) that are displayed in Sections 5.3.3 and 5.3.4.

Time τ can be eliminated from Equation 5.3-6 by simply dividing the equations; the result is

$$\frac{d\dot{\phi}}{d\phi} = -(b' + \cos\phi) + \frac{b'\omega'_\Delta - a'\sin\phi}{\dot{\phi}} .$$

(5.3-7)

This equation directly describes trajectories on the phase plane. It is the counterpart of Equation 5.2-16 which describes trajectories for the perfect integrator case discussed in Section 5.2.

5.3.1 Stable and Unstable Equilibrium Points

By using the vector

$$\vec{X}(\tau) \equiv \begin{bmatrix} x_1(\tau) \\ x_2(\tau) \end{bmatrix} = \begin{bmatrix} \phi(\tau) \\ \dot{\phi}(\tau) \end{bmatrix},$$

(5.3-8)

the system of equations given by Equation 5.3-6 can be expressed in state variable form as

$$\frac{d\vec{X}}{d\tau} = \vec{F}(\vec{X})$$

$$\vec{F}(\vec{X}) = \begin{bmatrix} f_1(\vec{X}) \\ f_2(\vec{X}) \end{bmatrix} \equiv \begin{bmatrix} x_2 \\ -(b' + \cos x_1)x_2 - a'\sin x_1 + b'\omega'_\Delta \end{bmatrix}.$$

(5.3-9)

The equilibrium points of Equation 5.3-9 are those values of $\vec{X}$ that satisfy $\vec{F}(\vec{X}) = \vec{0}$. For

$$0 \le \omega_\Delta' < \Omega_h' \equiv \left[\frac{a'}{b'}\right], \qquad (5.3\text{-}10)$$

they can be divided into two sets given by

$$\text{SET \#1:} \qquad \vec{X}_{2k-1} = \begin{bmatrix} (2k-1)\pi - \phi_0 \\ 0 \end{bmatrix} \qquad (5.3\text{-}11)$$

$$\text{SET \#2:} \qquad \vec{X}_{2k} = \begin{bmatrix} 2k\pi + \phi_0 \\ 0 \end{bmatrix}, \qquad (5.3\text{-}12)$$

where

$$\phi_0 \equiv \sin^{-1}(\omega_\Delta'/\Omega_h'), \qquad (5.3\text{-}13)$$

and k is an integer. For values of ω_Δ' that satisfy Equation 5.3-10, Equation 5.3-13 implies that ϕ_0 satisfies $0 \le \phi_0 < \pi/2$.

Two features distinguish these equilibrium points from those given by Equation 5.2-9 for the PLL based on a perfect integrator loop filter. First, the equilibrium points given by Equations 5.3-11 and 5.3-12 depend on PLL parameters. Secondly, they disappear as detuning ω_Δ' increases through Ω_h'. When this happens, structural changes occur in the phase plane. When no equilibrium points exist, the PLL eventually reaches a steady-state, out-of-lock condition characterized by a periodic beat note in the phase comparator output. Under this condition, the phase error has a magnitude which increases without bound as time becomes large. For $\omega_\Delta' > \Omega_h'$, all phase plane trajectories approach a periodic limit cycle.

Equilibrium points Equations 5.3-11 and 5.3-12 have a stability that can be determined by using the variational equation approach developed in Section 5.2. With respect to $\vec{X} = [x_1 \ x_2]^T$, the first variation Equation 5.2-13 has, when written for the System 5.3-9, a coefficient matrix given by

$$F_x(\vec{X}) = \begin{bmatrix} 0 & 1 \\ x_2 \sin(x_1) - a'\cos(x_1) & -b' - \cos(x_1) \end{bmatrix}. \qquad (5.3\text{-}14)$$

Now, substitute Equation 5.3-11 for $\vec{X}$ in Equation 5.3-14 to obtain a matrix with eigenvalues given by

$$\lambda_\pm = -\frac{1}{2}[b' - \cos\phi_0] \pm \sqrt{\frac{1}{4}[b' - \cos\phi_0]^2 + a'\cos\phi_0}, \qquad (5.3\text{-}15)$$

where ϕ_0 is given by Equation 5.3-13. Note that both the right- and left-half planes contain an eigenvalue since $\cos\phi_0 > 0$. Hence, the set of equilibrium points given by Equation 5.3-11 are unstable, and they are saddle points. In a similar manner, substitute Equation 5.3-12 for $\vec{X}$ in Equation 5.3-14 and obtain a matrix that has the eigenvalues

$$\lambda_\pm = -\frac{1}{2}\left[b' + \cos\phi_0\right] \pm \sqrt{\frac{1}{4}\left[b' + \cos\phi_0\right]^2 - a'\cos\phi_0}. \quad (5.3\text{-}16)$$

Both of these eigenvalues are in the left-half plane. This implies that the equilibrium points given by Equation 5.3-12 are asymptotically stable; they are stable nodes (foci) if the eigenvalues are real valued (complex valued).

The existence of stable equilibrium points *is necessary* in order for the PLL to pull into a stable phase-locked state. Equivalently, the nonexistence of equilibrium points implies that pull-in cannot occur. However, the existence of stable equilibrium points is *not sufficient* to guarantee that pull-in occurs. It is possible for a stable phase-lock state to exist and pull-in not occur. Under this condition, $|\phi|$ becomes unbounded with time, there is a periodic beat note in the phase comparator output, and the PLL is said to be *false locked*. As demonstrated in Section 5.3.3, a false lock state corresponds to a limit cycle on the PLL phase plane.

5.3.2 Phase Plane Structure Dependent on ω_Δ' — the Values Ω_p', Ω_2', and Ω_h'

For a second-order PLL, the phase plane can be subdivided into a number of separate regions, or cells (for a complete discussion of this topic, see Chapter 6 of Andronov et al.).[32] As $\tau \to \infty$, in any given cell, all trajectories have the same limiting behavior, and different cells are associated with different limiting behavior. That is, cells are distinguished by the limiting behavior of the trajectories that they contain. On a phase plane where phase is interpreted in a modulo-2π manner, a cell may not be a physically connected region. Instead, a cell may be the union of disjoint regions. Collectively, these cells define a cellular structure, or simply *structure*, for the phase plane. These ideas are made clear by the examples discussed in the two sections that follow.

As illustrated by the examples discussed below, equilibrium points, limit cycles, and separatrices are used to define the boundaries of the above-mentioned cells. In general, these quantities depend on PLL parameters; in what follows, their dependence on ω_Δ' is of interest. This implies that the cells and their boundaries vary as functions of ω_Δ'. At most values of ω_Δ', cell features and boundaries vary continuously, and the number of distinct cells does not change. However, there are discrete values of ω_Δ' where equilibrium points and limit cycles appear or disappear, and discontinuous changes occur in cell features and boundaries. At these discrete values of ω_Δ', the cellular structure of the phase plane changes. In the two sections that follow, the quantities Ω_p', Ω_2', and Ω_h' are used to denote specific values of ω_Δ' where the phase plane (for the PLL considered in this section) undergoes structural changes. But first, in the remainder of this section, these three values are assigned names that are based on simple qualitative characteristics.

Inspection of Equations 5.3-11 through 5.3-13 leads to the conclusion that equilibrium points exist (do not exist) for values of ω_Δ' that satisfy $0 \leq |\omega_\Delta'| \leq \Omega_h'$ ($|\omega_\Delta'| > \Omega_h'$), where Ω_h' is defined by Equation 5.3-10. A locked loop will become unlocked as $|\omega_\Delta'|$ increases through Ω_h'; for this reason, the quantity Ω_h' is denoted as the gain-normalized *hold-in range* for the imperfect integrator PLL (the hold-in range is $\Omega_h \equiv G\Omega_h'$). For values of ω_Δ' that satisfy $|\omega_\Delta'| > \Omega_h'$, all phase plane trajectories lead to a stable limit cycle; the whole phase plane belongs to a single cell.

As discussed above, $|\omega_\Delta'|$ can be less than the gain-normalized hold-in range Ω_h', so that a stable phase-lock state exists, but the PLL may be false locked (i.e., trapped by a limit cycle). However, there exists a value Ω_p', $0 < \Omega_p' < \Omega_h'$, with an important property: if ω_Δ' satisfies $|\omega_\Delta'| < \Omega_p'$, then no false lock states (i.e., limit cycles) exist, and phase lock is achieved regardless of initial conditions when the loop is closed (again, the whole phase plane belongs to a single cell). Because of this physical characteristic, the value Ω_p' is known as the gain-normalized *pull-in range*, and it is an important parameter in practical PLL design (Ω_p denotes the pull-in range and $\Omega_p' \equiv \Omega_p/G$). Unfortunately, an exact formula for Ω_p' cannot be obtained. However, Ω_p' can be computed numerically (Stensby).[57] Also, an approximate formula for Ω_p' is available; it is developed in Appendix 5.3.1, and it is discussed in Section 5.3.3.

Like the parameter Ω_h', the quantity Ω_p' denotes a value of $|\omega_\Delta'|$ where the phase plane structure changes (this is illustrated in the next section). However, instead of equilibrium points disappearing, either one or two limit cycles appear (they *bifurcate*) on the phase plane as $|\omega_\Delta'|$ increases through Ω_p'. The number of limit cycles that appear depends on the value of G (a fixed loop filter is assumed). As discussed in Section 5.3.3, when $|\omega_\Delta'|$ increases through Ω_p', exactly two limit cycles appear for the *high-gain case*. This case is most prevalent in practical applications. Alternatively, the *low-gain case* is discussed in Section 5.3.4. For this case, only one limit cycle appears as $|\omega_\Delta'|$ increases through Ω_p'.

For the high-gain case only, the *half-plane pull-in frequency* is denoted as Ω_2', where $\Omega_h' > \Omega_2' > \Omega_p'$ (to compute Ω_2' see Stensby and

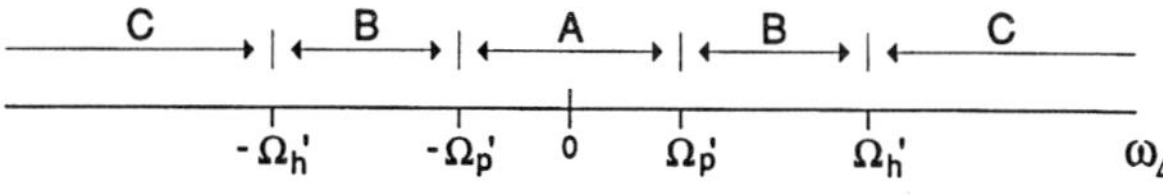

```
        C        B        A        B        C
   ──────→|←────→|←────→|←────→|←────→|←──────
   ───────┼──────┼──────┼──────┼──────┼──────────
        -Ω_h'  -Ω_p'    0     Ω_p'   Ω_h'         ω_Δ'
```

Region A: $|\omega_\Delta'| < \Omega_p'$
Stable equilibrium points exist. Limit cycles which hinder the pull-in process do not exist (pull-in always occurs).

Region B: $\Omega_p' < |\omega_\Delta'| < \Omega_h'$
Some trajectories lead to a stable equilibrium point while others lead to a stable periodic limit cycle (false lock).

Region C: $|\omega_\Delta'| > \Omega_h'$
No equilibrium points exist, and all trajectories lead to a stable limit cycle.

FIGURE 5.9
The phase plane structure depends on ω_Δ'.

Harb).[61] As discussed in Section 5.3.3, for values of ω_Δ' that satisfy $0 \leq \omega_\Delta' \leq \Omega_2'$ ($-\Omega_2' \leq \omega_\Delta' \leq 0$), all trajectories on the lower (upper) half of the phase plane lead, as $\tau \to \infty$, to a phase-locked state. That is, for the ranges of ω_Δ' that are given in the previous sentence, the PLL always pulls in from any point on half of the phase plane. For the low-gain case, there is no counterpart to the half-plane pull-in frequency parameter Ω_2'.

Figure 5.9 summarizes general properties of Ω_p' and Ω_h' that hold for both the high- and low-gain cases. In considering Figure 5.9, assume that the PLL parameters G, a, and b are held fixed; only detuning ω_Δ is allowed to vary. For ω_Δ' in Region A on the figure, the phase plane contains no limit cycles. Under this condition, all phase plane trajectories which are not separatrices lead to a stable equilibrium state (pull-in is successful). As illustrated in the two sections that follow, for ω_Δ' in Region B, some points on the phase plane have trajectories passing through them that lead to a stable equilibrium point; that is, pull-in occurs for some initial conditions. But, there are other points on the phase plane for which this is not true; passing through these points are trajectories that lead to a stable, periodic limit cycle. Pull-in fails when this happens, and the PLL false locks. Finally, phase lock is impossible for ω_Δ' in Region C on the figure; all phase plane trajectories lead to a stable, periodic limit cycle.

With the introduction of high- and low-gain cases, the gain parameter G has been reintroduced (it does not appear as an explicit parameter in Equations 5.3-4 and 5.3-6). In the discussion of the high- and low-gain cases, the loop filter is considered fixed (the parameters a = 100 and b = 20 are used throughout Section 5.3), and the parameters G and ω_Δ are allowed to vary. This approach is preferred over one that develops two cases based on large and small values of the gain-normalized pair (a′, b′) of loop filter constants. Also, the approach utilized here is consistent with the PLL literature that references extensively high-gain PLLs and the high-gain case.

5.3.3 The High-Gain Case

The high-gain case is discussed most frequently in the PLL literature, and it is the most important from an application standpoint. This section discusses important properties that are characteristic of the high-gain case. Sections 5.3.4 and 5.3.5 detail techniques that are useful for determining the lower limit on gain G for the high-gain case to apply.

The high-gain case is characterized by the occurrence of *saddle node bifurcation* at $\omega_\Delta' = \Omega_p'$ (also known as bifurcation from a limit cycle of the second kind). As illustrated by the phase plane diagrams in this section, saddle-node bifurcation is distinguished by the creation of a limit cycle on the phase plane for $\omega_\Delta' = \Omega_p'$; furthermore, as ω_Δ' increases beyond Ω_p', this limit cycle splits apart into two limit cycles on the phase plane. One of these limit cycles is stable; it is denoted here as Γ_s. With increasing ω_Δ', Γ_s moves upward on the phase plane; it approaches infinity as $\omega_\Delta' \to \infty$. The second limit cycle is unstable; it is denoted here as Γ_u. As ω_Δ' increases, the unstable limit cycle moves downward on the phase plane until a point $\omega_\Delta' = \Omega_2'$, $\Omega_p' < \Omega_2' < \Omega_h'$, is reached where it terminates on an unstable separatrix cycle (at $\omega_\Delta' = \Omega_2'$, Γ_u bifurcates from an unstable separatrix cycle). The unstable limit cycle does not exist for $\omega_\Delta' > \Omega_2'$. In this section the behavior just summarized is illustrated by using phase plane plots.

For the high-gain case, an exact formula does not exist for the PLL normalized pull-in range Ω_p'. However, by using numerical techniques, it is possible to compute an accurate value for Ω_p' (Stensby).[57] Alternatively, as shown in Appendix 5.3.1, Ω_p' can be approximated as

$$\Omega_p' \approx \sqrt{2\frac{a'}{b'}} \qquad (5.3\text{-}17)$$

for the high-gain case. As is demonstrated in Appendix 5.3.1, this simple formula can produce accurate results for the high-gain case with $a' \gg b'$.

Figure 5.10a depicts a phase plane plot for the case $a' = 0.5$, $b' = 0.1$, and $\omega_\Delta' = 2$ (or $a = 100$, $b = 20$, $G = 200$, and $\omega_\Delta = 400$). For these parameters, a value of 3.007 was computed for Ω_p' by using the computer-based method referenced above. Hence, ω_Δ' lies in Region A that is defined on Figure 5.9, and no limit cycles exist on the phase plane to hinder the pull-in process. Successful pull-in is achieved by this PLL

NONLINEAR PLL BEHAVIOR IN THE ABSENCE OF NOISE

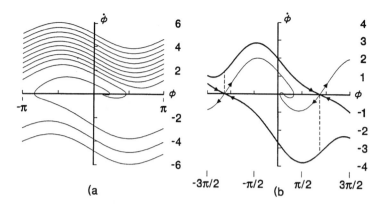

FIGURE 5.10
(a) Phase plane plot and (b) separatrices for a Type I PLL with a' = 0.5, b' = 0.1, and $\omega_\Delta' = 2$.

after it is started from any ordinary point on the phase plane (excluding the theoretical possibility of hang-up). Hence, the complete phase plane can be considered as a single cell. As $\tau \to \infty$, all trajectories (that are not separatrices) in this cell have the same limiting behavior; they lead to a stable equilibrium point.

Note the similarities and differences between this phase plane plot and those given in Section 5.2 for the perfect integrator case. Like the phase plane plots given for the perfect integrator case, Figure 5.10a has upper-half (lower-half) plane trajectories which traverse from left to right (right to left). Unlike the phase plane plots given for the perfect integrator case, the trajectories on Figure 5.10a lead to the stable equilibrium point $\phi_0 = \sin^{-1}(\omega_\Delta'/\Omega_h) \approx 0.41$, and there is no symmetry between upper- and lower-half plane trajectories (since $\omega_\Delta' \neq 0$ here).

Figure 5.10b depicts saddle points, separatrices, and the lock-in range for this case. Separatrices that approach a saddle point as $\tau \to \infty$ ($\tau \to -\infty$) are depicted by thick (thin) lines. The lock-in range is bounded on the left and right by the vertical dashed lines, and it is bounded above and below by the thick-lined separatrices. If started at a point within the lock-in range, a trajectory will lead to the stable phase-lock point shown, and it will do this without slipping cycles.

Figure 5.11a illustrates what happens when ω_Δ' increases to $\omega_\Delta' = 2.9$; this value of normalized detuning is slightly less than the computed $\Omega_p' \approx 3.007$ (a' and b' are unchanged from Figure 5.10). Hence, ω_Δ' still lies in Region A that is defined on Figure 5.9. Pull-in occurs from any ordinary point on the phase plane (excluding the theoretical possibility of hang-up); there are no limit cycles on the phase plane to hinder the pull-in process. Figure 5.11a has the same structure as Figure 5.10a (i.e., each phase plane defines a single cell). However, for ϕ in the vicinity of 2, the average value of $\dot\phi$ (when averaged over a 2π cycle of ϕ)

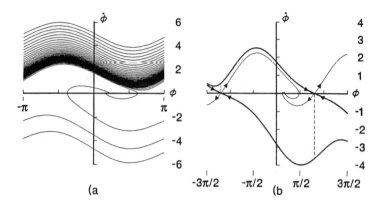

FIGURE 5.11
(a) Phase plane plot and (b) separatrices for a Type I PLL with a' = 0.5, b' = 0.1, and ω_Δ' = 2.9.

decreases very little from one 2π cycle to the next, and pull-in proceeds slowly (note the dark band in the upper-half plane where successive traces appear to "pile up").

Figure 5.11b depicts separatrices and the lock-in range for the case ω_Δ' = 2.9. A comparison of Figures 5.10b and 5.11b shows that an increase in ω_Δ' causes a notable decrease in the upper-half plane portion of the lock-in region. Much less affected is the lower-half plane portion of the lock-in region.

As ω_Δ increases, so that ω_Δ' passes through the value Ω_p', a limit cycle is created (at $\omega_\Delta' = \Omega_p'$), and it splits apart into a pair of limit cycles. That is, saddle node bifurcation occurs at $\omega_\Delta' = \Omega_p'$. Neither limit cycle exists for $\omega_\Delta' < \Omega_p'$. Figure 5.12a depicts this phenomenon for ω_Δ' = 3.1, a value slightly larger than the computed $\Omega_p' \approx 3.007$ (a' and b' are unchanged from the previous two figures). On Figure 5.12a, the limit cycles are displayed as dashed lines, and they are lettered as Γ_s and Γ_u. Note that Figure 5.12a displays only enough traces to serve the present discussion (some traces were "thinned out" to prevent the appearance of black "clumps of traces" that would hinder the visibility of individual traces).

On Figure 5.12a, the upper limit cycle Γ_s is stable. Trajectories that lie between the limit cycles, and trajectories that lie above Γ_s, converge to this stable limit cycle. This is indicated on Figure 5.12a by the pair of arrows that point towards Γ_s. On the phase plane, Γ_s moves monotonically upward with increasing ω_Δ'. It approaches infinity as $\omega_\Delta' \to \infty$.

On Figure 5.12a, the lower limit cycle Γ_u is unstable. Trajectories that lie between the limit cycles, and trajectories that lie immediately below Γ_u, diverge from this unstable limit cycle. This is indicated on Figure 5.12a by the pair of arrows that point away from Γ_u. On the

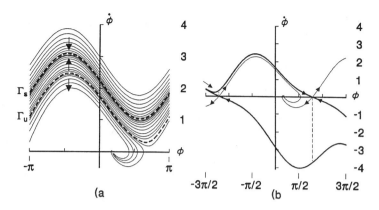

FIGURE 5.12
(a) Phase plane plot and (b) separatrices for a Type I PLL with $a' = 0.5$, $b' = 0.1$, and $\omega_\Delta' = 3.1$.

phase plane, Γ_u moves monotonically downward with increasing ω_Δ' (until it disappears as discussed below).

Unstable limit cycle Γ_u splits the phase plane into two cells. The trajectories that lie above Γ_u are in one cell; as $\tau \to \infty$ these trajectories converge to Γ_s. The trajectories that lie below Γ_u are in the second cell; as $\tau \to \infty$ these trajectories converge to an equlibrium point. As ω_Δ' increases through Ω_p', the phase plane splits into two cells, and the structure of the phase plane changes. For the loop filter pole and zero values that are employed in this section, this two-cell phase plane structure remains in effect for ω_Δ' in the range $\Omega_p' \leq \omega_\Delta' \leq \Omega_2' \approx 3.319$, where Ω_2' is discussed below.

Once again, detuning ω_Δ is increased, and Figure 5.13a depicts the phase plane for $\omega_\Delta' = 3.319$ (a' and b' are unchanged from the previous three figures). Note that, on the phase plane, Γ_s has moved upward from its position on Figure 5.12a. More importantly, as ω_Δ' approaches $\Omega_2' \approx 3.319$ from the left (i.e., $\omega_\Delta' \to (\Omega_2')^-$, $\Omega_2' \approx 3.319$), Γ_u moves downward on the phase plane, and it approaches a special type of periodic path that connects saddle points.

By using a thick dashed line, Figure 5.13b illustrates more clearly this saddle-to-saddle, periodic path. This dashed-line path is no longer classified as a limit cycle (it no longer has all of the qualitative properties of a limit cycle). Instead, it is known as a *saddle-to-saddle separatrix* or a *separatrix cycle*. Unstable limit cycle Γ_u is said to *bifurcate* from this separatrix cycle that exists at $\omega_\Delta' = \Omega_2'$, and Γ_u does not exist for $\omega_\Delta' > \Omega_2'$.

The separatrix cycle, from which Γ_u bifurcates, exists for only one value of ω_Δ', denoted here as Ω_2' (given that a' and b' are fixed). As discussed below and in Appendix 5.3.3, this separatrix cycle is externally

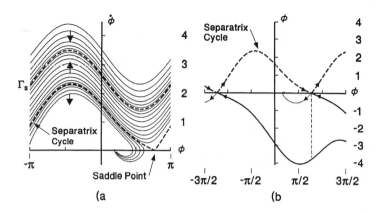

FIGURE 5.13
(a) Phase plane plot and (b) separatrices for a Type I PLL with $a' = 0.5$, $b' = 0.1$, and $\omega_\Delta' = 3.319$.

unstable for the high-gain case. Hence, the quantity Ω_2' is defined as the value of ω_Δ' at which the unstable limit cycle Γ_u bifurcates from an externally unstable separatrix cycle.

The separatrix cycle can be constructed from separatrices that emanate from adjacent saddle points; these separatrices "match up" at $\omega_\Delta' = \Omega_2'$. This is illustrated by Figures 5.14a through 5.14c, each of which shows parts of two (and only two) separatrices. For $\omega_\Delta' = 3.000 < \Omega_2'$ (Figure 5.14a), the separatrix with a positive slope lies below the separatrix with a negative slope (slopes are calculated at the saddle points). For $\omega_\Delta' = 3.319 = \Omega_2'$ (Figure 5.14b), the separatrices match up and form a separatrix cycle. Finally, for $\omega_\Delta' = 3.6000 > \Omega_2'$ (Figure 5.14c), the separatrix with a positive slope lies above the separatrix with a negative slope. This separatrix cycle is said to be a *structurally unstable path* on the phase plane since it only exists for $\omega_\Delta' = \Omega_2'$ (see p. 100 of Andronov et al.).[31]

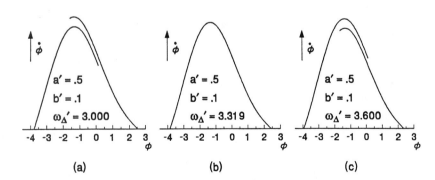

FIGURE 5.14
Formation of a separatrix cycle at $\omega_\Delta' = \Omega_2' = 3.319$ (high-gain case).

NONLINEAR PLL BEHAVIOR IN THE ABSENCE OF NOISE 147

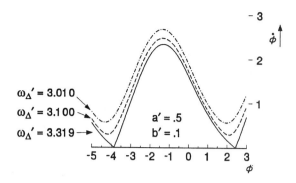

FIGURE 5.15
Bifurcation from an unstable separatrix cycle (high-gain case).

As shown on Figure 5.13a for the case $\omega_\Delta' = \Omega_2'$, trajectories converge to the stable limit cycle Γ_s if they lie above the separatrix cycle (Γ_s lies above the separatrix cycle). No matter how close a trajectory comes to the top side of the separatrix cycle, it will diverge from the separatrix cycle. The separatrix cycle (that exists at $\omega_\Delta' = \Omega_2'$) is said to be *externally unstable* since all trajectories diverge from its top side. Hence, the separatrix cycle that exists at $\omega_\Delta' = \Omega_2'$ is both externally unstable and structurally unstable (these are two distinct concepts).

From Figure 5.13b, note that the separatrix cycle forms the upper boundary of the lock-in region. On the figure, the lock-in region lies below the separatrix cycle, above the lower-half plane separatrix (the one drawn with the thick line), and to the left of the vertical dashed line (that lies in the lower-half plane). A trajectory cannot leave the lock-in region; it will converge, without slipping cycles, to the stable equilibrium point shown on the figure.

As suggested by Figures 5.12a and 5.13a, unstable limit cycle Γ_u bifurcates on an externally unstable separatrix cycle as ω_Δ' approaches Ω_2' from the left. This is depicted in more detail by Figure 5.15. This figure shows Γ_u for $\omega_\Delta' = 3.010$ and $\omega_\Delta' = 3.100$, and it shows the separatrix cycle at $\omega_\Delta' = 3.319$. As ω_Δ' approaches $\Omega_2' = 3.319$ from the left, the figure suggests that Γ_u decreases towards the separatrix cycle, and separatrix cycle bifurcation occurs at $\omega_\Delta' = \Omega_2'$.

Once again, detuning ω_Δ is increased, and Figure 5.16 depicts the phase plane for $\omega_\Delta' = 4.9$ (a' and b' are unchanged from the previous six figures). This phase plane is typical for values of ω_Δ' in the range $\Omega_2' < \omega_\Delta' < \Omega_h'$ where only stable Γ_s and equilibrium points exist. Two features distinguish Figure 5.16b from the previous plots of separatrices. First, consider the separatrices that connect saddle points and stable equilibrium points (each appears to leave a saddle point and terminate on a stable equilibrium point). These separatrices are short (the relevant saddle and stable equilibrium points are very close

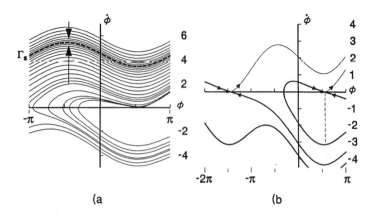

FIGURE 5.16
(a) Phase plane plot and (b) separatrices for a Type I PLL with a′ = 0.5, b′ = 0.1, and ω_Δ' = 4.9.

together for ω_Δ' = 4.9), they are close to the ϕ axis, and they are not visible on the figure. The second feature that distinguishes Figure 5.16b involves the separatrices that have a negative slope as they make their upper-half plane approach (as $\tau \to \infty$) to their saddle points. In the lower-half plane, these separatrices flow in a direction of decreasing ϕ only to reverse this direction when they enter the upper-half plane and terminate on saddle points. This behavior allows lower-half plane trajectories to "whip around" the origin and make their way to the stable limit cycle Γ_s.

This behavior is illustrated on Figures 5.16a and b; these figures show lower-half plane trajectories entering the upper-half plane and converging to the stable limit cycle Γ_s. This type of behavior is not evident on Figures 5.10 through 5.13, and it does not occur for $0 \leq \omega_\Delta' \leq \Omega_2'$. For this reason, the quantity Ω_2' is termed the *half-plane pull-in frequency*. More generally, all lower-half (upper-half) plane trajectories lead to a stable phase-lock state as long as $0 \leq \omega_\Delta' \leq \Omega_2'$ ($-\Omega_2' \leq \omega_\Delta' \leq 0$). The value of Ω_2' can be computed by using the algorithm outlined in Appendix 5.3.3 (also see Stensby and Harb).[61] However, no formula has been found that accurately approximates Ω_2'. Finally, the concept of half-plane pull-in frequency is applicable for only the high-gain case.

From inspection of Figures 5.16a and 5.16b, note that no upper-half plane trajectories lead to a phase-lock point unless they lie within the lock-in region. For Figure 5.16b, the lock-in region has a boundary composed of the vertical dashed line and the thick-line separatrix that reverses its direction and makes an upper-half plane approach (as $\tau \to \infty$) to the saddle point on the positive ϕ axis. Upper-half plane trajectories converge to the stable limit cycle Γ_s if they do not lie within the lock-in region.

NONLINEAR PLL BEHAVIOR IN THE ABSENCE OF NOISE

Figure 5.16 depicts a phase plane which exhibits a two-cell structure. Through points of the first cell pass trajectories that lead, as $\tau \to \infty$, to a stable equilibrium state. Through points of the second cell pass trajectories that lead to the stable limit cycle Γ_s. On a phase plane where phase is interpreted in a modulo-2π manner, the cells are not connected regions for the case depicted by Figure 5.16. Instead, each cell is the union of disjoint regions. A comparison of Figures 5.12 and 5.13 with Figure 5.16 suggests that the phase plane undergoes a structural change as ω_Δ' increases through Ω_2'.

The final case that must be considered is $|\omega_\Delta'| > \Omega_h'$ (Region C on Figure 5.9). As discussed previously, no equilibrium points exist for these values of ω_Δ', and phase lock is impossible. On the phase plane, all trajectories lead to the stable limit cycle Γ_s. As was the case for $|\omega_\Delta'| < \Omega_p'$, the phase plane describes only one cell for values of ω_Δ' in Region C on Figure 5.9. Finally, Γ_s expands monotonically upward towards infinity as $\omega_\Delta' \to \infty$.

The essential features of the high-gain case can be summarized by a *bifurcation diagram*. On the phase plane, limit cycles Γ_s and Γ_u are 2π-periodic functions $\dot{\phi}(\phi)$ that depend on the parameter ω_Δ'. The essential features of this dependence can be illustrated by plotting $\dot{\phi}(0)$ as a function of ω_Δ' (the point $(\phi, \dot{\phi}(\phi)) = (0, \dot{\phi}(0))$ lies on the limit cycle). Plots of this nature are known as bifurcation diagrams, and they are used frequently in the analysis of nonlinear systems.

Figure 5.17 depicts a bifurcation diagram that is generic for the high-gain case discussed in this section. The plot illustrates the fact that no limit cycles exist for values of ω_Δ' that satisfy $0 \le \omega_\Delta' < \Omega_p' \approx 3.007$ (computed for a' = 0.5 and b' = 0.1). Then, as ω_Δ' increases through Ω_p', Γ_s and Γ_u form and split apart (saddle node bifurcation occurs at Ω_p'). As ω_Δ' increases, stable Γ_s moves upward on the phase plane (the solid-line plot has a positive slope), and the stable limit cycle approaches infinity with ω_Δ' ($\dot{\phi}(0) \to \infty$ as $\omega_\Delta' \to \infty$). On the other hand, with increasing ω_Δ', unstable Γ_u moves downward (the dashed-line plot has a negative slope) until the point $\omega_\Delta' = \Omega_2' \approx 3.319$ is reached where the unstable limit cycle bifurcates from a separatrix cycle. Qualitative behavior of the type depicted on Figure 5.17 is characteristic of the high-gain case. With its saddle node bifurcation at pull-in frequency Ω_p' and separatrix cycle bifurcation at half-plane pull-in frquency Ω_2', Figure 5.17 is generic for the high-gain case.

5.3.4 The Low-Gain Case

As mentioned previously, the high-gain case applies to properly designed and operated second-order PLLs that contain an imperfect integrator, and it is this case that is covered most frequently in the PLL

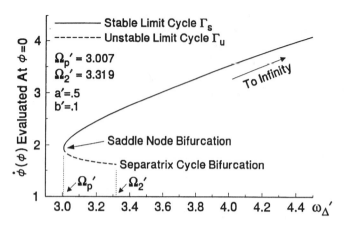

FIGURE 5.17
A typical bifurcation diagram for the high-gain case. Saddle node bifurcation occurs at the pull-in frequency Ω_p'. Separatrix cycle bifurcation occurs at the half-plane pull-in frequency Ω_2'.

literature. However, the high-gain case, with its saddle-node bifurcation at Ω_p' and separatrix cycle bifurcation at Ω_2', does not cover all possibilities that may occur in practical applications.

Besides Figure 5.17, a second characteristic bifurcation diagram is possible, and it is generic for the *low-gain case*. The existence of two cases, each described by a generic bifurcation diagram, was pointed out by Greenstein[62] and others; see also the two bifurcation diagrams described by Meyr and Ascheid (see Figure 4.4-2 of Reference 8). This section provides a qualitative description of the low-gain case.

The low-gain case is important since it can apply in practical applications. Recall that gain parameter $G = AK_1K_mK_v$ has the amplitude of the input reference signal (i.e., the parameter A) as a multiplicative factor. The low-gain case applies when this reference signal level drops sufficiently low (as it might during a system anomaly). In this section the low-gain case is shown to apply for values of G in a set that includes the interval $0 < G < b$.

Figure 5.18 depicts a bifurcation diagram that illustrates generic properties of the low-gain case. The diagram describes the PLL with $a' = a/G = 10$ and $b' = b/G = 2$. Note that Figures 5.17 and 5.18 describe results which were computed using the same values for the loop filter parameters a and b. However, the data plotted on Figure 5.17 was computed by using a value of G that is 20 times larger than that used to compute the data plotted on Figure 5.18.

The low-gain case is characterized by the fact that the stable limit cycle Γ_s bifurcates from an externally stable separatrix cycle at the pull-in frequency Ω_p'. Unlike the high-gain case, saddle node bifurcation

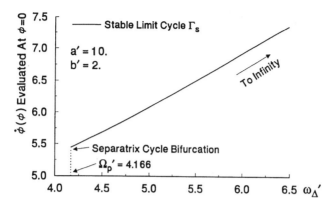

FIGURE 5.18
A typical bifurcation diagram for the low-gain case. Separatrix cycle bifurcation occurs at the pull-in frequency Ω_p'.

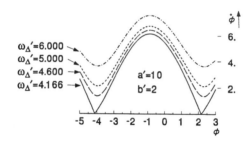

FIGURE 5.19
Bifurcation from a stable separatrix cycle (low-gain case).

plays no role here. However, both cases have an important feature in common; Figures 5.17 and 5.18 describe limit cycles which contract downward on the phase plane until they terminate in a separatrix cycle bifurcation.

For the low-gain case, Figure 5.19 illustrates bifurcation of Γ_s from the separatrix cycle. It shows what happens at the separatrix cycle bifurcation point that is marked on Figure 5.18. For $\omega_\Delta' = \Omega_p' = 4.166$, it shows the computed separatrix cycle. For values of ω_Δ' equal to 4.6, 5.0, and 6.0, Figure 5.19 shows the computed limit cycle Γ_s that bifurcates from the separatrix cycle.

As can be seen from Figure 5.19, the separatrix cycle for the low-gain case is a 2π-periodic path connecting adjacent saddle points. In this respect, it is similar to the separatrix cycle encountered during discussion of the high-gain case (compare the appearance of Figures 5.15 and 5.19). However, the two separatrix cycles have some important differences. The separatrix cycle for the low-gain (high-gain) case is externally stable (unstable). This difference can be described in terms of phase plane trajectories that lie sufficiently close to the upper side of the separatrix cycle. These trajectories will converge to (diverge from) the upper side of the externally stable (unstable) separatrix cycle.

Fortunately, a simple formula exists for determining the stability of the separatrix cycle. As discussed in Appendix 5.3.3 (see also Theorem 44, p. 304 of Andronov et al.),[31] external stability can be determined by the simple test

$$s(\phi_{sp}) \equiv -(b + G\cos\phi_{sp}) \begin{array}{c} \text{stable} \\ \text{(low-gain case)} \\ < \\ \\ > \\ \text{unstable} \\ \text{(high-gain case)} \end{array} 0, \qquad (5.3\text{-}18)$$

where $\phi_{sp} \equiv \pi - \sin^{-1}(b\omega_\Delta/aG)$ is a saddle point on the path of the separatrix cycle. In computing ϕ_{sp} for use in Equation 5.3-18, the quantity ω_Δ is taken as the value of loop detuning for which the separatrix cycle exists, given values of G, a, and b. This value of ω_Δ can be computed by using the algorithm outlined in Appendix 5.3.3.

Simple inspection of Equation 5.3-18 reveals that $s(\phi_{sp})$ must be negative for values of G that satisfy $0 \le G < b$. For this range of gain, the separatrix cycle is externally stable, and the low-gain case applies. Hence, the low-gain case applies for a set of G values which includes the interval $0 \le G < b$.

5.3.5 General Phase Plane Characteristics for the Low-Gain Case

Figures 5.10 through 5.13 and Figure 5.16 detail the general phase plane behavior for the important high-gain case. For the low-gain case, a similar phase plane analysis is given in this section. This analysis is relatively simple since the low-gain case is characterized by a single separatrix cycle bifurcation at the pull-in frequency Ω_p' (see Figure 5.18).

For the low-gain case the phase plane structure is very simple. First, for values of ω_Δ' that satisfy $|\omega_\Delta'| < \Omega_p'$, the phase plane contains no limit cycles, and pull-in occurs from any initial starting point (after excluding the theoretical possibility of hang-up). That is, the entire phase plane can be considered as one cell that contains trajectories that lead to an equilibrium point. Next, for values of ω_Δ' that satisfy $|\omega_\Delta'| > \Omega_h'$, the phase plane contains no equilibrium points, and pull-in to phase lock is not possible. Instead, all trajectories lead to a stable limit cycle. Again, the phase plane is one cell that contains trajectories that lead to a stable limit cycle. In this section, neither of these simple cases are illustrated by phase plane plots.

Figure 5.20 depicts a phase plane plot that is typical for the low-gain case with ω_Δ in the range $\Omega_p < |\omega_\Delta| < \Omega_h$. For this example, a = 100, b = 20, ω_Δ = 46, and G = 10 (so that a' = 10, b' = 2, ω_Δ' = 4.6). Figure 5.20a shows the stable limit cycle Γ_s as a dashed-line plot. Note that all trajectories that lie above Γ_s converge to this limit cycle. Also, all other upper-half plane trajectories, which are not within the lock-in region, converge to Γ_s. As displayed on Figure 5.20b, the lock-in region lies to the left of the vertical dashed line; the remainder of its boundary is the separatrix that changes direction and approaches (with ϕ increasing as $\tau \to \infty$) the saddle point $x_s = \pi - \sin^{-1}(b\omega_\Delta/aG)$. Some lower-half plane trajectories enter the lock-in region and terminate on a stable equilibrium point. But other lower-half plane trajectories enter the corridor, shown on Figure 5.20b, between the two separatrices that cross the negative $\dot\phi$ axis. These trajectories "whip around" the origin and make their way into the upper-half plane. From there, they converge to the stable limit cycle Γ_s shown on Figure 5.20a. Finally, note that this summary of the low-gain case with $\Omega_p < \omega_\Delta < \Omega_h$ is similar to the high-gain case with $\Omega_2 < \omega_\Delta < \Omega_h$ (compare Figures 5.16 and 5.20). In both cases, the phase planes contain a single stable, upper-half plane limit cycle that can attract some trajectories that start in the lower-half plane.

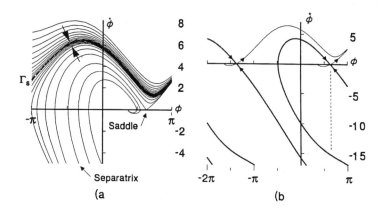

FIGURE 5.20
(a) Phase plane plot and (b) separatrices for a Type I PLL with a' = 10, b' = 2, and ω_Δ' = 4.6. This illustrates typical behavior for the low-gain case with $\Omega_p' < \omega_\Delta' < \Omega_h'$.

5.3.6 Computing the Separatrix Cycle and Determining its Stability

As shown in Appendix 5.3.3, the separatrix cycle can be computed easily. The algorithm discussed in the appendix accepts parameters a, b, and G as user-supplied input, and it computes the value of ω_Δ where the separatrices (from adjacent saddle points) match up and form the separatrix cycle (see also Stensby and Harb[61] for a discussion of this algorithm). This value of ω_Δ can be used to compute ϕ_{sp} in Equation 5.3-18 and determine the stability of the computed separatrix cycle.

Loop filter parameters a = 100 and b = 20 were used to compute the separatrix cycles depicted on Figures 5.21 and 5.22. Note that the same ratio a/b = 5 was used to compute the data that appears on Figures 5.10 through 5.23. This feature is valuable when comparing results. For example, the separatrix cycle that appears on Figure 5.15 (Figure 5.19) also appears as one of the choices on Figure 5.21 (Figure 5.22).

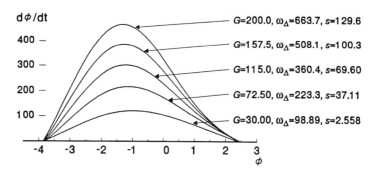

FIGURE 5.21
Unstable separatrix cycles for a Type I PLL with a = 100 and b = 20. Recall that $d\phi/dt = G\dot{\phi}$.

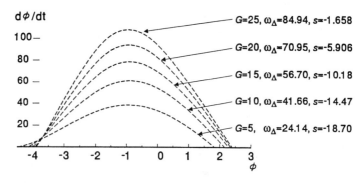

FIGURE 5.22
Stable separatrix cycles for a Type I PLL with a = 100 and b = 20. Recall that $d\phi/dt = G\dot{\phi}$.

NONLINEAR PLL BEHAVIOR IN THE ABSENCE OF NOISE 155

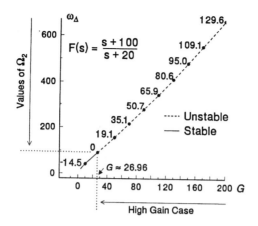

FIGURE 5.23
Values of G and ω_Δ for which a separatrix cycle occurs.

On Figures 5.21 and 5.22, only one 2π period of each separatrix cycle is shown. Of course, each cycle is 2π-periodic, and it spans its own set of saddle points (which are different for each plot shown). Also, note that instantaneous frequency error is displayed along the vertical axes of these figures (recall that $d\phi/dt = Gd\phi/d\tau = G\dot\phi$). Each graph is annotated with the value of G used with the algorithm described in Appendix 5.3.3; also given is the value of ω_Δ computed by the algorithm and the value of s computed by using Equation 5.3-18.

Figure 5.21 depicts separatrix cycles for values of G ranging from 30 to 200. Note that each of these cycles is externally unstable since it is associated with a positive value of s. Hence, the high-gain case applies for these separatrix cycles. Note that for $(G, \omega_\Delta) = (200, 663.7)$, the separatrix cycle depicted is, when gain normalized, the separatrix cycle shown on Figures 5.13 and 5.15.

Figure 5.22 depicts separatrix cycles for values of G ranging from 5 to 25. Note that each of these cycles is externally stable since it is associated with a negative value of s. The low-gain case applies for these values of G. Note that for $(G, \omega_\Delta) = (10, 41.66)$, the separatrix cycle depicted on Figure 5.22 is, when gain normalized, the separatrix cycle shown on Figure 5.19.

Figure 5.23 depicts a plot of points (G, ω_Δ) at which a separatrix cycle was computed for the PLL with $a = 100$ and $b = 20$. Values of s, ranging from -14.5 to 129.6, appear next to solid dots on the plot (s appears parametrically on the plot). The dashed (solid)-line segment of the plot, where s is positive (negative), corresponds to an unstable (stable) separatrix cycle. By computing a large number of closely spaced points on the plot, parameter s was found to change sign at $G \approx 26.96$; this value approximates the lower bound on G for the high-gain case when a = 100 and b = 20. For the high-gain case, recall that Ω_2 denotes that value of ω_Δ where the externally unstable separatrix cycle exists. For this reason, Ω_2 appears as a label on the ω_Δ axis of Figure 5.23.

5.4 EFFECTS OF IF FILTERING ON THE LONG LOOP

The long loop is modeled in Section 2.7, and its block diagram is given by Figure 2.13. First, the VCO output is used by a down converter to hetrodyne the input signal into a bandpass signal at an intermediate frequency ω_{if}. This IF signal is passed through a bandpass IF filter before it is compared in phase to the output of a crystal oscillator that oscillates at ω_{if}. This phase comparison operation produces a lowpass signal that drives the loop filter and VCO combination. As discussed in Section 2.7, an IF filter within the loop reduces problems associated with noise rectification in an imperfect phase comparator.

An IF filter within the loop can cause a stability problem, and it can reduce the pull-in range. The stability problem can be analyzed by applying standard linear stability theory to the baseband model depicted by Figure 2.15, and it is not discussed here. The problem of pull-in range reduction is discussed in this section. In most cases, these two problems decrease in severity as simpler (lower-order) IF filters are used. To minimize these problems, the IF filter should have a bandwidth which is large compared to the desired closed-loop bandwidth, and it should introduce only a minimal amount of phase shift over its bandwidth (equivalently, it should introduce little additional time delay). A conservative design approach is to use only enough IF filtering to overcome the phase comparator threshold problem discussed in Section 2.7. Other problems, such as significant intermodulation products in the down converter output, can be dealt with by choosing ω_{if} properly and by down converter design.

The problem of an IF filter reducing the pull-in range may be traced to the fact that new limit cycles may be introduced into the dynamics of the loop when the IF filter is placed within the loop. In the PLL literature, these limit cycles are called false-lock states, and they can be stable or unstable. They can be modeled as

$$\phi = \omega_f t + \psi(t), \tag{5.4-1}$$

where ω_f denotes the apparent frequency error in the false-locked loop, and

$$\psi(t) \equiv \sum_{k=1}^{\infty} a_k \cos(k\omega_f t) + b_k \sin(k\omega_f t) \tag{5.4-2}$$

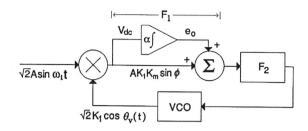

FIGURE 5.24
Baseband equivalent of a PLL containing an IF signal path.

is periodic with a fundamental frequency of ω_f. A rigorous mathematical treatment of false lock is not given here; instead, the approach used in this section is similar to that given by Gardner,[6] and it illustrates the problem in a simple manner.

The false-lock problem can be very perplexing from an operational standpoint. The pull-in process halts when a false-lock state is reached, and the PLL appears to lock at a frequency that is incorrect. Common lock detectors can indicate incorrectly that the loop is phase locked and operating properly. The false-lock state may be detected by observing a low-frequency periodic beat note in the phase detector output. However, the beat note may be hidden by noise, and this method may fail to detect the false-lock state. In fact, the false-lock state may go undetected until overall system failure is noted.

Figure 5.24 depicts the PLL under consideration in this section. The loop filter in this PLL is a cascade of filters $F_1(s)$ and $F_2(s)$. Filter $F_1 = 1 + \alpha/s$ is the standard perfect integrator loop filter, and F_2 represents the lowpass equivalent of a symmetrical IF filter. When $F_2 = 1$, the PLL has an infinite pull-in range; there are no limit cycles to impede the pull-in process. However, as discussed below, new limit cycles appear for some filters $F_2 \neq 1$.

In general, the problem of false lock is difficult to analyze without making simplifying assumptions. An assumption that greatly simplifies the analysis is that $\omega_f \gg G$; that is, in the false-locked loop, the apparent frequency error is large compared to G, so that the beat note lies outside of the closed-loop bandwidth. This assumption holds in many applications where the additional filtering F_2 introduces significant phase shift only for frequencies that are large compared to G. The assumption simplifies the analysis since it implies that ψ has a small peak value, and this periodic function is almost sinusoidal.

The assumption $\omega_f \gg G$ permits simple approximations to be made for both ϕ and $\sin\phi$. In terms of Equations 5.4-1 and 5.4-2, the closed-loop phase error can be approximated as

$$\phi = \omega_f t + a_1 \cos\omega_f t + b_1 \sin\omega_f t, \tag{5.4-3}$$

where a_1 and b_1 have small magnitudes. This approximation can be used to simplify the phase comparator output. This output can be expressed as

$$AK_1K_m \sin\phi$$
$$= AK_1K_m \sin(\omega_f t + a_1 \cos\omega_f t + b_1 \sin\omega_f t) \qquad (5.4\text{-}4)$$
$$\approx AK_1K_m[\sin\omega_f t + \cos\omega_f t(a_1 \cos\omega_f t + b_1 \sin\omega_f t)]$$

since a_1 and b_1 are small.

The nominal value of α is on the order of $G/2$ in properly designed and operated loops. Hence, the above-mentioned assumption implies that $\omega_f \gg \alpha$, and the sinusoidal component in the integrator output e_o is small compared to the sinusoidal component delivered to the input of F_2 (see Figure 5.24). The integrator output e_o is slowly changing in time (or it is DC), and it is assumed to contain no component at ω_f.

Based on the VCO model discussed in Section 2.1.3, Figure 5.24 depicts a loop described by the functional differential equation

$$\frac{d\phi}{dt} = \omega_\Delta - K_v F_2[e_o + AK_1K_m \sin\phi]. \qquad (5.4\text{-}5)$$

The notation F_2 is used in Equation 5.4-5 as a linear operator that maps periodic input $e_o + AK_1K_m\sin\phi$ into a unique periodic output $y \equiv F_2[e_o + AK_1K_m\sin\phi]$. In what follows, assume that F_2 has a unity DC gain, and denote as $|F_2|$ and $\angle F_2$ the amplitude and phase responses, respectively, of F_2 at the frequency ω_f.

False lock in the PLL depicted by Figure 5.24 can be analyzed easily in terms of the equations listed above. First, substitute Equations 5.4-3 and 5.4-4 into Equation 5.4-5 to obtain

$$\omega_f - \omega_f(a_1 \sin\omega_f t - b_1 \cos\omega_f t)$$
$$= \omega_\Delta - K_v F_2[e_o] \qquad (5.4\text{-}6)$$
$$- GF_2[\sin\omega_f t + \cos\omega_f t(a_1 \sin\omega_f t + b_1 \cos\omega_f t)].$$

The fundamental frequency components in this equation are of interest. Equate the fundamental components on both sides of Equation 5.4-6 to obtain

$$-\omega_f(a_1 \sin\omega_f t - b_1 \cos\omega_f t) = -GF_2[\sin\omega_f t]. \qquad (5.4\text{-}7)$$

NONLINEAR PLL BEHAVIOR IN THE ABSENCE OF NOISE

Now, sinusoid $\sin\omega_f t$ is scaled in magnitude by $|F_2|$, and it is shifted in phase by $\angle F_2$, as it passes through filter F_2. This observation leads to

$$-\omega_f(a_1 \sin\omega_f t - b_1 \cos\omega_f t)$$
$$= -G|F_2|\sin(\omega_f t + \angle F_2) \qquad (5.4\text{-}8)$$
$$= -G|F_2|(\sin\angle F_2 \cos\omega_f t + \cos\angle F_2 \sin\omega_f t).$$

Finally, approximations for a_1 and b_1 can be obtained by equating sin and cos terms on both sides of Equation 5.4-8. This procedure leads to

$$a_1 = \frac{G|F_2|}{\omega_f}\cos\angle F_2$$
$$b_1 = -\frac{G|F_2|}{\omega_f}\sin\angle F_2. \qquad (5.4\text{-}9)$$

The DC component in the phase detector output is known as the *pull-in voltage*. This quantity, denoted here as V_{dc}, is responsible for slewing the integrator output towards the value of voltage required for phase lock. The simple approximation

$$V_{dc} = AK_1 K_m \frac{G|F_2|}{2\omega_f}\cos\angle F_2 \qquad (5.4\text{-}10)$$

results from computing the DC component in Equation 5.4-4 with a_1 given by Equation 5.4-9.

As an example, consider modeling the lowpass equivalent circuit of the IF filter by $|F_2| = 1$ and $\angle F_2 = (\pi/3)(\omega_f/G)$. This simple model assumes that the IF filter has a unity-gain passband which is wide enough to pass all frequencies in the IF signal. Also, the filter is assumed to introduce a linear phase shift over its passband. A simple model similar to this may serve to analyze adequately the effects of including within the loop a high-order crystal or ceramic filter.

Figure 5.25 compares the case of no additional filtering to the simple example mentioned above. Figure 5.25a depicts the normalized pull-in voltage when there is no additional filtering ($F_2 = 1$). Figure 5.25b is a plot of the normalized pull-in voltage when F_2 is modeled by the above-mentioned unity-gain, linear-phase filter.

Note that nulls in V_{dc} can result from the $\cos\angle F_2$ term in Equation 5.4-10; Figure 5.25b illustrates this phenomenon. Consider

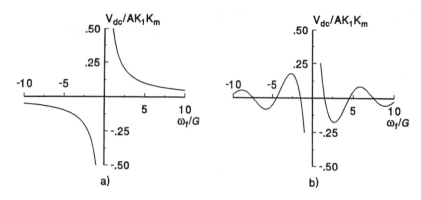

FIGURE 5.25
Normalized pull-in voltage for (a) no excess phase and (b) excess phase $\zeta = (\pi/3)(\omega_f/G)$.

values of normalized ω_f that are smaller than the first null (which is an unstable false-lock point). For these values of ω_f, V_{dc} has the same polarity on Figures 5.25a and 5.25b. However, for these frequencies, V_{dc} on Figure 5.25b is smaller in magnitude than it is on Figure 5.25a. This implies that the additional filtering causes the loop to pull in more slowly.

For values of normalized ω_f slightly larger than the first null on Figure 5.25b, V_{dc} has an opposite polarity than the voltage shown on Figure 5.25a for the standard loop. For these values of normalized ω_f, the polarity is incorrect, and pull-in does not occur in the loop containing the additional filtering. Instead, the reverse polarity causes the loop to push away from the correct lock frequency; when this happens, the closed-loop frequency error increases. *Frequency pushing* continues until ω_f coincides with the second null on Figure 5.25b which is a stable false-lock point. The loop cannot leave this false-lock point, and the pull-in mechanism fails.

A conservative design approach would utilize an IF filter for which there are no stable false-lock points in the loop response. For example, if the IF filter is based on a single LC-tuned circuit, then its lowpass equivalent contains only one pole. Such a lowpass equivalent contributes less than $\pi/2$ radians of phase shift, and no nulls exist in the loop pull-in voltage V_{dc}. If two resonant LC circuits are used in the IF filter, then lowpass equivalent F_2 introduces less than π radians of phase shift, and the only null in V_{dc} is unstable. From Equation 5.4-10 it is easy to see that F_2 must have four or more poles in order for stable false-lock points to exist.

Additional filtering, beyond what is contributed by the loop and IF filters, is always present in practical PLL. For example, at least one pole is introduced by the operational amplifier used in the loop filter.

The VCO, and other loop components, may have significant dynamics which were not accounted for in the original design. This additional filtering, which may be difficult to identify and model, shows up as part of F_2. Hence, a conservative design would employ an IF filter whose lowpass equivalent has, at most, two poles.

APPENDIX 5.2.1 PULL-IN TIME FOR A SECOND-ORDER PLL

A second-order PLL containing a perfect integrator in its loop filter and a constant frequency reference is discussed in Section 5.2. As part of the discussion, it is shown that the PLL pull-in range is infinite. That is, phase lock is achieved regardless of the initial conditions in effect when the loop is closed or the difference ω_Δ between the reference and VCO center frequencies.

However, pull-in may occur too slowly and be too unreliable for many applications. A quantitative measure of the effectiveness of the pull-in phenomenon is the *pull-in time* parameter. This parameter is a function of the initial phase and frequency errors, and it is defined as the time required for the PLL to go from these initial errors to a frequency-locked state. The nonlinear nature of the pull-in phenomenon prevents the determination of an exact formula for the pull-in time.

In this appendix, a simple approximation for pull-in time is developed for the second-order PLL containing a perfect integrator in its loop filter. Like the equivalent result developed by Richman,[63] the result developed here shows that pull-in time grows at a rate that is proportional to ω_Δ^2 for large values of loop detuning ω_Δ.

A number of related assumptions must be made to approximate the pull-in time, and they are explained here with the aid of Figure 5.26. First, the assumption is made that the initial value of frequency error is $d\phi/dt \approx \omega_\Delta$, and this value is large compared to the frequency error in effect when the lock-in region is entered. This means that the PLL

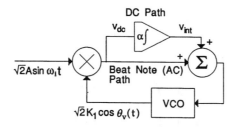

FIGURE 5.26
Diagram used to develop an approximation for the pull-in time.

will slip a large number of cycles before reaching a frequency-locked state. Also, prior to entering the lock-in region, the phase plane frequency error function $\dot{\phi}(\phi)$, when averaged over a 2π cycle of phase, changes relatively little from one 2π cycle to the next. This implies that the integrator output v_{int} (see Figure 5.26) changes little from one period of the beat note to the next. A second assumption is that the phase comparator output contains an "AC-like" component at the beat note frequency and a "DC-like" component v_{dc}. Furthermore, the AC component is filtered out by the integrator, and it has no *direct* influence on v_{int}. Instead, the AC component passes through the direct connection in the loop filter. The DC component v_{dc} has a negligible *direct* effect on the VCO frequency; it is assumed that the VCO frequency, when averaged over a beat note cycle, is given by $\omega_0 + K_v v_{int}$ (the VCO is tuned by v_{int} only). These assumptions are interrelated, and they are justified for large initial beat note frequencies.

The difference between the reference frequency and the average VCO frequency (averaged over one cycle of the beat note) is of interest in what follows. In terms of the assumptions outlined above, this difference can be expressed as

$$\tilde{\omega} \equiv \omega_i - [\omega_0 + K_v v_{int}]. \tag{5.2.1-1}$$

The pull-in time, denoted as T_d, is approximated here as the time required for $\tilde{\omega}$ to go from its initial value ω_Δ (i.e., the initial value of v_{int} is zero) to its value when the PLL trajectory enters the lock-in region.

A relationship between $\tilde{\omega}$ and v_{dc} must be obtained in order to approximate T_d. To obtain this relationship, note that

$$\tilde{\omega} = \omega_\Delta - \alpha K_v \int_0^t v_{dc}\, du. \tag{5.2.1-2}$$

Now, differentiate this last equation and obtain

$$\frac{d\tilde{\omega}}{dt} = -\alpha K_v v_{dc}. \tag{5.2.1-3}$$

An algebraic relationship for v_{dc} in terms of $\tilde{\omega}$ must be obtained in order to eliminate v_{dc} from Equation 5.2.1-3 and obtain a differential equation in $\tilde{\omega}$.

An algebraic equation relating v_{dc} to $\tilde{\omega}$ can be obtained once an approximation for ϕ is found. From Figure 5.26, Equation 2.2-2, and the assumptions outlined above, the phase error must satisfy

NONLINEAR PLL BEHAVIOR IN THE ABSENCE OF NOISE 163

$$\frac{d\phi}{dt} = \omega_i - [\omega_o + K_v v_{int}] - G \sin\phi$$

$$= \tilde{\omega} - G \sin\phi.$$

(5.2.1-4)

As stated above, it is assumed that $\tilde{\omega}$ varies slowly compared to one cycle of the beat note; in what follows, $\tilde{\omega}$ is treated as a constant larger than G when Equation 5.2.1-4 is solved for its periodic solution. This solution is used to approximate v_{dc} as a function of $\tilde{\omega}$.

An approximation to the periodic beat note can be found by substituting

$$\phi \approx \omega_f t + a_1 \cos\omega_f t + b_1 \sin\omega_f t \qquad (5.2.1\text{-}5)$$

into Equation 5.2.1-4. Then, in the result that this produces, equate the DC, $\sin\omega_f t$, and $\cos\omega_f t$ terms. From this, obtain a system of algebraic equations that can be solved to obtain

$$a_1 = \frac{\tilde{\omega}}{G} - \sqrt{\left(\frac{\tilde{\omega}}{G}\right)^2 - 2}$$

$$b_1 = 0 \qquad (5.2.1\text{-}6)$$

$$\omega_f = \frac{\tilde{\omega}}{2} + \frac{G}{2}\sqrt{\left(\frac{\tilde{\omega}}{G}\right)^2 - 2}.$$

Note that a_1 and ω_f approach zero and $\tilde{\omega}$, respectively, as $\tilde{\omega}$ becomes large compared to G.

The quantity v_{dc} for use in Equation 5.2.1-3 must be approximated. This voltage can be approximated by using Equations 5.2.1-5 and 5.2.1-6 in the phase comparator output to obtain

$$AK_1 K_m \sin\phi = AK_1 K_m \sin(\omega_f t + a_1 \cos\omega_f t)$$

$$= AK_1 K_m \big[\sin\omega_f t \cdot \cos(a_1 \cos\omega_f t)$$

$$+ \cos\omega_f t \cdot \sin(a_1 \cos\omega_f t)\big]$$

$$\approx AK_1 K_m \big[\sin(\omega_f t) + a_1 \cos^2(\omega_f t)\big].$$

(5.2.1-7)

This last approximation is based on the assumption that a_1 is small. Finally, note that the DC component in the phase comparator output can be approximated as

$$V_{dc} \approx AK_1K_m \frac{a_1}{2} = \frac{1}{2}AK_1K_m\left[\frac{\tilde{\omega}}{G} - \sqrt{\left(\frac{\tilde{\omega}}{G}\right)^2 - 2}\right]. \quad (5.2.1\text{-}8)$$

Equations 5.2.1-3 and 5.2.1-8 can be used to approximate the pull-in time. Combine these two equations to obtain

$$dt \approx \frac{-\alpha^{-1}(2/\tilde{\omega})}{1-\sqrt{1-2(G/\tilde{\omega})^2}}d\tilde{\omega}. \quad (5.2.1\text{-}9)$$

However, for $\tilde{\omega} \gg G$ the approximation

$$\sqrt{1-2(G/\tilde{\omega})^2} \approx 1-(G/\tilde{\omega})^2 \quad (5.2.1\text{-}10)$$

can be made so that

$$dt \approx -\frac{2\tilde{\omega}}{\alpha G^2}d\tilde{\omega}. \quad (5.2.1\text{-}11)$$

The desired result follows by integrating Equation 5.2.1-11 to obtain

$$T_d \approx -\int_{\omega_\Delta}^{\omega_{lok}} \frac{2\tilde{\omega}}{\alpha G^2}d\tilde{\omega} = \frac{\omega_\Delta^2 - \omega_{lok}^2}{\alpha G^2}, \quad (5.2.1\text{-}12)$$

where ω_{loc} is the value of $\tilde{\omega}$ in effect when the PLL trajectory enters the lock-in region (so that frequency lock is achieved). Now, recall the assumption that the initial value of frequency error is $d\phi/dt \approx \omega_\Delta$, and this value is large compared to ω_{loc}. This assumption leads to the approximation

$$T_d \approx \frac{\omega_\Delta^2}{\alpha G^2}. \quad (5.2.1\text{-}13)$$

For the second-order PLL considered here, the theory predicts an unlimited pull-in range. However, it predicts that the time required to pull in grows at a rate proportional to the square of the initial frequency error when this value is large. Hence, the natural pull-in mechanism may be too slow for some applications. For these cases, the PLL can be augmented by circuitry that aids the acquisition process (see Chapter 5 of Gardner).[6]

APPENDIX 5.3.1 PULL-IN RANGE OF A SECOND-ORDER PLL

The second-order PLL with an imperfect integrator in its loop filter has phase plane trajectories that satisfy

$$\frac{d\dot\phi}{d\phi} = -(b' + \cos\phi) + \frac{b'\omega_\Delta' - a'\sin\phi}{\dot\phi} \qquad (5.3.1\text{-}1)$$

(see Section 5.3). The parameters a' and b' are the gain-normalized zero and pole, respectively, of the loop filter. The quantity ω_Δ' denotes the gain-normalized detuning.

A value Ω_p' exists such that pull-in is guaranteed if $|\omega_\Delta'| < \Omega_p'$ (Lindsey).[4] The reason for this is that the phase plane contains no limit cycles for $|\omega_\Delta'| < \Omega_p'$. The quantity Ω_p' is known in the PLL literature as the pull-in range. In the analysis that follows, it is assumed that $\omega_\Delta' > 0$ and $a' > b' > 0$.

For the high-gain case discussed in Section 5.3, two limit cycles bifurcate (they "split apart") at $\omega_\Delta' = \Omega_p'$; one is stable and the other is unstable (see Figure 5.17). As ω_Δ' increases from Ω_p', the stable limit cycle expands upward on the phase plane; it expands upward towards infinity as $\omega_\Delta' \to \infty$. As ω_Δ' increases from Ω_p', the unstable limit cycle moves downward on the phase plane. This continues until $\omega_\Delta' = \Omega_2'$, $\Omega_p' < \Omega_2' < \Omega_h'$, is reached; at this point, the unstable limit cycle bifurcates from an unstable separatrix cycle. This behavior is described by the bifurcation diagram in Figure 5.17. These limit cycles prohibit some trajectories from reaching a stable equilibrium point. That is, they limit the pull-in range of the PLL.

For the high-gain case, a simple approximation for the pull-in range Ω_p' is developed in this appendix. First, a truncated Fourier series expansion is used to model a limit cycle, and this expansion is substituted into Equation 5.3.1-1. Next, a harmonic balance approach is used to obtain a system of algebraic equations for the unknown Fourier coefficients in the limit cycle expansion. Finally, the desired approximation is obtained by examining the roots of the algebraic system; Ω_p' is chosen as the value of ω_Δ' where two roots merge together.

In many cases of practical importance, a good approximation of Ω_p' can be obtained from the method outlined above. This method models the limit cycles with an expansion of the form

$$\dot\phi \approx \alpha_0 + \alpha_1 \cos\phi + \beta_1 \sin\phi, \qquad (5.3.1\text{-}2)$$

where α_0, α_1, and β_1 are unknown Fourier coefficients. Of course, more terms can be employed in the approximation of $\dot\phi$; however, the amount of work required to approximate Ω_p' grows rapidly with the number of terms in ϕ. Now, substitute Equation 5.3.1-2 into Equation 5.3.1-1 and obtain

$$[\alpha_o + \alpha_1 \cos\phi + \beta_1 \sin\phi][-\alpha_1 \sin\phi + \beta_1 \cos\phi]$$
$$= -(b' + \cos\phi)[\alpha_o + \alpha_1 \cos\phi + \beta_1 \sin\phi] + b'\omega_\Delta' - a'\sin\phi .$$
(5.3.1-3)

Next, obtain

$$-b'\alpha_o - \frac{\alpha_1}{2} + b'\omega_\Delta' = 0 \qquad (5.3.1\text{-}4)$$

by equating the DC component on both sides of Equation 5.3.1-3. In a similar manner obtain

$$\alpha_o \beta_1 = -b'\alpha_1 - \alpha_o \qquad (5.3.1\text{-}5)$$

and

$$\alpha_o \alpha_1 = b'\beta_1 + a' \qquad (5.3.1\text{-}6)$$

by equating the $\cos\phi$ and $\sin\phi$ components, respectively. Finally, eliminate α_1 and β_1 in Equation 5.3.1-4 through Equation 5.3.1-6 and obtain

$$F(\alpha_o) \equiv \alpha_o^3 - \omega_\Delta' \alpha_o^2 + \left(b'^2 + \frac{1}{2}\left[\frac{a'}{b'} - 1\right]\right)\alpha_o - b'^2 \omega_\Delta' = 0 . \quad (5.3.1\text{-}7)$$

The roots of Equation 5.3.1-7 are of interest. Note that there is one real root that approaches zero as $b' \to 0^+$ (positive ω_Δ' is held fixed). In general, this root is very small for the high-gain case; it is a result of the approximations used to obtain Equation 5.3.1-7, and it has no physical significance. When real valued, the remaining two roots correspond to limit cycles on the phase plane.

Before using algebraic techniques to study the roots of Equation 5.3.1-7, it is instructive to examine plots of $F(\alpha_o)$. For the values $a' = 0.5$ and $b' = 0.1$ considered in Section 5.3, Figure 5.27 displays plots of $F(\alpha_o)$ for several values of ω_Δ'. The graph for $\omega_\Delta' = 2$ shows no root pair corresponding to the limit cycles discussed in Section 5.3. This suggests that no limit cycles exist for $\omega_\Delta' = 2$, and pull-in occurs without any problems (see Figure 5.10). The case of two

NONLINEAR PLL BEHAVIOR IN THE ABSENCE OF NOISE

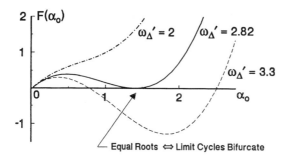

FIGURE 5.27
$F(\alpha_o)$ for the large gain case with $a' = 0.5$ and $b' = 0.1$.

real equal roots is depicted on Figure 5.27 by the graph labeled $\omega_\Delta' = 2.82$. This value of ω_Δ' is near the point Ω_p' where bifurcation occurs and the pair of limit cycles is created.

Note from Figure 5.27 that the real roots move in opposite directions as ω_Δ' increases past 2.82. This suggests that after splitting apart, one limit cycle moves up on the phase plane while the other moves down. This is supported by Figures 5.11 and 5.12 which show the limit cycles moving in opposite directions as ω_Δ' increases. Also, it is supported by the qualitative theory of rotated vector fields (Section 4.5 of Perko).[30] Unfortunately, Figure 5.27 and this simple theory do not provide any information on the fact that one of the limit cycles bifurcates from a separatrix cycle at $\omega_\Delta' = \Omega_2'$ as is discussed in Section 5.3.

The roots of Equation 5.3.1-7 must sum to ω_Δ', a value equal to minus the coefficient of α_o^2. As discussed in the paragraph following Equation 5.3.1-7, only two of these roots are significant. Let Ω_p' denote the value of ω_Δ' at which the significant roots become real and equal. When real and equal, the roots can be approximated as

$$\alpha_o \approx \Omega_p'/2. \qquad (5.3.1\text{-}8)$$

However, the equal roots of Equation 5.3.1-7 must satisfy

$$\left.\frac{dF}{d\alpha_o}\right|_{\alpha_o = \Omega_p'/2} = -\frac{1}{4}\left(\Omega_p'\right)^2 + (b')^2 + \frac{1}{2}\left[\frac{a'}{b'} - 1\right] \qquad (5.3.1\text{-}9)$$

$$= 0$$

Hence, an approximation to the pull-in frequency (i.e., the frequency at which the significant roots become real and equal) is

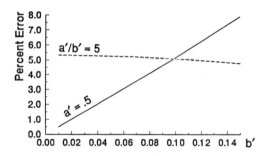

FIGURE 5.28
Percent difference between computed Ω_p' and the approximation $\Omega_p' \approx \sqrt{2a'/(b')}$.

$$\Omega_p' \approx 2\sqrt{(b')^2 + \frac{1}{2}\left(\frac{a'}{b'}-1\right)}. \qquad (5.3.1\text{-}10)$$

Now, the inequality

$$\left|\frac{a'}{b'}\right| \gg \left|1 - 2(b')^2\right| \qquad (5.3.1\text{-}11)$$

holds in most applications of high-gain loops. Under this condition, the approximation for the pull-in frequency can be written as

$$\Omega_p' \approx \sqrt{2\frac{a'}{b'}}. \qquad (5.3.1\text{-}12)$$

Approximation Equation 5.3.1-12 is equivalent to a widely used pull-in frequency approximation given in the PLL literature (after differences in notation are accounted for, Equation 5.3.1-12 is identical to the results given by Example 1, p. 469 of Lindsey).[4]

Despite its simplicity, Equation 5.3.1-12 can provide accurate results. Figure 5.28 shows the percent difference between Equation 5.3.1-12 and results obtained from an accurate numerical calculation of Ω_p' (using the algorithm outlined in Stensby).[57] On Figure 5.28, the solid line plot results from using the fixed value a' = 0.5, and the dashed-line graph was produced by using the fixed ratio a'/b' = 5.

APPENDIX 5.3.2 COMPUTATION OF SEPARATRICES FOR A SECOND-ORDER PLL

Saddle points and separatrices play an important role in the dynamics of second-order PLLs. As discussed in Chapter 5, saddle

NONLINEAR PLL BEHAVIOR IN THE ABSENCE OF NOISE 169

points are unstable equilibrium points; on the phase plane, they are approached by four trajectories known as separatrices. Two of these trajectories approach each saddle point as $\tau \to \infty$ (τ is the independent time variable); the remaining two trajectories approach the saddle point as $\tau \to -\infty$. These features are illustrated by Figure 5.4 which depicts a saddle point with its separatrices; also, some trajectories are displayed that lie close to the separatrices.

An algorithm is given below for computing the separatrices of a second-order PLL containing an imperfect integrator in its loop filter (see Section 5.3). It can be modified easily to handle the second-order PLL containing a perfect integrator (the PLL discussed in Section 5.2). The algorithm uses a two-part approach. First, the selected separatrix is approximated by an asymptotic expansion of the form

$$\dot{\phi}(\phi) = \sum_k \Gamma_k (\phi - \phi_s)^k, \qquad (5.3.2\text{-}1)$$

where ϕ_s is the saddle point, and algebraic expressions for Γ_k, $1 \leq k \leq 4$, are computed. An approximation based on these four terms is used to compute the point $(\phi_s + \Delta, \dot{\phi}(\phi_s + \Delta))$, where step size Δ is user supplied. Starting at this point, the remainder of the separatrix is computed by integrating the time-dependent differential Equation 5.3-6 that describes the PLL.

The asymptotic expansion representation of $\dot{\phi}$ is computed by substituting Equation 5.3.2-1 and

$$\cos\phi = \cos\bigl([\phi-\phi_s]+\phi_s\bigr) = \cos(\phi-\phi_s)\cos\phi_s - \sin(\phi-\phi_s)\sin\phi_s$$
$$\sin\phi = \sin\bigl([\phi-\phi_s]+\phi_s\bigr) = \sin(\phi-\phi_s)\cos\phi_s + \cos(\phi-\phi_s)\sin\phi_s \qquad (5.3.2\text{-}2)$$

into Equation 5.3-7 to obtain

$$\sum_k \Gamma_k(\phi-\phi_s)^k \sum_k k\Gamma_k(\phi-\phi_s)^{k-1}$$
$$= -[\,b' + \cos(\phi-\phi_s)\cos\phi_s - \sin(\phi-\phi_s)\sin\phi_s\,]\sum_k \Gamma_k(\phi-\phi_s)^k \qquad (5.3.2\text{-}3)$$
$$+ b'\omega'_\Delta - a'[\,\sin(\phi-\phi_s)\cos\phi_s + \cos(\phi-\phi_s)\sin\phi_s\,].$$

Next, on both sides of this equation, powers of $(\phi - \phi_s)$ are equated, and a system of algebraic equations is obtained for the coefficients Γ_k.

When first-order terms in Equation 5.3.2-3 are equated the result is

$$\Gamma_1^2 + (b' + \cos\phi_s)\Gamma_1 + a'\cos\phi_s = 0. \qquad (5.3.2\text{-}4)$$

Note that $\cos\phi_s$ in Equation 5.3.2-4 is negative since ϕ_s, taken modulo-2π, is in the interval $(\pi/2, \pi)$ (this is shown in Section 5.3). Hence, the two separatrices that approach ϕ_s as $\tau \to -\infty$ have a positive slope of

$$\Gamma_1^+ = -\frac{(b'+\cos\phi_s)}{2} + \left[\frac{(b'+\cos\phi_s)^2}{4} - a'\cos\phi_s\right]^{\frac{1}{2}}. \qquad (5.3.2\text{-}5)$$

Likewise, the separatrices that approach the saddle as $\tau \to \infty$ do so with the negative slope

$$\Gamma_1^- = -\frac{(b'+\cos\phi_s)}{2} - \left[\frac{(b'+\cos\phi_s)^2}{4} - a'\cos\phi_s\right]^{\frac{1}{2}}. \qquad (5.3.2\text{-}6)$$

One of these two values of slope must be selected during the computation of a separatrix.

This process can be continued, and additional coefficients can be found. Equating terms in Equation 5.3.2-3 that are second order in $(\phi - \phi_s)$ yields a result that can be solved for

$$\Gamma_2^\pm = \frac{[\Gamma_1^\pm + a'/2]\sin\phi_s}{3\Gamma_1^\pm + b' + \cos\phi_s}. \qquad (5.3.2\text{-}7)$$

In a similar manner, third- and fourth-order terms can be equated to solve for

$$\Gamma_3^\pm = \frac{-2(\Gamma_2^\pm)^2 + \Gamma_2^\pm \sin\phi_s + (\Gamma_1^\pm/2 + a'/6)\cos\phi_s}{4\Gamma_1^\pm + b' + \cos\phi_s} \qquad (5.3.2\text{-}8)$$

and

$$\Gamma_4^\pm = \frac{-5\Gamma_2^\pm \Gamma_3^\pm + (\Gamma_3^\pm - \Gamma_1^\pm/6 - a'/24)\sin\phi_s + (\Gamma_2^\pm/2)\cos\phi_s}{5\Gamma_1^\pm + b' + \cos\phi_s}, \qquad (5.3.2\text{-}9)$$

respectively. Note the ± signs in Equations 5.3.2-7 through 5.3.2-9; these signs imply that two quantities are defined by each of these formulas. If Γ_1^+, given by Equation 5.3.2-5, is used to compute a positive slope, then label all coefficients with the plus superscript. In a similar manner, use the negative superscript if Equation 5.3.2-6 is used to compute the slope.

The results given above are used to approximate the value $\dot{\phi}(\phi_s + \Delta)$ of the separatrix cycle at the point $\phi_s + \Delta$, where Δ is a user-supplied small increment. Then, starting from these initial conditions, the remainder of the separatrix is obtained by numerically integrating the time-dependent differential Equation 5.3-6 that describes the PLL. The separatrices displayed on Figures 5.9 through 5.13 are typical of those produced by this method.

APPENDIX 5.3.3 THE SEPARATRIX CYCLE OF A SECOND-ORDER PLL

For values of positive ω_Δ' that are less than the hold-in range Ω_h', a second-order PLL with an imperfect integrator in its loop filter has saddle points that are spaced 2π radians apart on the phase plane. As discussed in Appendix 5.3.2, each of these saddle points is approached by four separatrices. As $\tau \to \infty$, two separatrices approach each saddle point, and they do this with a negative slope on the phase plane (τ is the independent time variable in the differential equation describing the PLL). And, as $\tau \to -\infty$, two other separatrices approach each saddle point, and they do this with a positive slope on the phase plane.

For certain values of loop parameters, there is a periodic phase plane path, an example of which is given by Figure 5.29, that connects the above-mentioned sequence of saddle points. Such a periodic phase plane trajectory is called a *separatrix cycle* (or *separatrix loop* or *saddle-to-saddle separatrix*). As discussed below (see also Chapter 5), the external stability of the separatrix cycle determines whether the high- or low-gain case applies. Also, it bifurcates into a limit cycle which profoundly changes the pull-in behavior of the PLL.

Intuitively, it is helpful to think of the separatrix cycle as being constructed from separatrices that "match up" at certain values of loop parameters. This is illustrated by Figure 5.30 which shows parts of separatrices from adjacent saddle points. For a precise value of ω_Δ', the two separatrices overlie one another as is depicted by Figure 5.30b. For slightly smaller (larger) values of ω_Δ', the separatrix from the smaller saddle point always lies below (above) the separatrix from the larger saddle point as is shown by Figure 5.30a (Figure 5.30c).

The separatrix cycle is an example of a *structurally unstable path* on the phase plane (see p. 100 of Andronov et al.).[31] Suppose this saddle-to-saddle path exists for a given set of PLL parameters a', b', and ω_Δ'.

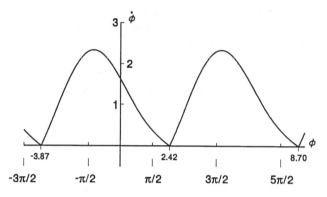

FIGURE 5.29
Separatrix cycle for a' = 0.5, b' = 0.1, and $\omega_\Delta' = 3.319$.

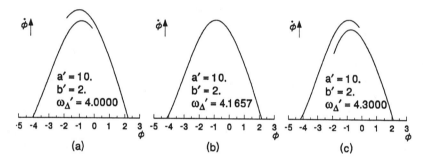

FIGURE 5.30
Separatrices from adjacent saddle points match up to form a separatrix cycle (low-gain case).

Then the saddle-to-saddle nature of the path no longer exists for arbitrarily small changes in ω_Δ' (the path "breaks"). More importantly, if the separatrix cycle is externally stable, it bifurcates into a stable limit cycle for arbitrarily small increases in positive ω_Δ'; this is illustrated by Figures 5.18 and 5.19. In a similar manner, if the separatrix cycle is externally unstable, it bifurcates into an unstable limit cycle for arbitrarily small decreases in positive ω_Δ'; this is illustrated by Figures 5.15 and 5.17. This limit cycle has a major influence on the pull-in properties of the PLL.

In the second-order PLL containing an imperfect integrator in its loop filter, a detailed and rigorous theory can be developed that qualitatively describes the global nonlinear behavior of the loop. This theory proves the existence of the separatrix cycle in both the low- and high-gain cases, and it details the two types of bifurcations that are possible. It is based on the theory of *rotated vector fields* (see Section 4.5 of Perko).[30]

NONLINEAR PLL BEHAVIOR IN THE ABSENCE OF NOISE 173

This theory is not given here; instead, this appendix details simple methods for computing and determining the external stability of the separatrix cycle.

5.3.3.1 Computing the Separatrix Cycle

One period of this cycle can be computed by setting up boundary-value problems at successive saddle points of Equation 5.3-6 and computing separatrices as discussed in Appendix 5.3.2. The boundary value problem that starts at the smaller saddle is integrated in a forward direction to produce one separatrix; the boundary value problem starting at the larger saddle is integrated in a reverse direction to produce a second separatrix. The algorithm described in this section finds the value of ω_Δ' (for given constants a' and b') so that the two separatrices meet halfway between the saddle points and describe the separatrix cycle.

On the phase plane, Equation 5.3-6, with $0 < \omega_\Delta' < a'/b'$, has a saddle point at

$$\phi_k = \left[\pi - \sin^{-1}(b'\omega_\Delta'/a')\right] - 2\pi k \qquad (5.3.3\text{-}1)$$

$$\dot{\phi}_k = 0 \qquad (5.3.3\text{-}2)$$

for each integer k. As discussed in Appendix 5.3.2, a separatrix with the positive slope

$$\frac{d\dot{\phi}}{d\phi}(\phi_1) = -\frac{b' + \cos(\phi_1)}{2} + \sqrt{\frac{(b' + \cos(\phi_1))^2}{4} - a'\cos(\phi_1)} \qquad (5.3.3\text{-}3)$$

departs from the saddle point (ϕ_1, 0), and it traverses through the upper half of the phase plane for increasing ϕ, $\phi > \phi_1$. In a similar manner, a separatrix with the negative slope

$$\frac{d\dot{\phi}}{d\phi}(\phi_0) = -\frac{b' + \cos(\phi_0)}{2} - \sqrt{\frac{(b' + \cos(\phi_0))^2}{4} - a'\cos(\phi_0)} \qquad (5.3.3\text{-}4)$$

departs from the saddle point (ϕ_0, 0), and it traverses through the upper half of the phase plane for decreasing ϕ, $\phi < \phi_0$. These separatrices can be computed by using the algorithm discussed in Appendix 5.3.2. Examples of these separatrices are depicted on Figure 5.30.

The algorithm for computing the separatrix cycle must find the value of ω_Δ' where the above-mentioned separatrices form a cycle by meeting halfway between ϕ_1 and ϕ_0; Figure 5.30b shows an example of separatrices that meet to form a complete cycle. It must be remembered that Slopes 5.3.3-3 and 5.3.3-4, as well as Equation 5.3-6, depend on ω_Δ'. Hence, PLL detuning ω_Δ' has an influence on the boundary conditions utilized when integrating Equation 5.3-6; also, it enters this differential equation in an explicit manner.

The algorithm for calculating the separatrix cycle starts integrating Equation 5.3-7 subject to Boundary Condition 5.3.3-3 at $(\phi_1, \dot\phi_1)$. The integration is done in a forward direction until $\dot\phi(\phi_1 + \pi)$ is reached; denote this numerical value as $\dot\phi_{\text{fwd}}(\omega_\Delta')$. Next, Equation 5.3-7 is started at $(\phi_0, \dot\phi_0)$ with Boundary Condition 5.3.3-4. The integration is done in a backward direction until $\dot\phi(\phi_0 - \pi)$ is reached; denote this numerical value as $\dot\phi_{\text{rev}}(\omega_\Delta')$. The algorithm strives to find the value of ω_Δ' that satisfies the nonlinear algebraic equation

$$g(\omega_\Delta') \equiv \dot\phi_{\text{fwd}}(\omega_\Delta') - \dot\phi_{\text{rev}}(\omega_\Delta') = 0. \qquad (5.3.3\text{-}5)$$

Equation 5.3.3-5 can be solved for ω_Δ' by numerical integration coupled with standard routines for solving algebraic equations. The MINPACK routine HYBRD1 was used to obtain the results described here (Garbow et al.).[81] A user-supplied Jacobian is not required to use HYBRD1; the routine computes one numerically by using finite-difference techniques.

Values of a = 100 and b = 20 were used in the algorithm outlined above; also, integral multiples of 5 up to and including 200 were used for G. The results are depicted by Figure 5.23. For every point (G, ω_Δ) on this graph, Equation 5.3-6 has a separatrix cycle. The separatrix cycle vanishes ("breaks") as (G, ω_Δ) moves off this graph. Next, the stability of this cycle is discussed.

5.3.3.2 Stability of the Separatrix Cycle

The separatrix cycle may be externally stable or externally unstable (see p. 301 of Andronov et al.).[31] If the upper-half plane separatrix cycle is externally stable (unstable), then trajectories will converge to (diverge from) the separatrix cycle upper side if they come sufficiently close to the cycle. Regardless of the separatrix cycle external stability, trajectories that start below the cycle do not converge to the cycle. The qualitative property of external stability is different from the fact that the separatrix cycle is a structurally unstable phase plane path.

Fortunately, the question of separatrix cycle external stability is answered easily (p. 304 of Andronov et al.).[31] Let ϕ_{sp} denote a saddle point approached by the separatrix cycle. The separatrix cycle is externally stable (unstable) if $s' \equiv -(b' + \cos\phi_{sp})$ is negative (positive). If gain normalization is removed, the test for stability can be stated as

$$s = -\left(b + G\cos\phi_{sp}\right) \underset{>}{\overset{<}{\underset{\text{unstable}}{\overset{\text{stable}}{}}}} 0. \qquad (5.3.3\text{-}6)$$

This simple formula was used to test the data plotted on Figure 5.23. The solid (dashed) line portion of this graph describes points (G, ω_Δ) where the separatrix cycle is stable (unstable). As discussed next, the low (high)-gain case applies when G is sufficiently small (large) so that the separatrix cycle is externally stable (unstable).

5.3.3.3 Bifurcation of a Periodic Limit Cycle from the Separatrix Cycle

Suppose that an externally stable separatrix cycle exists for $\omega_\Delta = \Omega_p > 0$ (the low-gain case). Then, this separatrix cycle breaks, and the PLL phase locks, as ω_Δ *decreases* from Ω_p (Ω_p is the pull-in range parameter). This separatrix cycle breaks and bifurcates into a stable upper-half plane limit cycle Γ_s as ω_Δ *increases* from Ω_p.

Figure 5.19 illustrates bifurcation results that are typical for the low-gain case where the separatrix cycle is stable. As computed by the algorithm outlined above, a stable separatrix cycle exists for a = 100, b = 20, G = 10, and $\omega_\Delta = \Omega_p \approx 41.657$. It breaks, and a stable periodic limit cycle bifurcates, as ω_Δ increases from Ω_p. Figure 5.19 shows examples of this limit cycle for values of ω_Δ equal to 46, 50, and 60.

Suppose that an unstable separatrix cycle exists for $\omega_\Delta = \Omega_2 > 0$ (the high-gain case). As ω_Δ *decreases* from Ω_2, this separatrix cycle breaks and bifurcates into an unstable limit cycle Γ_u that expands upward on the phase plane. An example of this is depicted by Figure 5.15 which shows the separatrix cycle for the values a = 100, b = 20, G = 200, and $\omega_\Delta = \Omega_2 \approx 663.8$ radians per second. Also, the figure illustrates Γ_u for values of ω_Δ equal to 620.0 and 602.0 radians per second.

Chapter 6

STOCHASTIC METHODS FOR THE NONLINEAR PLL MODEL

In this chapter, fundamental results and techniques are introduced which are useful for analyzing a PLL driven by a noisy reference signal. They build upon the modeling methods for noise that are introduced in Chapter 2. However, the methods introduced here are more general than those covered in Chapters 2 and 3. In these earlier chapters, a linear PLL model is employed; here, fundamentals and techniques are covered that are useful in analyzing the PLL nonlinear model.

For most PLL modeling tasks, linear techniques can be used productively *if* the phase error variance is sufficiently small (so that the actual value of the phase error remains small, at least most of the time). However, as input reference noise and phase error variance increase, a point is reached where these requirements no longer hold, and the linear theory breaks down. When this happens, the methods introduced in this chapter must be applied to analyze the problem since all linear methods produce inaccurate analytical and numerical results. In addition, the methods introduced in Chapters 6 through 8 allow the study and characterization of phenomena that have no counterpart in the linear theory. For example, only nonlinear methods can be used to analyze the important problem of noise-induced cycle slips. Roughly stated, a noise-induced cycle slip occurs when input noise causes the closed-loop phase error to advance or retard by an integer multiple of 2π radians. This phenomenon must be analyzed by using a nonlinear PLL model; it has no counterpart in the linear PLL theory.

As provided in Chapters 2 and 3, linear noise analysis techniques model the PLL state vector as a Gaussian random process. This Gaussian modeling assumption is an approximation that deteriorates as the input reference noise and phase error variance increase. Eventually, it becomes unacceptable and must be abandoned in favor of a non-Gaussian state vector model.

Fortunately, the PLL model described in Section 2.5 has a state vector that can be modeled as a Markov process. This is a *key* modeling assumption that is independent of input noise level and the value of the phase error variance. By using a Markov model, it is possible to develop theory and techniques for analyzing the nonlinear nature of the PLL. Use of a PLL model that is both Markov and nonlinear can lead to results that are valid for the important case of high phase error variance.

This chapter is devoted to laying the foundation for the analysis of this Markov, nonlinear PLL model. First, a simple example of a discrete Markov process is described in Section 6.1. Limiting conditions are given under which this discrete process approaches a continuous Markov process that has important attributes in common with the phase error in a PLL. Next, Section 6.2 introduces the theory of first-order Markov processes. This section includes the development of the one-dimensional Fokker–Planck equation that describes the time-evolution of the first-order PLL phase error density function. From studying Section 6.2, the reader should note that many important closed-form results can be given for one-dimensional Markov processes. Finally, Section 6.3 provides an extension of the Fokker–Planck equation to the n-dimensional case.

6.1 THE RANDOM WALK — A SIMPLE MARKOV PROCESS

Before launching the formal study of Markov processes in Section 6.2, the reader is introduced in this section to a simple, one-dimensional example that has some important attributes in common with the phase error process in a PLL. This commonly discussed example is the classical "random walk", and many forms of it (i.e., the "gambler's ruin") have been studied over the years. At first, a discrete random walk is introduced. Then, a limiting form of this random walk is shown to be the well-known continuous Wiener process. Finally, simple equations are developed that provide a complete statistical description of the discrete and limiting form of the random walk. The approach taken here is similar to that discussed in the paper by Chandrasekhar (see the collection of papers edited by Wax).[33]

Suppose a man takes a random walk by starting at a designated origin on a straight line path. With probability p ($q \equiv 1 - p$), he takes a step to the right (left). Suppose that each step is of length ℓ meters, and each step is completed in τ_s seconds. After N steps (completed in $N\tau_s$ seconds), the man is located $X_d(N)$ steps from the origin; note that $-N \le X_d(N) \le N$ since the man starts at the origin. If $X_d(N)$ is positive (negative), the man is located to the right (left) of the origin. The

STOCHASTIC METHODS FOR THE NONLINEAR PLL MODEL 179

quantity $P[X_d(N) = n]$, $-N \leq n \leq N$ denotes the probability that the man's location is n steps from the origin after he has taken N steps.

The calculation of $P[X_d(N) = n]$ is simplified greatly by the assumption, implied in the previous paragraph, that the man takes independent steps. That is, the direction taken at the Nth step is independent of $X_d(k)$, $0 \leq k \leq N-1$ and the directions taken at all previous steps. Also simplifying the development is the assumption that p does not depend on step index N. Under these conditions, it is possible to write

$$P[X_d(N+1) = X_d(N)+1] = p$$
$$P[X_d(N+1) = X_d(N)-1] = 1-p = q. \qquad (6.1\text{-}1)$$

Let R_{n0} and L_{n0} denote the number of steps to the right and left, respectively, that will place the man n, $-N \leq n \leq N$, steps from the origin after he has completed a total of N steps. Integers R_{n0} and L_{n0} depend on integers N and n; the relationship is given by

$$R_{n0} - L_{n0} = n$$
$$R_{n0} + L_{n0} = N \qquad (6.1\text{-}2)$$

since $-N \leq n \leq N$. Integer values for R_{n0} and L_{n0} exist when n and N are both even or odd. When a solution of Equation 6.1-2 exists, it is given by

$$R_{n0} = \frac{N+n}{2}$$
$$L_{n0} = \frac{N-n}{2}. \qquad (6.1\text{-}3)$$

Of course, there are multiple sequences of N steps, R_{n0} to the right and L_{n0} to the left, that the man can take to ensure that he is n steps from the origin. In fact, the number of such sequences is given by

$$\binom{N}{R_{n0}} = \frac{N!}{R_{n0}!\, L_{n0}!} \qquad (6.1\text{-}4)$$

This quantity represents the number of subsets of size R_{n0} that can be formed from N distinct objects. These sequences are mutually exclusive events. Furthermore, they are equally probable, and the probability of each of them is

$$P[R_{no}(L_{no}) \text{ Steps To Right (Left) in a Specific Sequence}]$$
$$= p^{R_{no}} q^{L_{no}}.$$
(6.1-5)

The desired probability $P[X_d(N) = n]$ can be computed easily with the use of Equations 6.1-3, 6.1-4, and 6.1-5. From the theory of independent Bernoulli trials, the result

$$P[X_d(N) = n] = \frac{N!}{R_{no}! \, L_{no}!} p^{R_{no}} q^{L_{no}} \quad (6.1\text{-}6)$$

follows easily. If there are no integer solutions to Equation 6.1-2 for given values of n and N, then it is not possible to arrive at n steps from the origin after taking N steps and $P[X_d(N) = n] = 0$. Note that Equation 6.1-6 is just the probability that the man takes R_{no} steps to the right given that he takes N independent steps.

The analysis leading to Equation 6.1-6 can be generalized to include a nonzero starting location. Instead of starting at the origin, assume that the man starts his random walk at m steps from the origin, where $-N \le m \le N$. Then, after the man has completed N independent steps, $P[X_d(N) = n \mid X_d(0) = m]$ denotes the probability that he is n steps from the origin given that he started m steps from the origin.

A simple modification of Equation 6.1-6 leads to $P[X_d(N) = n \mid X_d(0) = m]$. Let $v \equiv n - m$, so that v denotes the man's net increase in the number of steps to the right after he has completed N steps. Also, R_{nm} (L_{nm}) denotes the number of steps to the right (left) that are required if the man starts and finishes m and n, respectively, steps from the origin. Note that

$$R_{nm} = \frac{N+v}{2}$$
$$L_{nm} = \frac{N-v}{2}$$
(6.1-7)

if $|v| \le N$ and $N + v$, $N - v$ are even; otherwise, integers R_{nm} and L_{nm} do not exist. Finally, the desired result follows by substituting R_{nm} and L_{nm} for R_{no} and L_{no} in Equation 6.1-6; this procedure leads to

$$P[X_d(N) = n \mid X_d(0) = m] = P[R_{nm} \text{ steps to the right out of N steps}]$$
(6.1-8)
$$= \frac{N!}{R_{nm}! \, L_{nm}!} p^{R_{nm}} q^{L_{nm}}$$

if integers R_{nm} and L_{nm} exist, and

$$P[X_d(N) = n | X_d(0) = m] = 0 \qquad (6.1\text{-}9)$$

if R_{nm} and L_{nm} do not exist.

For $Npq \gg 1$, an asymptotic approximation is available for Equation 6.1-8. In the development that follows, it is assumed that $p = q = 1/2$. According to the DeMoivre–Laplace theorem (see Papoulis),[20] for $N/4 \gg 1$ and $|R_{nm} - N/2| < \sqrt{N/4}$, the approximation

$$P[X_d(N) = n | X_d(0) = m] = \frac{N!}{R_{nm}! L_{nm}!} \left(\frac{1}{2}\right)^{R_{nm}} \left(\frac{1}{2}\right)^{L_{nm}}$$

$$\approx \frac{1}{\sqrt{2\pi(N/4)}} \exp\left[-\frac{(R_{nm} - N/2)^2}{2(N/4)}\right] \qquad (6.1\text{-}10)$$

can be made.

6.1.1 The Wiener Process as a Limit of the Random Walk

Recall that each step corresponds to a distance of ℓ meters, and each step is completed in τ_s seconds. At time $t = N\tau_s$, let $X(N\tau_s)$ denote the man's physical displacement from the origin. Then $X(N\tau_s)$ is a random process given by $X(N\tau_s) \equiv \ell X_d(N)$, since $X_d(N)$ denotes the number of steps the man is from the origin after he takes N steps. Note that $X(N\tau_s)$ is a discrete-time random process that takes on only discrete values.

For large N and small ℓ and τ_s, the probabilistic nature of $X(N\tau_s)$ is of interest. First, note that $P[X(N\tau_s) = \ell n | X(0) = \ell m] = P[X_d(N) = n | X_d(0) = m]$; this observation and the binomial distribution function leads to the result

$$P[X(N\tau_s) \leq \ell n | X(0) = \ell m] = P[\text{Number of Steps to Right} \leq R_{nm}]$$

$$= \sum_{k=0}^{R_{nm}} \binom{N}{k} \left(\tfrac{1}{2}\right)^k \left(\tfrac{1}{2}\right)^{N-k}. \qquad (6.1\text{-}11)$$

For large N, the DeMoivre–Laplace theorem (see pp. 50–52 of Papoulis)[20] leads to the approximation

$$P[X(N\tau_s) \le \ell n | X(0) \approx \ell m] = G\left[\frac{R_{nm} - N/2}{\sqrt{N/4}}\right]$$

$$= G\left[\frac{v}{\sqrt{N}}\right], \tag{6.1-12}$$

where G is the distribution function for a zero-mean, unit-variance Gaussian random variable.

The discrete random walk process outlined above has the continuous Wiener process as a formal limit. To see this, let $\ell \to 0$, $\tau_s \to 0$, and $N \to \infty$ in such a manner that

$$\frac{\ell^2}{2\tau_s} \to D$$

$$t = N\tau_s$$

$$x = n\ell \tag{6.1-13}$$

$$x_o = m\ell$$

$$X(t) = X(N\tau_s),$$

where D is known as the *diffusion constant*. In terms of D, x, x_o, and t, the results of Equation 6.1-13 can be used to write

$$\frac{v}{\sqrt{N}} = \frac{(x - x_o)/\ell}{\sqrt{t/\tau_s}}$$

$$= \frac{(x - x_o)}{\sqrt{2Dt}}. \tag{6.1-14}$$

The probabilistic nature of the limiting form of X(t) is seen from Equations 6.1-12 and 6.1-14. In the limit, the process X(t) is described by the first-order conditional density function

$$f(x, t | x_o) = \frac{1}{\sqrt{4\pi Dt}} \exp\left[-\frac{(x - x_o)^2}{4Dt}\right]. \tag{6.1-15}$$

When $X(0) = x_o = 0$, this result describes the conditional probability density function of a continuous-time Wiener process. Clearly, process

STOCHASTIC METHODS FOR THE NONLINEAR PLL MODEL

$X(t)$ is Gaussian, and it is nonstationary since it has a variance that grows with time.

6.1.2 The Diffusion Equation for the Transition Density Function

In terms of physical displacement (from the origin) X, the conditional probability $P[X(N\tau_s) = \ell n \,|\, X(0) = \ell m]$ describes the probabilistic nature of the discrete-time random walk problem outlined above. In what follows, this conditional probability is denoted by the short hand notation $P[\ell n, N\tau_s \,|\, \ell m]$. For the case $p = q = \frac{1}{2}$, it is easy to see that it satisfies the difference equation

$$P\big[\ell n, (N+1)\tau_s \,\big|\, \ell m\big] = \frac{1}{2} P\big[\ell(n-1), N\tau_s \,\big|\, \ell m\big] \\ + \frac{1}{2} P\big[\ell(n+1), N\tau_s \,\big|\, \ell m\big]. \qquad (6.1\text{-}16)$$

The initial condition on the above difference equation is

$$P\big[\ell n, 0 \,\big|\, \ell m\big] = 1 \quad \text{if } n = m \\ = 0 \quad \text{if } n \neq m. \qquad (6.1\text{-}17)$$

The continuous conditional density $f(x, t \,|\, x_o)$ given by Equation 6.1-15 satisfies a partial differential equation. Note that the difference Equation 6.1-16 can be used to write

$$\frac{P\big[\ell n, (N+1)\tau_s \,\big|\, \ell m\big] - P\big[\ell n, N\tau_s \,\big|\, \ell m\big]}{\tau_s} \\ = \frac{\ell^2}{2\tau_s}\left[\frac{P\big[\ell(n+1), N\tau_s \,\big|\, \ell m\big] - 2P\big[\ell n, N\tau_s \,\big|\, \ell m\big] + P\big[\ell(n-1), N\tau_s \,\big|\, \ell m\big]}{\ell^2}\right]. \qquad (6.1\text{-}18)$$

Now, in the sense described by Equation 6.1-13, the formal limit of Equation 6.1-18 is

$$\frac{\partial}{\partial t} f(x, t \,|\, x_o) = D \frac{\partial^2}{\partial x^2} f(x, t \,|\, x_o), \qquad (6.1\text{-}19)$$

where $f(x,t \,|\, x_o)$ denotes the conditional probability density function given by Equation 6.1-15.

Equation 6.1-19 is a one-dimensional *diffusion equation*. It describes how probability diffuses (or flows) with time. It implies that probability is conserved in much the same way that the well-known continuity equation implies the conservation of electric charge. To draw this analogy, note that f describes the density of probability (or density of *probability particles*) on the one-dimensional real line. That is, f can be assigned units of particles per meter. Since D has units of square meters per second, a unit check on both sides of Equation 6.1-19 produces

$$\left(\frac{1}{\text{second}}\right)\left(\frac{\text{particles}}{\text{meter}}\right) = \left(\frac{\text{meter}^2}{\text{second}}\right)\left(\frac{1}{\text{meter}}\right)^2\left(\frac{\text{particles}}{\text{meter}}\right). \quad (6.1\text{-}20)$$

Now, write Equation 6.1-19 as

$$\frac{\partial}{\partial t}f = -\nabla \cdot \mathfrak{I}, \quad (6.1\text{-}21)$$

where

$$\mathfrak{I} \equiv -D\frac{\partial}{\partial x}f, \quad (6.1\text{-}22)$$

and ∇ is the divergence operator. The quantity $\mathfrak{I}$ is a one-dimensional *probability current*, and it has units of particles per second. Note the similarity between Equation 6.1-21 and the well-known continuity equation for electrical charge.

Probability current $\mathfrak{I}(x,t \mid x_o)$ indicates the rate of particle flow past point x at time t. Let (x_1, x_2) denote an interval; integrate Equation 6.1-21 over this interval to obtain

$$\frac{\partial}{\partial t}P[x_1 < X(t) \leq x_2 \mid x_o] = \frac{\partial}{\partial t}\int_{x_1}^{x_2} f(x,t \mid x_o)dx$$
$$= -[\mathfrak{I}(x_2,t \mid x_o) - \mathfrak{I}(x_1,t \mid x_o)]. \quad (6.1\text{-}23)$$

As illustrated by Figure 6.1, the left-hand side of this equation represents the time rate of probability buildup on (x_1, x_2). That is, between the limits of x_1 and x_2, the area under f is changing at a rate equal to the left-hand side of Equation 6.1-23. As depicted by Figure 6.1, the right-hand side of Equation 6.1-23 represents the probability currents entering the ends of the interval (x_1, x_2).

STOCHASTIC METHODS FOR THE NONLINEAR PLL MODEL

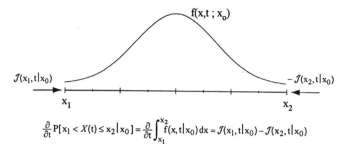

$$\frac{\partial}{\partial t}P[x_1 < X(t) \le x_2 | x_0] = \frac{\partial}{\partial t}\int_{x_1}^{x_2} f(x,t|x_0)\,dx = \mathcal{J}(x_1,t|x_0) - \mathcal{J}(x_2,t|x_0)$$

FIGURE 6.1
Probability buildup on (x_1, x_2) expressed in terms of net current entering the interval.

In Sections 6.2.3 and 6.3.2, generalized versions of Equation 6.1-21 are introduced. Also, in Chapter 7, an equation having the form of Equation 6.1-21 is used to describe the flow of probability current in nonlinear PLL models. Finally, it must be pointed out that the diffusion phenomenon is a transport mechanism that describes flow in many important applications (heat, electric current, molecular, etc.).

6.1.3 An Absorbing Boundary on the Random Walk

The quantity $X_d(N)$ is unconstrained in the discrete random walk described by Equation 6.1-8. Now, consider placing an absorbing boundary condition at n_1. No further displacements are possible after X_d reaches the boundary at n_1; the man stops his random walk the instant he arrives at the boundary (he is absorbed). Such a boundary condition has applications in many problems of practical importance. As discussed in Section 7.2, it plays a significant role in determining the average rate at which a first-order PLL slips cycles due to a noisy reference signal.

As before, assume that the man starts his random walk at m steps from the origin where $m < n_1$. This initial condition implies that $X_d(0) = m$ since random process X_d denotes the man's displacement (in steps) from the origin. He takes random steps; either he completes N of them, or he is absorbed at the boundary before completing N steps. If he manages to complete N steps, $X_d(N)$ denotes his displacement (in steps) from the origin. For the random walk with absorbing boundary conditions, the quantity $P[n, N \mid m; n_1]$ denotes the probability that $X_d(N) = n$ given that $X_d(0) = m$ and an absorbing boundary exists at n_1. In what follows, an expression is developed for this probability.

It is helpful to trace the man's movement by using a plane as shown by Figures 6.2a and b. On these diagrams, the horizontal axis denotes displacement, in steps, from the origin; the vertical axis denotes the

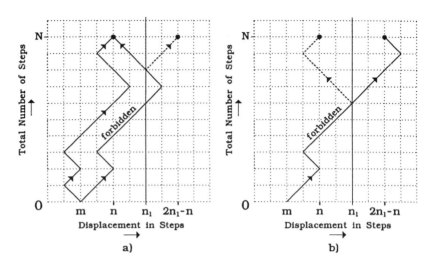

FIGURE 6.2
Random walk with an absorbing boundary at n_1.

total number of steps taken by the man. Every time the man takes a step, he moves upward on the diagram; also, he moves laterally to the right or left. The absorbing boundary is depicted on these figures by a solid vertical line at n_1. In the remainder of this section, these diagrams are used to illustrate the *reflection principle* for dealing with random processes that hit absorbing boundaries (also see Papoulis).[20]

Figure 6.2a depicts two N-step sequences (the solid-line paths) that start at m and arrive at n. One of these is "forbidden" since it intersects the boundary. A "forbidden" N-step sequence is one that intersects the boundary one or more times; for the present argument, assume that a "forbidden" sequence is not stopped by the boundary. For all steps above the *last* point of contact with the boundary, the "forbidden" sequence on Figure 6.2a has been reflected across the boundary to produce a dashed-line path that leads to the point $2n_1 - n$, the reflected (across the boundary) image of the point n. In this same manner, every "forbidden" path that reaches n can be partially reflected to produce a unique path that leads to the image point $2n_1 - n$.

The solid-line path on Figure 6.2b is an N-step sequence that arrives at the point $2n_1 - n$; as was the case on Figure 6.2a, this point is the mirror image across the boundary of point n. For all steps above the *last* point of contact with the boundary, the "forbidden" sequence on Figure 6.2b has been reflected across the boundary to produce a dashed-line path that leads to the point n. In this same manner, every "forbidden" path that reaches $2n_1 - n$ can be partially reflected to produce a unique path that leads to n.

STOCHASTIC METHODS FOR THE NONLINEAR PLL MODEL

From the observations outlined in the last two paragraphs, it can be concluded that a one-to-one correspondence exists between N-step "forbidden" sequences that reach point n and N-step sequences that reach the image point $2n_1 - n$. That is, for every "forbidden" sequence that reaches n (that reaches $2n_1 - n$), there is a sequence that reaches $2n_1 - n$ (that reaches n). Out of all N-step sequences that start at m, the proportion that are forbidden and reach n is exactly equal to the proportion that reach the image point. This observation is crucial in the development of $P[n, N \mid m; n_1]$.

Without the boundary in place, the computation of $P[n, N \mid m]$ involves computing the relative frequency of the man arriving at $X_d = n$ after N steps. That is, to compute the probability $P[n, N \mid m]$, the number of N step sequences that lead to n must be normalized by the total number of distinct N-step sequences. Now, after the boundary at n_1 is put in place, the number of "forbidden" sequences that reach n must be subtracted from the total number (i.e., the number without a boundary) of N-step sequences that lead to n. But, this difference is exactly equal to the number of N-step sequences that lead to n *minus* the number that lead to $2n_1 - n$ (both numbers are computed without a boundary in place). This difference must be normalized by the number of distinct N-step sequences to compute the probability $P[n, N \mid m; n_1]$. Hence, it is easy to conclude that

$$P[n, N \mid m; n_1] = P[n, N \mid m] - P[2n_1 - n, N \mid m], \quad (6.1-24)$$

where $P[n, N \mid m]$ is given by Equation 6.1-8. For the absorbing boundary case, the probability of reaching n can be expressed in terms of probabilities that are calculated for the boundary-free case.

6.1.4 An Absorbing Boundary on the Wiener Process

As in Section 6.1.1, suppose that each step corresponds to a distance of ℓ meters, and it takes τ_s seconds to take a step. Furthermore, $X(N\tau_s) = \ell X_d(N)$ denotes the man's physical distance (in meters) from the origin after completing N steps. Also, for the case where an absorbing boundary exists at $\ell n_1 > \ell m$, $P[\ell n, N\tau_s \mid \ell m; \ell n_1]$ denotes the probability that the man is ℓn meters from the origin at $t = N\tau_s$, given that he starts at ℓm when $t = 0$. Using the argument which led to Equation 6.1-24, it is possible to write

$$P[\ell n, N\tau_s \mid \ell m; \ell n_1] = P[\ell n, N\tau_s \mid \ell m] - P[2n_1\ell - n\ell, N\tau_s \mid \ell m]. \quad (6.1-25)$$

Both terms on the right-hand side of Equation 6.1-25 can be approximated by Equation 6.1-13 for large N. Now, let $\ell \to 0$, $\tau_s \to 0$, and $N \to \infty$ in the manner described by Equation 6.1-14 to obtain (with $\ell n_1 \to x_1$)

$$f(x,t|x_o;x_1) = \frac{1}{\sqrt{4\pi Dt}}\left[\text{Exp}\left[-\frac{(x-x_o)^2}{4Dt}\right] - \text{Exp}\left[-\frac{(2x_1-x-x_o)^2}{4Dt}\right]\right]. \quad (6.1\text{-}26)$$

This conditional probability density function for the absorbing boundary case is the counterpart of Equation 6.1-15 that describes the conditional probability density function for the boundary-free case.

From Equation 6.1-26, note that

$$f(x_1,t|x_o;x_1) = 0 \quad (6.1\text{-}27)$$

for all $t \geq 0$. That is, the placement of an absorbing boundary at x_1 forces the density function to vanish at the boundary. If $X(t)$ represents the position of a random particle, then Equation 6.1-27 implies that, only rarely, can the particle be found in the vicinity of the boundary. In Section 7.2, this boundary condition plays an important role in determining the rate of cycle slippage in a first-order PLL driven by a noisy reference.

6.1.5 Gaussian White Noise as the Formal Derivative of the Wiener Process

Gaussian white noise is a mathematical concept that is widely used in PLL analysis and other areas of applied science. Generally, it is understood to be a stationary, zero-mean, Gaussian process that has a covariance function consisting of a delta function. Often, it can be used to model an existing, bandlimited Gaussian noise source in a practical system. Usually, this is the case when the bandwidth of the existing source is large compared to the bandwidth of the system. In most cases, use of a white noise model simplifies the analysis of the system that contains the model.

As is widely known, white noise serves only as a mathematical model; in reality, it does not exist as a physical process. However, it is common to relate the mathematical concept of Gaussian white noise to the Wiener process. Such a relationship is discussed in Section 3.2 of Arnold[34] and Section 3.8 of Jazwinski.[35] As discussed by Arnold,[34]

STOCHASTIC METHODS FOR THE NONLINEAR PLL MODEL 189

Gaussian white noise is a *generalized stochastic process*, and it is the formal derivative of the Wiener process. To obtain an understanding of this, let X(t) denote a zero-mean Wiener process with

$$\text{Variance}\left[X(t+\Delta t) - X(t)\right] = E\left[\{X(t+\Delta t) - X(t)\}^2\right]$$
$$= \frac{N_o}{2}\Delta t, \quad (6.1\text{-}28)$$

where $\Delta t > 0$ is a small time increment. Use this Wiener process to define

$$\eta(t;\Delta t) \equiv \frac{X(t+\Delta t) - X(t)}{\Delta t}. \quad (6.1\text{-}29)$$

As Δt approaches zero, the limit of $\eta(t; \Delta t)$ does not exist in the classical Calculus sense. However, in the relationship between Gaussian white noise and the Wiener process, insight can be gained by computing the covariance function for $\eta(t; \Delta t)$. By using the fact that the Wiener process has independent increments, process $\eta(t; \Delta t)$ can be shown to have the covariance function

$$R(\tau_d) = \frac{N_o}{2}\left[\frac{\Delta t - |\tau_d|}{(\Delta t)^2}\right] \quad |\tau_d| \leq \Delta t$$
$$= 0 \quad |\tau_d| > \Delta t, \quad (6.1\text{-}30)$$

and Figure 6.3 is an illustration of this result. Now, the area under $R(\tau_d)$ is the constant $N_0/2$, and it is independent of Δt. As Δt approaches zero, $R(\tau_d)$ approaches the generalized function $(N_0/2)\delta(\tau_d)$, the correlation function of white noise. For this reason, in the engineering literature, it is stated that Gaussian white noise is the formal derivative of the Wiener process.

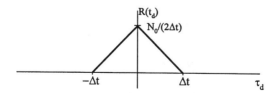

FIGURE 6.3
Autocorrelation of $[X(t + \Delta t) - X(t)]/\Delta t$ for small Δt.

6.2 THE FIRST-ORDER MARKOV PROCESS

Usually, simplifying assumptions are made when performing an engineering analysis of a nonlinear system driven by noise. Often, assumptions are made that place limits on the amount of information that is required to analyze the system. For example, it is commonly assumed that a finite dimensional model can be used to describe the system. The model is described by a finite number of state variables which are modeled as random processes. An analysis of the system might involve the determination of the probability density function that describes these state variables. A second common assumption has to do with how this probability density evolves with time, and what kind of initial data it depends on. This assumption states that the future evolution of the density function can be expressed in terms of the present values of the state variables; knowledge of past values of the state variables is not necessary. As discussed in this chapter, this second assumption means that the state vector can be modeled as a continuous Markov process. Furthermore, the process has a density function that satisfies a parabolic partial differential equation known as the Fokker–Planck equation.

The one-dimensional Markov process and Fokker–Planck equation are discussed in this section. Unlike the situation in multidimensional problems, a number of exact closed-form results can be obtained for the one-dimensional case, and this justifies treating the case separately. Also, the one-dimensional case is simpler to deal with from a notational standpoint. The more general N-dimensional case is discussed in Section 6.3.

A random process has the *Markov property* if its distribution function at any future instant, conditioned on present and past values of the process, does not depend on the past values. Consider increasing, but arbitrary, values of time $t_1 < t_2 < \ldots < t_n$, where n is an arbitrary positive integer. A random process X(t) has the Markov property if its conditional probability distribution function satisfies

$$F(x_n, t_n | x_{n-1}, t_{n-1}; \ldots; x_1, t_1)$$
$$= P[X(t_n) \leq x_n | X(t_{n-1}) = x_{n-1}, X(t_{n-2}) = x_{n-2}, \ldots, X(t_1) = x_1]$$
$$= P[X(t_n) \leq x_n | X(t_{n-1}) = x_{n-1}]$$
$$= F(x_n, t_n | x_{n-1}, t_{n-1})$$

(6.2-1)

for all values $x_1, x_2, \ldots, x_n$ and all sequences $t_1 < t_2 < \ldots < t_n$.

The Wiener and random walk processes discussed in Section 6.1 are examples of Markov processes. In the development that produced Equation 6.1-15, the initial condition x_o was specified at $t = 0$. Now, suppose the initial condition is changed so that x_o is specified at $t = t_o$. In this case, transitions in the displacement random process X can be described by the conditional probability density function

$$f(x, t \mid x_o, t_o) = \frac{1}{\sqrt{4\pi D(t-t_o)}} \operatorname{Exp}\left[-\frac{(x-x_o)^2}{4D(t-t_o)}\right] \quad (6.2-2)$$

for $t > t_o$. Note that the displacement random process X is Markov since, for t greater than t_o, Density 6.2-2 can be expressed in terms of the displacement value x_o at time t_o; prior to time t_o, the history of the displacement is not relevant.

If random process $X(t)$ is a Markov process, the joint density of $X(t_1), X(t_2), \ldots, X(t_n)$, where $t_1 < t_2 < \ldots < t_n$, has a simple representation. First, recall the general formula

$$f(x_n, t_n; x_{n-1}, t_{n-1}; \cdots ; x_1, t_1)$$
$$= f(x_n, t_n \mid x_{n-1}, t_{n-1}; \cdots ; x_1, t_1) f(x_{n-1}, t_{n-1} \mid x_{n-2}, t_{n-2}; \cdots ; x_1, t_1) \cdots \quad (6.2-3)$$
$$\cdots f(x_2, t_2 \mid x_1, t_1) f(x_1, t_1)$$

Now, utilize the Markov property to write Equation 6.2-3 as

$$f(x_n, t_n; x_{n-1}, t_{n-1}; \cdots ; x_1, t_1)$$
$$= f(x_n, t_n \mid x_{n-1}, t_{n-1}) f(x_{n-1}, t_{n-1} \mid x_{n-2}, t_{n-2}) \cdots \quad (6.2-4)$$
$$\cdots f(x_2, t_2 \mid x_1, t_1) f(x_1, t_1).$$

Equation 6.2-4 states that a Markov process $X(t)$, $t \geq t_1$, is completely specified by the initial marginal density $f(x_1, t_1)$ and the set of first-order conditional densities

$$f(x_p, t_p \mid x_q, t_q), \ t_p > t_q > t_1. \quad (6.2-5)$$

For this reason, conditional densities of the form of Equation 6.2-5 are known as *transition densities*.

Some important special cases arise regarding the time dependence of the marginal and transitional densities of a Markov process. A

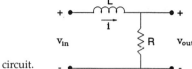

FIGURE 6.4
A simple RL circuit.

Markov process is said to be *homogeneous* if $f(x_2, t_2 | x_1, t_1)$ is invariant to a shift in the time origin. In this case, the transition density depends only on the time difference $t_2 - t_1$. Now, recall that stationarity implies that both $f(x,t)$ and $f(x_2, t_2 | x_1, t_1)$ are invariant to a shift in the time origin. Hence, stationary Markov processes are homogeneous. However, the converse of this last statement is not generally true.

6.2.1 An Important Application of Markov Processes

Consider a physical problem, such as a first-order PLL, that can be modeled as a first-order system driven by white Gaussian noise. Let $X(t)$, $t \geq t_0$ denote the state of this system; the statistical properties of state X are of interest here. Suppose that the initial condition $X(t_0)$ is a random variable that is independent of the white noise excitation of the system. Then state $X(t)$ belongs to a special class of Markov processes known as *diffusion processes*. The advanced reader should consult Chapter 9 of Arnold[34] for a proof of the claim that state X is Markov. As is characteristic of diffusion processes, almost all sample functions of X are continuous, but they are differentiable nowhere. Finally, these statements are generalized easily to nth-order systems driven by white Gaussian noise.

As an example of a first-order system driven by white Gaussian noise, consider the RL circuit illustrated by Figure 6.4. In this circuit, current $i(t)$, $t \geq t_0$ is the state, and white Gaussian noise $v_{in}(t)$ is assumed to have a mean of zero. The initial condition $i(t_0)$ is assumed to be a zero-mean Gaussian random variable, and it is independent of the input noise $v_{in}(t)$ for all t.

The formal differential equation that describes state $i(t)$ is

$$\frac{di}{dt} = -\frac{R}{L}i + \frac{1}{L}v_{in}. \qquad (6.2\text{-}6)$$

However, sample functions of current i are not differentiable, so Equation 6.2-6 only serves as a symbolic representation of the circuit dynamical model. Now, recall from Section 6.1.5 that white noise v_{in} can be represented as a generalized derivative of a Wiener process. If W_t denotes this Wiener process, and $i_t \equiv i(t)$ denotes the circuit current (in these representations, the independent time variable is depicted as a subscript), then it is possible to say that

$$di_t = -\frac{R}{L}i_t\, dt + \frac{1}{L}dW_t \qquad (6.2\text{-}7)$$

is formally equivalent to Equation 6.2-6.

Equations 6.2-6 and 6.2-7 are *stochastic differential equations*, and they should be considered as nothing but formal symbolic representations for the integral equation

$$i_t - i_{t_k} = -\frac{R}{L}\int_{t_k}^{t} i_t\, dt + \frac{1}{L}\int_{t_k}^{t} dW_t, \qquad (6.2\text{-}8)$$

where $t > t_k \geq t_0$. On the right-hand side of Equation 6.2-8, the first integral can be interpreted in the classical Rieman sense. However, sample functions of W_t are not of bounded variation, so the second integral cannot be a Rieman–Stieltjes integral. Instead, it can be interpreted as a *stochastic Itô integral*, and a major field of mathematical analysis exists to support this effort.

On the right-hand side of Equation 6.2-8, the stochastic differential dW_t does not depend on i_t, $t < t_k$. Based on this observation, it is possible to conjecture that any probabilistic description of future values of i_t, $t > t_k$, when conditioned on the present value i_{t_k} *and* past values $i_{t_{k-1}}$, $i_{t_{k-2}}$, ..., does not depend on the past current values. That is, the structure of Equation 6.2-8 suggests that i_t is Markov. The proof of this conjecture is a major result in the theory of stochastic differential equations (see Chapter 9 of Arnold).[34]

Of great practical interest are methods for characterizing the statistical properties of diffusion processes that represent the state of a dynamical system driven by white Gaussian noise. At least two schools of thought exist for characterizing these processes. The first espouses direct numerical simulation of the system dynamical model. The second school is adhered to here, and it utilizes indirect analysis tools based on some form of diffusion equation (such as the Fokker–Planck equation).

6.2.2 The Chapman–Kolmogorov Equation

Suppose that $X(t)$ is a random process described by the conditional density function $f(x_3, t_3 | x_1, t_1)$. Clearly, this density function must satisfy

$$f(x_3, t_3 \mid x_1, t_1) = \int_{-\infty}^{\infty} f(x_3, t_3; x_2, t_2 \mid x_1, t_1) dx_2, \quad (6.2\text{-}9)$$

where $t_1 < t_2 < t_3$. Now, a standard result from probability theory can be used here; substitute

$$\begin{aligned} f(x_3, t_3 &; x_2, t_2 \mid x_1, t_1) \\ &= f(x_3, t_3 \mid x_2, t_2; x_1, t_1) f(x_2, t_2 \mid x_1, t_1) \end{aligned} \quad (6.2\text{-}10)$$

into Equation 6.2-9 and obtain

$$\begin{aligned} f(x_3, t_3 &\mid x_1, t_1) \\ &= \int_{-\infty}^{\infty} f(x_3, t_3 \mid x_2, t_2; x_1, t_1) f(x_2\ t_2 \mid x_1, t_1) dx_2. \end{aligned} \quad (6.2\text{-}11)$$

Equation 6.2-11 can be simplified if X is a Markov process. By using the Markov property, this last equation can be simplified to obtain

$$f(x_3, t_3 \mid x_1, t_1) = \int_{-\infty}^{\infty} f(x_3, t_3 \mid x_2, t_2) f(x_2, t_2 \mid x_1, t_1) dx_2. \quad (6.2\text{-}12)$$

This is the well-known *Chapman–Kolmogorov* equation for Markov processes (it is also known as the *Smoluchowski* equation). It provides a useful formula for the transition probability from x_1 at time t_1 to x_3 at time t_3 in terms of an intermediate step x_2 at time t_2, where t_2 lies between t_1 and t_3. In Section 6.3, a version of Equation 6.2-12 is used in the development of the N-dimensional Fokker–Planck equation.

6.2.3 The One-Dimensional Kramers–Moyal Expansion

As discussed in Section 6.1.1, a limiting form of the random walk is a Markov process described by the transition Density 6.1-15. This density function satisfies the diffusion Equation 6.1-19. These results are generalized in this section where it is shown that a first-order Markov process is described by a transition density that satisfies an equation known as the *Kramers–Moyal expansion*. When the Markov process is the state of a dynamical system driven by white Gaussian noise, this equation simplifies to what is known as the Fokker–Planck equation. Equation 6.1-19 is a simple example of a Fokker–Planck equation.

Consider the random increment

$$\Delta X_{t_1} \equiv X(t_1 + \Delta t) - X(t_1), \qquad (6.2\text{-}13)$$

where Δt is a small, positive time increment. Given that $X(t_1) = x_1$, the conditional characteristic function Θ of ΔX_{t_1} is given by

$$\Theta(\omega; x_1, t_1, \Delta t) = E\left[\exp(j\omega \Delta X_{t_1}) \mid X(t_1) = x_1\right]$$

$$= \int_{-\infty}^{\infty} e^{j\omega(x-x_1)} f(x, t_1 + \Delta t \mid x_1, t_1) dx. \qquad (6.2\text{-}14)$$

If the Markov process is homogeneous, then Θ depends on the time difference Δt, but not the absolute value of t_1. The inverse of Equation 6.2-14 is

$$f(x, t_1 + \Delta t \mid x_1, t_1) = \frac{1}{2\pi} \int_{-\infty}^{\infty} e^{-j\omega(x-x_1)} \Theta(\omega; x_1, t_1, \Delta t) d\omega, \qquad (6.2\text{-}15)$$

which is an expression for the transition density in terms of the characteristic function of the random increment. Now, use Equation 6.2-15 in

$$f(x, t_1 + \Delta t) = \int_{-\infty}^{\infty} f(x, t_1 + \Delta t; x_1, t_1) \, dx_1$$

$$= \int_{-\infty}^{\infty} f(x, t_1 + \Delta t \mid x_1, t_1) \, f(x_1, t_1) \, dx_1 \qquad (6.2\text{-}16)$$

to obtain

$$f(x, t_1 + \Delta t)$$

$$= \int_{-\infty}^{\infty} \left[\frac{1}{2\pi} \int_{-\infty}^{\infty} e^{-j\omega(x-x_1)} \Theta(\omega; x_1, t_1, \Delta t) \, d\omega \right] f(x_1, t_1) \, dx_1. \qquad (6.2\text{-}17)$$

The characteristic function Θ can be expressed in terms of the moments of the process X. To obtain such a relationship for use in Equation 6.2-17, expand the exponential in Equation 6.2-14 and obtain

$$\Theta(\omega; x_1, t_1, \Delta t) = E\left[\exp(j\omega\Delta X_{t_1}) \mid X(t_1) = x_1\right]$$

$$= E\left[\sum_{q=0}^{\infty} \frac{(j\omega)^q}{q!} (\Delta X_{t_1})^q \mid X(t_1) = x_1\right] \qquad (6.2\text{-}18)$$

$$= \sum_{q=0}^{\infty} \frac{(j\omega)^q}{q!} m^{(q)}(x_1, t_1, \Delta t)$$

where

$$m^{(q)}(x_1, t_1, \Delta t) = E\left[(\Delta X_{t_1})^q \mid X(t_1) = x_1\right]$$
$$= E\left[(X(t_1 + \Delta t) - X(t_1))^q \mid X(t_1) = x_1\right] \qquad (6.2\text{-}19)$$

is the qth conditional moment of the random increment ΔX_{t_1}.

This expansion of the characteristic function can be used in Equation 6.2-17. Substitute Equation 6.2-18 into Equation 6.2-17 and obtain

$$f(x, t_1 + \Delta t) \qquad (6.2\text{-}20)$$

$$= \int_{-\infty}^{\infty} \left[\frac{1}{2\pi} \int_{-\infty}^{\infty} e^{-j\omega(x-x_1)} \sum_{q=0}^{\infty} \frac{(j\omega)^q}{q!} m^{(q)}(x_1, t_1, \Delta t) \, d\omega\right] f(x_1, t_1) \, dx_1$$

$$= \sum_{q=0}^{\infty} \frac{1}{q!} \int_{-\infty}^{\infty} \left[\frac{1}{2\pi} \int_{-\infty}^{\infty} (j\omega)^q e^{-j\omega(x-x_1)} \, d\omega\right] m^{(q)}(x_1, t_1, \Delta t) \, f(x_1, t_1) \, dx_1$$

This result can be simplified by using the identity

$$\frac{1}{2\pi} \int_{-\infty}^{\infty} (j\omega)^q e^{-j\omega(x-x_1)} \, d\omega = \frac{1}{2\pi}\left(-\frac{\partial}{\partial x}\right)^q \int_{-\infty}^{\infty} e^{-j\omega(x-x_1)} \, d\omega$$
$$= \left(-\frac{\partial}{\partial x}\right)^q \delta(x - x_1). \qquad (6.2\text{-}21)$$

The use of Identity 6.2-21 in Equation 6.2-20 results in

$$f(x, t_1 + \Delta t)$$
$$= \sum_{q=0}^{\infty} \frac{1}{q!} \int_{-\infty}^{\infty} \left[\left(-\frac{\partial}{\partial x} \right)^q \delta(x - x_1) \right] m^{(q)}(x_1, t_1, \Delta t) f(x_1, t_1) dx_1$$
$$= \sum_{q=0}^{\infty} \frac{1}{q!} \left(-\frac{\partial}{\partial x} \right)^q \int_{-\infty}^{\infty} \delta(x - x_1) m^{(q)}(x_1, t_1, \Delta t) f(x_1, t_1) dx_1 \quad (6.2\text{-}22)$$
$$= f(x, t_1) + \sum_{q=1}^{\infty} \frac{1}{q!} \left(-\frac{\partial}{\partial x} \right)^q m^{(q)}(x_1, t_1, \Delta t) f(x_1, t_1)$$

since $m^{(0)} = 1$. Now, no special significance is attached to time variable t_1 in Equation 6.2-22; hence, substitute t for t_1 and write

$$f(x, t + \Delta t) - f(x, t) = \sum_{q=1}^{\infty} \frac{1}{q!} \left(-\frac{\partial}{\partial x} \right)^q m^{(q)}(x, t, \Delta t) f(x, t). \quad (6.2\text{-}23)$$

Finally, divide both sides of this last result by Δt, and let $\Delta t \to 0$ to obtain the formal limit

$$\frac{\partial}{\partial t} f(x, t) = \sum_{q=1}^{\infty} \frac{1}{q!} \left(-\frac{\partial}{\partial x} \right)^q K^{(q)}(x, t) f(x, t), \quad (6.2\text{-}24)$$

where

$$K^{(q)}(x, t) \equiv \lim_{\Delta t \to 0} \frac{E\left[\{X(t + \Delta t) - X(t)\}^q \mid X(t) = x \right]}{\Delta t}, \quad (6.2\text{-}25)$$

$q \geq 1$, are called the *intensity coefficients*. Integer q denotes the order of the coefficient. Equation 6.2-24 is called the *Kramers–Moyal expansion*. In general, the coefficients given by Equation 6.2-25 depend on time. However, the intensity coefficients are time invariant in cases where the underlying process is homogeneous. In what follows, all of the examples involve homogeneous processes. Hence, in the results developed in this chapter, the intensity coefficients are assumed to be independent of time.

6.2.4 The One-Dimensional Fokker–Planck Equation

The intensity coefficients $K^{(q)}$ vanish for $q \geq 3$ in applications where X is the state of a first-order system driven by white Gaussian noise (see Risken).[36] This means that incremental changes $\Delta X_t \equiv [X(t + \Delta t) - X(t)]$ in the process occur slowly enough so that their third- and higher-order moments vanish more rapidly than Δt. Under these conditions, Equation 6.2-24 reduces to the one-dimensional *Fokker–Planck* equation

$$\frac{\partial}{\partial t}f(x,t) = -\frac{\partial}{\partial x}\left[K^{(1)}(x)\,f(x,t)\right] + \frac{1}{2}\frac{\partial^2}{\partial x^2}\left[K^{(2)}(x)\,f(x,t)\right]. \quad (6.2\text{-}26)$$

When $K^{(q)} = 0$ for $q \geq 3$, random process X is known as a *diffusion process*, and its sample functions are continuous (see Risken).[36] Applications of this type are of interest here where the theory is applied to the PLL. Apart from initial and boundary conditions, Equation 6.2-26 shows that the probability density function for a diffusion process is determined completely by only first- and second-order moments of the process increments.

As a simple example, consider the RL circuit depicted by Figure 6.4. This circuit is driven by v_{in}, a zero-mean, white Gaussian noise process with a double-sided spectral density of $N_o/2$ watts/Hz. This white noise process is the formal derivative of a Wiener process W_t; the variance of an increment of this Wiener process is $(N_o/2)\Delta t$ (see Equation 6.1-28). The RL circuit is described by Equations 6.2-7 and 6.2-8 which can be used to write

$$\Delta i_t = -\frac{R}{L}i_t\,\Delta t + \frac{1}{L}\int_t^{t+\Delta t} dW_t. \quad (6.2\text{-}27)$$

The commonly used notation $i_t \equiv i(t)$ and $\Delta i_t \equiv i(t + \Delta t) - i(t)$ is used in Equation 6.2-27, and the differential dW_t is formally equivalent to $v_{in}dt$. This current increment can be used in Equation 6.2-25 to obtain

$$\begin{aligned}K^{(1)} &= \lim_{\Delta t \to 0}\frac{E\left[\Delta i_t \mid i_t = i\right]}{\Delta t} \\ &= -\frac{R}{L}i\end{aligned} \quad (6.2\text{-}28)$$

In a similar manner, the second intensity coefficient is

$$K^{(2)} = \lim_{\Delta t \to 0} \frac{E\left[(\Delta i_t)^2 \mid i_t = i\right]}{\Delta t}$$

$$= \lim_{\Delta t \to 0} \frac{\frac{1}{L^2} \int_t^{t+\Delta t} \int_t^{t+\Delta t} E\left[v_{in}(t_1) v_{in}(t_2)\right] dt_1\, dt_2}{\Delta t} \quad (6.2\text{-}29)$$

$$= \frac{1}{L^2} \frac{N_0}{2}.$$

Hence, the Fokker–Planck equation that describes the RL circuit is

$$\frac{\partial f}{\partial t} = \frac{\partial}{\partial i}\left(\frac{R}{L} i\right) f + \frac{N_0}{4L^2} \frac{\partial^2}{\partial i^2} f. \quad (6.2\text{-}30)$$

Finally, as can be shown by direct substitution, a steady-state solution of this equation is

$$f(i) = \frac{1}{\sqrt{2\pi(N_0/4RL)}} \exp\left[-\frac{1}{2} i^2 \Big/ (N_0/4RL)\right]. \quad (6.2\text{-}31)$$

The intensity coefficients used in Equation 6.2-26 can be given physical interpretations when process X(t) denotes the time-dependent displacement of a particle. Consider a particle on one of an ensemble of paths (sample functions) which pass through point x at time t. As the particle passes through point x at time t, its velocity is dependent on which path it is on. Inspection of Equation 6.2-25 shows that $K^{(1)}(x)$ is the average of these velocities. In a similar manner, coefficient $K^{(2)}$ can be given a physical interpretation. In Δt seconds after time t, the particle has undergone a displacement of $\Delta X_t \equiv X(t + \Delta t) - X(t)$ from point x. Of course, ΔX_t depends on which path the particle is on, so there is uncertainty in the magnitude of ΔX_t. That is, after leaving point x, there is uncertainty in how far the process has gone during the time increment Δt. For small Δt, a measure of this uncertainty is given by $\Delta t K^{(2)}(x)$, a first-order-in-Δt approximation for the variance of the displacement increment ΔX_t.

In many applications, $K^{(2)}$ is constant (independent of x); this is the case for the first-order PLL considered in Chapter 7. If $K^{(2)}(x)$ is not constant, then Equation 6.2-26 can be transformed into a new Fokker–Planck equation where the new coefficient $\tilde{K}^{(2)}$ is a positive constant (see pg. 97 of Risken[36] for details of the transformation). For this reason, coefficient $K^{(2)}$ in Equation 6.2-26 is assumed to be a positive constant in what follows.

Note that Equation 6.2-26 can be written as

$$\frac{\partial}{\partial t}f(x,t) = -\frac{\partial}{\partial x}\left[K^{(1)}(x)f(x,t) - \frac{1}{2}\frac{\partial}{\partial x}\left[K^{(2)}f(x,t)\right]\right]$$

$$= -\frac{\partial}{\partial x}\left[K^{(1)}(x) - \frac{1}{2}\frac{\partial}{\partial x}K^{(2)}\right]f(x,t) \qquad (6.2\text{-}32)$$

$$= -\nabla \cdot \Im(x,t).$$

where

$$\Im(x,t) \equiv \left[K^{(1)}(x) - \frac{1}{2}\frac{\partial}{\partial x}K^{(2)}\right]f(x,t), \qquad (6.2\text{-}33)$$

and $\nabla \cdot \Im$ denotes the divergence of $\Im$. Notice the similarity of Equation 6.2-32 with the well-known continuity equation

$$-\frac{\partial \rho}{\partial t} = \nabla \cdot J \qquad (6.2\text{-}34)$$

from electromagnetic theory. In this analogy, f (particles per meter) and $\Im$ (particles per second) in Equation 6.2-32 are analogous to one-dimensional charge density ρ (electrons per meter) and one-dimensional current J (electrons per second), respectively, in Equation 6.2-34. $\Im$ is the probability current; in an abstract sense, it can be thought of as describing the "amount" of probability crossing a point x in the positive direction per unit time. In the literature, it is common to see $\Im$ cited as a flow rate of "probability particles". That is, $\Im(x,t)$ can be thought of as the rate of particle flow at point x and time t (see Section 6.5 of Lindsey,[4] Section 4.3 of Stratonovich,[37] and Section 5.2 of Gardiner).[38]

The $K^{(1)}f$ term in $\Im$ is analogous to a *drift current* component. To see the current aspect, recall that $K^{(1)}$ has units of velocity if X is analogous to particle displacement and f has units of particles per meter. Hence, the product $K^{(1)}f$ is analogous to a one-dimensional current since

$$(\text{meters/second})(\text{particles/meter}) = \text{particles/second}. \qquad (6.2\text{-}35)$$

It is a drift current (i.e., a current that results from an external force or potential acting on particles) since, in applications, $K^{(1)}$ is due to the

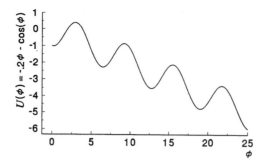

FIGURE 6.5
Potential function with periodically spaced local minima.

presence of an external force. This external force acts on the particles, and

$$\mathfrak{I}_s \equiv K^{(1)} \, f \qquad (6.2\text{-}36)$$

can be thought of as a drift current that results from movement of the forced particles.

In fact, drift coefficient $K^{(1)}$ is used in

$$U(x) \equiv -2 \int^x \frac{K^{(1)}(\alpha)}{K^{(2)}} d\alpha \qquad (6.2\text{-}37)$$

to define the *one-dimensional potential function* for Equation 6.2-32. An important conceptual role can be developed for U; it is more likely for probability (probability particles) to flow in the direction of a lower potential.

As shown in Chapter 7, Figure 6.5 depicts a potential function $U(x)$ which is similar to what is encountered in a first-order PLL. The significant feature of this potential function is the sequence of periodically spaced potential wells. A particle can move from one well to the next, and it is more likely to move to a well of lower potential than to a well of higher potential. In the applications discussed in Chapters 7 and 8, noise-induced cycle slips are associated with the movement of particles between the wells.

The component

$$\mathfrak{I}_d \equiv -\frac{1}{2} \frac{\partial}{\partial x} \Big[K^{(2)} \, f(x, t) \Big] \qquad (6.2\text{-}38)$$

in Equation 6.2-33 is analogous to a *diffusion current* (diffusion currents result in nature from a nonzero gradient of charged particles which undergo random motion — see p. 339 of Smith and Dorf).[39] As in Equation 6.2 35, $\mathfrak{I}_d$ is analogous to a current since it has units of particles per second. That $\mathfrak{I}_d$ is analogous to a diffusion current is easy to see when $K^{(2)}$ is a constant. In this case, $\mathfrak{I}_d$ is proportional to the gradient of the analogous charge density f, and $K^{(2)}$ is the constant diffusion coefficient. The negative sign associated with Equation 6.2-38 is due to the fact that particles tend to diffuse in the direction of lower probability concentrations.

Initial and boundary conditions on f(x, t) must be supplied when a solution is sought for the Fokker–Planck equation. An initial condition is a constraint that f(x, t) must satisfy at some instant of time; initial condition $f(x, t_1)$, where t_1 is the initial time, is specified as a function of the variable x. A boundary condition is a constraint that f(x, t) must satisfy at some displacement x_1 (x_1 may be infinite); boundary condition $f(x_1, t)$ is specified as a function of the variable t. Usually, initial and boundary conditions are determined by the physical properties of the application under consideration.

6.2.5 Transition Density Function

Often, the process X(t) is known to assume a value x_1 at time t_1. In this case, the initial condition is given by

$$f(x, t_1) = \delta(x - x_1), \qquad (6.2\text{-}39)$$

a delta function at x_1. When Equation 6.2-39 is used as the initial condition, the solution of Equation 6.2-26 is the *transition density function* for the Markov process. That is, the transition density $f(x, t \mid x_1, t_1)$ for the Markov process X(t) can be found by solving

$$\frac{\partial}{\partial t} f(x, t \mid x_1, t_1)$$
$$= -\frac{\partial}{\partial x}\left[K^{(1)}(x) f(x, t \mid x_1, t_1)\right] + \frac{1}{2}\frac{\partial^2}{\partial x^2}\left[K^{(2)} f(x, t \mid x_1, t_1)\right] \qquad (6.2\text{-}40)$$

subject to

$$f(x, t_1 \mid x_1, t_1) = \delta(x - x_1). \qquad (6.2\text{-}41)$$

STOCHASTIC METHODS FOR THE NONLINEAR PLL MODEL 203

In computing the transition density, the boundary conditions that must be imposed on Equation 6.2-40 are problem dependent.

6.2.6 Natural, Periodic, and Absorbing Boundary Conditions

Boundary conditions must be specified when looking for a solution of Equation 6.2-26. In general, the specification of boundary conditions can be a complicated task that requires the consideration of subtle issues. Fortunately, only three types of simple boundary conditions are required for the PLL applications considered in Chapters 7 and 8, and they are described in this section.

The first type of boundary conditions to be considered arise naturally in many applications where sample functions of X(t) are unconstrained in value. In these applications, the Fokker–Planck equation applies over the whole real line, and the boundary conditions specify what happens as x approaches ±∞. With respect to x, integrate Equation 6.2-32 over the real line to obtain

$$-\frac{\partial}{\partial t}\int_{-\infty}^{\infty} f(x,t)dx = \lim_{x \to \infty} \Im(x,t) - \lim_{x \to -\infty} \Im(x,t). \quad (6.2\text{-}42)$$

Now, combine this result with the normalization condition

$$\int_{-\infty}^{\infty} f(x,t)dx = 1, \quad (6.2\text{-}43)$$

which must hold for all t, to obtain

$$\lim_{x \to \infty} \Im(x,t) = \lim_{x \to -\infty} \Im(x,t). \quad (6.2\text{-}44)$$

While this must be true in general, the stronger requirement

$$\lim_{x \to \infty} \Im(x,t) = \lim_{x \to -\infty} \Im(x,t) = 0 \quad (6.2\text{-}45)$$

holds in all applications of the theory to physical problems (where probability buildup at infinity cannot occur). Furthermore, the requirement

$$\lim_{x \to \infty} f(x,t) = \lim_{x \to -\infty} f(x,t) = 0 \quad (6.2\text{-}46)$$

is necessary for the normalization Condition 6.2-43 to hold. As $x \to \pm\infty$, the density function must satisfy the requirement that $f(x,t) \to 0$ on the order of $1/x^{1+\varepsilon}$, for some $\varepsilon > 0$; this requirement is necessary for convergence of the integral in Equation 6.2-43. Equations 6.2-45 and 6.2-46 constitute what is called a set of *natural boundary conditions*.

Boundary conditions of a periodic nature are used in PLL analysis. They require a constraint of the form

$$f(x,t) = f(x+L_o,t)$$
$$\Im(x,t) = \Im(x+L_o,t),$$
(6.2-47)

where L_o is the period. These are called *periodic boundary conditions*, and they are used for certain types of analyses when the intensity coefficients $K^{(1)}$ and $K^{(2)}$ are periodic functions of x. The PLL applications discussed in Chapters 7 and 8 are of this nature. Periodic intensity coefficients occur in a class of problems generally referred to as Brownian motion in a periodic potential (see Chapter 11 of Risken).[36]

The last type of boundary conditions discussed in this chapter are of the *absorbing type*. Suppose that X(t) denotes the displacement of a particle that starts at $X(0) = x_o$. As is illustrated by Figure 6.6, the particle undergoes a random displacement X(t) until, at $t = t_a$, it makes first contact with the boundary at $x_b > x_o$. The particle is absorbed (it vanishes) at this point of first contact. Clearly, the time interval $[0, t_a]$ from start to absorption depends on the path (sample function) taken by the particle, and the length of this time interval is a random variable.

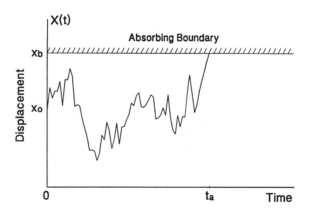

FIGURE 6.6
At time $t = t_a$, process X(t) hits an absorbing boundary placed at $x = x_b$.

For the case illustrated by Figure 6.6, an absorbing boundary at x_b requires that

$$\lim_{x \to x_b^-} f(x,t) = 0 \qquad (6.2\text{-}48)$$

for all t. Intuitively, this condition says that the particle can be found only rarely in a small neighborhood $(x_b - \Delta x, x_b)$, $\Delta x > 0$, of the boundary. In Section 6.1, the boundary Condition 6.2-48 was shown to hold for a Wiener process subjected to an absorbing boundary (see Equation 6.1-27). In the remainder of this section, this boundary condition is supported by an intuitive argument based on arriving at a contradiction; that is, Equation 6.2-48 is assumed to be false, and it is shown that this leads to a contradiction (see also pg. 219 of Cox and Miller).[40] The argument given requires that X(t) be homogeneous; however, the boundary condition holds in the more general nonhomogeneous case.

Suppose that Equation 6.2-48 is not true; suppose some time interval (t_1, t_2) and some displacement interval $(x_b - \Delta x, x_b)$ exist such that

$$f(x,t) \geq \epsilon > 0, \quad t_1 < t < t_2, \quad x_b - \Delta x < x < x_b, \qquad (6.2\text{-}49)$$

for some small $\epsilon > 0$. That is, suppose the density f(x,t) is bounded away from zero for x in some small neighborhood of the boundary and for t in some time interval. Then, on any infinitesimal time interval $(t, t + \Delta t) \subset (t_1, t_2)$, the probability $g(t)\Delta t$ that absorption occurs during $(t, t + \Delta t)$ is greater than or equal to the joint probability that the particle is near x_b at time t and the process increment $\Delta X_t \equiv X(t + \Delta t) - X(t)$ carries the particle into the boundary. That is, to first-order in Δt, the probability $g(t)\Delta t$ must satisfy

$$g(t)\Delta t \geq P[X(t+\Delta t) - X(t) \equiv \Delta X_t > \Delta x, \, x_b - \Delta x < X(t) < x_b]$$

$$= P[X(t+\Delta t) - X(t) \equiv \Delta X_t > \Delta x \,|\, x_b - \Delta x < X(t) < x_b] \qquad (6.2\text{-}50)$$

$$\cdot P[x_b - \Delta x < X(t) < x_b],$$

where Δx is a small and arbitrary positive increment. Note that g(t) is the probability density function of the absorption time.

As $\Delta t \to 0$, the right-hand side of Equation 6.2-50 approaches zero on the order of $\sqrt{\Delta t}$ or slower if Δx is allowed to approach zero on the order of $\sqrt{\Delta t}$. To see this, first note that Equation 6.2-49 implies

$$P[x_b - \Delta x < X(t) < x_b] \geq \epsilon \, \Delta x \tag{6.2-51}$$

so that

$$g(t)\Delta t \geq \epsilon \, \Delta x \, P[X(t+\Delta t) - X(t) \equiv \Delta X_t > \Delta x \mid x_b - \Delta x < X(t) < x_b]. \tag{6.2-52}$$

However, as pointed out in Section 6.2.3, a first-order-in-Δt approximation for the conditional variance of ΔX_t is

$$\text{Var}[\Delta X_t \mid X(t) = x] \approx \Delta t K^{(2)}, \tag{6.2-53}$$

where $K^{(2)}$ is a positive constant (see the paragraph preceding Equation 6.2-32). With a nonzero probability, the magnitude of a random variable exceeds its standard deviation. Hence, Equation 6.2-53 implies the existence of a $p_o > 0$ such that

$$P\left[\Delta X_t > \sqrt{K^{(2)} \Delta t} \mid x_b - \Delta x < X(t) < x_b\right] \geq p_o > 0 \tag{6.2-54}$$

for sufficiently small Δt. Set $\Delta x = \sqrt{K^{(2)} \Delta t}$ in Equation 6.2-52, and use Equation 6.2-54 to obtain

$$g(t)\Delta t \geq \epsilon \sqrt{K^{(2)} \Delta t} \, p_o, \tag{6.2-55}$$

or

$$g(t) \geq \epsilon \frac{\sqrt{K^{(2)}}}{\sqrt{\Delta t}} \, p_o. \tag{6.2-56}$$

Now, allow $\Delta t \to 0$ in Equation 6.2-56 to obtain the contradiction that the density function $g(t)$ is infinite on $t_1 < t < t_2$. This contradiction implies that Assumption 6.2-49 cannot be true; hence, boundary Condition 6.2-48 must hold for all $t \geq 0$.

6.2.7 Steady-State Solution to the Fokker–Planck Equation

Many applications are of a time-invariant nature, and they are characterized by Fokker–Planck equations that have time-invariant intensity coefficients. Usually, such an application is associated with a Markov process $X(t)$ that becomes stationary as $t \to \infty$. As t becomes large, the underlying density $f(x, t \mid x_0, t_0)$ that describes the process approaches a steady-state density function $f(x)$ that does not depend on t, t_0, or x_0. For example, this special case is relevant to the nonlinear PLL model with a noisy, sinusoidal reference. Often, the goal of system analysis in these cases is to find the steady-state density $f(x)$. Alternatively, in the steady state, the first and second moments of X may be all that are necessary in some applications.

The steady-state density $f(x)$ satisfies

$$0 = -\frac{d}{dx}\left[K^{(1)}(x)f(x) - \frac{1}{2}\frac{d}{dx}\left[K^{(2)} f(x)\right]\right], \quad (6.2\text{-}57)$$

the steady-state version of Equation 6.2-32. In this equation, the diffusion coefficient $K^{(2)}$ is assumed to be constant (see the paragraph before Equation 6.2-32). Integrate both sides of Equation 6.2-57 to obtain

$$\mathfrak{I}_{ss} = K^{(1)}(x) f(x) - \frac{1}{2}\frac{d}{dx}\left[K^{(2)} f(x)\right], \quad (6.2\text{-}58)$$

where constant $\mathfrak{I}_{ss}$ represents a steady-state value for the probability current discussed in Section 6.2.3.

The general solution of Equation 6.2-58 can be found by using standard techniques. First, simplify the equation by substituting $y(x) \equiv K^{(2)}f(x)$ to obtain

$$\frac{dy}{dx} - 2\left[\frac{K^{(1)}(x)}{K^{(2)}}\right] y = -2\mathfrak{I}_{ss}. \quad (6.2\text{-}59)$$

The integrating factor for Equation 6.2-59 is

$$\mu(x) = \exp\left[-2\int^x \left[\frac{K^{(1)}(\rho)}{K^{(2)}}\right] d\rho\right]. \quad (6.2\text{-}60)$$

This result and Equation 6.2-59 can be used to write

$$\mu\left[\frac{dy}{dx} - 2\left(\frac{K^{(1)}(x)}{K^{(2)}}\right)y\right] = \frac{d}{dx}[\mu y] = -2\mathfrak{I}_{ss}\mu \quad (6.2\text{-}61)$$

so that

$$\frac{d}{dx}\left[\mu(x)f(x)K^{(2)}\right] = -2\mathfrak{I}_{ss}\mu(x). \quad (6.2\text{-}62)$$

Finally, the general solution to Equation 6.2-58 can be written as

$$f(x) = \left(\mu(x) K^{(2)}\right)^{-1} \left[-2\Im_{ss} \int^x \mu(\rho) d\rho + C\right]. \quad (6.2\text{-}63)$$

Note that Equation 6.2-63 depends on the constants $\Im_{ss}$ and C. The steady-state value of probability current $\Im_{ss}$ and the constant C are chosen so that f(x) satisfies specified boundary conditions (which are application dependent) and the normalization condition

$$\int_{-\infty}^{\infty} f(x) \, dx = 1. \quad (6.2\text{-}64)$$

These results are used in Section 7.2.3 to compute the steady-state probability density which describes the phase error in a first-order PLL driven by a noisy reference sinusoid.

6.2.8 The One-Dimensional First-Passage Time Problem

Suppose Markov process X(t) denotes the instantaneous position of a particle that starts at x_o when t = 0. Assume that absorbing boundaries exist at b_1 and b_2 with $b_1 < x_o < b_2$. Let t_a denote the amount of time that is required for the particle to reach an absorbing boundary for the first time. Time t_a is called the *first-passage time*, and it varies from one path (sample function of X(t)) to the next. Hence, it is a random variable which depends on the initial value x_o. Figure 6.7 depicts the absorbing boundaries and two typical paths of the particle. The first-passage time problem plays a major role in Section 7.2 where it is used in the analysis of the cycle slip phenomenon.

As it evolves with time, process X(t) is described by the density $f(x,t \mid x_o, t_o)$, $t_o = 0$. This evolution is described here in a qualitative manner; Figure 6.8 is used in this effort, and it depicts the typical behavior of $f(x,t \mid x_o, 0)$. At t = 0, the process starts at x_o; this implies that all of the probability is massed at x_o as is shown by Figure 6.8a. As t increases, the location of the particle becomes more uncertain, and $f(x,t \mid x_o, 0)$ tends to "spread out" in the x variable as shown by Figure 6.8b (this figure depicts the density for some $t = t_1 > 0$). For t > 0, $f(x,t \mid x_o, 0)$ is represented as the sum of a continuous function $q(x,t \mid x_o, 0)$ and a pair of delta functions placed at b_1 and b_2. The delta functions represent the fact that probability will accumulate at the boundaries; this accumulation is due to the fact that sample functions will, sooner or later, terminate on a boundary and become absorbed. As shown by Figure 6.8c which depicts the case $t_2 > t_1$, q continues to "spread out" as time

STOCHASTIC METHODS FOR THE NONLINEAR PLL MODEL   209

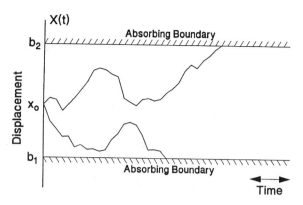

FIGURE 6.7
Two sample functions of X(t) and absorbing boundaries at b_1 and b_2.

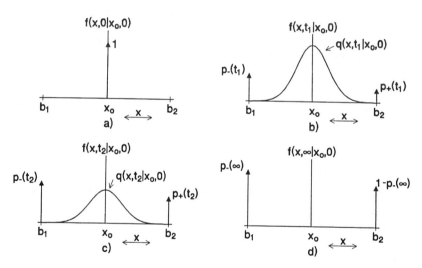

FIGURE 6.8
Density $f(x,t \mid x_0,0)$ at (a) $t = 0$, (b) $t = t_1$, (c) $t = t_2 > t_1$, and (d) $t = \infty$.

increases, and it is more likely that the particle impacts a boundary and is absorbed. The area under $q(x,t \mid x_0,0)$ decreases with time, and the delta function weights increase with time; however, the sum of the area under q and the delta function weights is unity. As $t \to \infty$, the probability that the particle is absorbed approaches unity; this requirement implies that

$$\lim_{t \to \infty} q(x,t \mid x_0,0) = 0. \qquad (6.2\text{-}65)$$

This time-limiting case is depicted by Figure 6.8d; on this figure, the quantity q is zero, and the delta function weights add to unity.

Function $q(x,t \mid x_o,0)$ is a solution of the one-dimensional Fokker–Planck equation given by Equation 6.2-26. The initial condition

$$q(x,0 \mid x_o,0) = \delta(x - x_o), \qquad (6.2\text{-}66)$$

and the absorbing boundary conditions

$$q(b_1,t \mid x_o,0) = 0$$
$$q(b_2,t \mid x_o,0) = 0, \qquad (6.2\text{-}67)$$

$t \geq 0$, should be used in this effort.

6.2.9 The Distribution and Density of the First-Passage Time Random Variable

Function $q(x,t \mid x_o,0)$ can be used to compute the distribution and density functions of the first-passage time random variable t_a. First, the quantity

$$\psi(t \mid x_o,0) = \int_{b_1}^{b_2} q(x,t \mid x_o,0) dx, \qquad (6.2\text{-}68)$$

$t \geq 0$ represents the probability that the particle has not been absorbed by time $t > 0$ given that the particle position is x_o, $b_1 < x_o < b_2$, at $t = 0$. In a similar manner,

$$F_{t_a}(t \mid x_o,0) = P[t_a \leq t \mid x_o,0]$$
$$= 1 - \psi(t \mid x_o,0) \qquad (6.2\text{-}69)$$
$$= 1 - \int_{b_1}^{b_2} q(x,t \mid x_o,0) dx$$

represents the probability that the first-passage time random variable t_a is not greater than t; that is, $F_{t_a}(t \mid x_o,0)$ is the distribution function for the first-passage time random variable t_a. Finally, the desired density function can be obtained by differentiating Equation 6.2-69 with respect to t; this procedure yields the formula

STOCHASTIC METHODS FOR THE NONLINEAR PLL MODEL

$$f_{t_a}(t \mid x_o, 0) = \frac{\partial}{\partial t} F_{t_a}(t \mid x_o, 0)$$
$$= -\frac{\partial}{\partial t} \psi(t \mid x_o, 0)$$
(6.2-70)

for the density function of t_a.

6.2.10 The Expected Value of the First-Passage Time Random Variable

The expected value of the first-passage time random variable is important in many applications; in Section 7.2.7, it is related to the frequency of slipping cycles in a first-order PLL. A simple expression for $E[t_a]$ is determined in this section for the case where diffusion coefficient $K^{(2)}$ is constant, exactly the case for the first-order PLL discussed in Chapter 7.

The average value of the first-passage time can be expressed in terms of $q(x, t \mid x_o, 0)$. To accomplish this, note that Equation 6.2-70 can be used to write

$$E[t_a] = \int_0^\infty t f_{t_a}(t \mid x_o, 0) \, dt$$
$$= -\int_0^\infty t \left[\frac{\partial}{\partial t} \psi(t \mid x_o, 0) \right] dt.$$
(6.2-71)

However, the integral in Equation 6.2-71 can be evaluated by parts to yield

$$E[t_a] = -t\psi(t \mid x_o, 0) \Big|_0^\infty + \int_0^\infty \psi(t \mid x_o, 0) \, dt.$$
(6.2-72)

Now, the integral in Equation 6.2-72 is assumed to converge. Hence, it is necessary that $\psi(t \mid x_o, 0)$ approach zero faster than $1/t$ as $t \to \infty$; this implies that the first term on the right of Equation 6.2-72 is zero and that

$$E[t_a] = \int_0^\infty \psi(t \mid x_o, 0) \, dt.$$
(6.2-73)

Finally, substitute Equation 6.2-68 into this and obtain

$$E[t_a] = \int_{b_1}^{b_2} Q(x \mid x_o, 0) \, dx, \qquad (6.2\text{-}74)$$

where

$$Q(x \mid x_o, 0) \equiv \int_0^\infty q(x, t \mid x_o, 0) \, dt. \qquad (6.2\text{-}75)$$

Note that Equation 6.2-67 implies that Q satisfies the condition

$$\begin{aligned} Q(b_1 \mid x_o, 0) &= 0 \\ Q(b_2 \mid x_o, 0) &= 0. \end{aligned} \qquad (6.2\text{-}76)$$

Equations 6.2-74 and 6.2-75 show that the expected value of the first-passage time can be expressed in terms of $q(x,t \mid x_o,0)$.

A simple first-order linear differential equation can be obtained that describes Q. To obtain this equation, first note that $q(x,t \mid x_o,0)$ satisfies the one-dimensional Fokker–Planck equation (see Section 6.2.7 above)

$$\begin{aligned} \frac{\partial}{\partial t} q(x,t \mid x_o, 0) &= -\frac{\partial}{\partial x}\left[K^{(1)}(x) q(x,t \mid x_o, 0)\right] \\ &\quad + \frac{1}{2} K^{(2)} \frac{\partial^2}{\partial x^2} q(x,t \mid x_o, 0), \end{aligned} \qquad (6.2\text{-}77)$$

where it has been assumed that $K^{(2)} > 0$ is constant. Now, integrate both sides of this last equation with respect to time and obtain

$$\begin{aligned} q(x, \infty \mid x_o, 0) - q(x, 0 \mid x_o, 0) &= -\frac{d}{dx}\left[K^{(1)}(x) Q(x \mid x_o, 0)\right] \\ &\quad + \frac{1}{2} K^{(2)} \frac{d^2}{dx^2} Q(x \mid x_o, 0), \end{aligned} \qquad (6.2\text{-}78)$$

where Q is given by Equation 6.2-75. Equations 6.2-65 and 6.2-66 can be used with Equation 6.2-78 to obtain

$$-\delta(x - x_o) = -\frac{d}{dx}\left[K^{(1)}(x) Q(x \mid x_o, 0)\right] + \frac{1}{2} K^{(2)} \frac{d^2}{dx^2} Q(x \mid x_o, 0). \qquad (6.2\text{-}79)$$

Now, integrate both sides of this result to obtain

$$\frac{d}{dx}Q(x|x_o,0) - 2\left[\frac{K^{(1)}(x)}{K^{(2)}}\right]Q(x|x_o,0) = \frac{2}{K^{(2)}}\left[\mathbb{C}_o - U(x-x_o)\right], \quad (6.2\text{-}80)$$

where $\mathbb{C}_o$ is a constant of integration, and $U(x)$ denotes the unit step function. Equation 6.2-80 is a first-order linear differential equation that describes Q. Due to Equation 6.2-76, it must be solved subject to the boundary conditions

$$Q(b_1|x_o,0) \equiv \int_0^\infty q(b_1,t|x_o,0)\,dt = 0$$
$$Q(b_2|x_o,0) \equiv \int_0^\infty q(b_2,t|x_o,0)\,dt = 0. \quad (6.2\text{-}81)$$

The integrating factor μ for Equation 6.2-80 is related to the potential function $U(x)$ (see Equation 6.2-37) by

$$\mu(x) = \exp\left[-\frac{2}{K^{(2)}}\int^x K^{(1)}(\alpha)\,d\alpha\right] = \exp[U(x)] \quad (6.2\text{-}82)$$

since

$$\frac{d}{dx}[\mu(x)Q(x|x_o,0)] = \mu(x)\left[\frac{2}{K^{(2)}}\right]\left[\mathbb{C}_o - U(x-x_o)\right]. \quad (6.2\text{-}83)$$

Now, integrate both sides of Equation 6.2-83 to obtain

$$Q(x|x_o,0) = \mu(x)^{-1}\left[\frac{2}{K^{(2)}}\int_{b_1}^x \mu(\alpha)[\mathbb{C}_o - U(\alpha-x_o)]\,d\alpha + \mathbb{C}_1\right], \quad (6.2\text{-}84)$$

where $\mathbb{C}_1$ is a constant of integration. Application of boundary Conditions 6.2-81 leads to the determination that $\mathbb{C}_1 = 0$ and

$$\mathbb{C}_o = \frac{\int_{b_1}^{b_2} \mu(\alpha)\,U(\alpha-x_o)\,d\alpha}{\int_{b_1}^{b_2} \mu(\alpha)\,d\alpha}. \quad (6.2\text{-}85)$$

A formula for the average first-passage time can be obtained by substituting Equation 6.2-84, with $\mathbb{C}_1 = 0$, into Equation 6.2-74. This effort yields

$$E[t_a] = \frac{2}{K^{(2)}} \int_{b_1}^{b_2} \mu(x)^{-1} \left[\int_{b_1}^{x} \mu(\alpha) [\mathbb{C}_o - U(\alpha - x_o)] \, d\alpha \right] dx, \quad (6.2\text{-}86)$$

where constant $\mathbb{C}_o$ is given by Equation 6.2-85. This result is used in Section 7.2.7 to compute the average rate at which a first-order PLL slips cycles due to noise on its reference signal.

6.2.11 Ratio of Boundary Absorption Rates

Probability current flows into both boundaries b_1 and b_2. On Figure 6.8, this is illustrated by the weights $p_-(t)$ and $p_+(t)$ increasing with time. However, in general, the flow is unequal, and one boundary may receive more probability current than the other. Hence, over a time interval $[0, T]$, the probability that flows into the boundaries may be unequal. This aspect of probability flow is analyzed in this section, and the results are used in Chapter 7 to analyze the phenomenon of noise-induced cycle slips in a first-order PLL.

Figure 6.1 illustrates the fact that the amount of probability in an interval changes at a rate that is equal to the current flowing into the interval. This implies that

$$\int_0^T \mathcal{I}(x_1, t \mid x_o, 0) \, dt$$

represents the total amount of probability that flows (in the direction of increasing x) past point x_1 during $[0, T]$. This result can be used to quantify the amount of probability that enters the boundaries depicted on Figure 6.8.

As is illustrated on Figure 6.8, probability flows into the boundaries at b_1 and b_2 as time increases. During the time interval $[0, T]$, the total amount of probability that flows into the boundaries b_1 and b_2 is

$$p_-(T) = -\int_0^T \mathcal{I}(b_1, t \mid x_o, 0) \, dt$$

$$p_+(T) = \int_0^T \mathcal{I}(b_2, t \mid x_o, 0) \, dt,$$

(6.2-87)

respectively. The minus sign on the first of these integrals results from the fact that probability must flow in a direction of decreasing x (to the left) if it enters boundary b_1. As T approaches infinity, the total probability that enters the boundaries b_1 and b_2 is $p_-(\infty)$ and $p_+(\infty) = 1 - p_-(\infty)$, respectively.

In terms of $q(x,t|x_0,0)$ introduced in Section 6.2.7, the probability current on the interval $b_1 \leq x \leq b_2$ can be expressed as (see Equation 6.2-33)

$$\mathcal{I}(x,t|x_0,0) = K^{(1)}(x)q(x,t|x_0,0) - \frac{1}{2}\frac{\partial}{\partial x}\left[K^{(2)}q(x,t|x_0,0)\right]. \quad (6.2\text{-}88)$$

Integrate this equation over $0 \leq t < \infty$, and use Equation 6.2-75 to write

$$\int_0^\infty \mathcal{I}(x,t|x_0,0)\,dt$$
$$= K^{(1)}(x)Q(x|x_0,0) - \frac{1}{2}\frac{d}{dx}\left[K^{(2)}Q(x|x_0,0)\right]. \quad (6.2\text{-}89)$$

This result describes the total probability that flows to the right (in the direction of increasing x) past point x during the time interval [0, ∞). Now, in this last result, use boundary Conditions 6.2-76 and p_+ given by Equation 6.2-87 to write

$$p_+(\infty) = -\frac{1}{2}\frac{d}{dx}\left[K^{(2)}Q(x|x_0,0)\right]\bigg|_{x=b_2} \quad (6.2\text{-}90)$$

for the total probability absorbed at boundary b_2. In a similar manner, the total probability absorbed at boundary b_1 is given as

$$p_-(\infty) = \frac{1}{2}\frac{d}{dx}\left[K^{(2)}Q(x|x_0,0)\right]\bigg|_{x=b_1}. \quad (6.2\text{-}91)$$

The sign difference in the last two equations results from the fact that current entering b_1 must flow in a direction that is opposite to the current flowing into boundary b_2.

The ratio of absorption probabilities $p_+(\infty)$ and $p_-(\infty)$ is needed in Chapter 7. By using Equations 6.2-90 and 6.2-91, this ratio can be expressed as

$$\frac{P_+(\infty)}{P_-(\infty)} = -\frac{\frac{d}{dx}\left[K^{(2)} Q(x|x_o,0)\right]\bigg|_{x=b_2}}{\frac{d}{dx}\left[K^{(2)} Q(x|x_o,0)\right]\bigg|_{x=b_1}}. \qquad (6.2\text{-}92)$$

The derivatives in this result can be supplied by Equation 6.2-80 when, as assumed here, $K^{(2)}$ is independent of x. Use boundary Conditions 6.2-81 in Equation 6.2-80 to obtain

$$\frac{d}{dx} Q(x|x_o,0)\bigg|_{x=b_1,b_2} = \frac{2}{K^{(2)}}\left[\mathbb{C}_o - U(x-x_o)\right]\bigg|_{x=b_1,b_2}, \qquad (6.2\text{-}93)$$

where $\mathbb{C}_o$ is given by Equation 6.2-85. Now, combine these last results to obtain

$$\frac{P_+(\infty)}{P_-(\infty)} = -\frac{\mathbb{C}_o - U(b_2 - x_o)}{\mathbb{C}_o - U(b_1 - x_o)} \qquad (6.2\text{-}94)$$

which becomes

$$\frac{P_+(\infty)}{P_-(\infty)} = -\frac{\mathbb{C}_o - 1}{\mathbb{C}_o} \qquad (6.2\text{-}95)$$

for the usual $b_1 < x_o < b_2$ case. This last result provides a ratio of the absorption probabilities in terms of the constant $\mathbb{C}_o$. It is used in Section 7.2.7 to determine the average number of cycle slips to the right vs. the average number of slips to the left for a first-order PLL driven by a noisy reference signal.

6.3 The Vector Markov Process

Let $\vec{\mathbf{X}}(t)$ denote an n-dimensional, vector-valued random process (boldfaced capitol letters are used to represent vectors). It has the Markov property if its distribution function at any future instant, conditioned on present and past values of the process, does not depend on the past values. In a manner similar to the first-order case described by Equation 6.2-1, for every integer m, a vector-valued Markov process has a conditional probability distribution that obeys

STOCHASTIC METHODS FOR THE NONLINEAR PLL MODEL 217

$$F(\vec{X}_m, t_m \mid \vec{X}_{m-1}, t_{m-1}; \ldots; \vec{X}_1, t_1) = F(\vec{X}_m, t_m \mid \vec{X}_{m-1}, t_{m-1}), \quad (6.3\text{-}1)$$

where $t_1 < t_2 < \ldots < t_m$. In Equation 6.3-1, the vectors $\vec{X}_1, \ldots, \vec{X}_m$ are n-dimensional, vector-valued variables.

6.3.1 The n-Dimensional Kramers–Moyal Expansion

The method used in Section 6.2 to derive the first-order Kramers–Moyal expansion can be generalized to handle the n-dimensional case considered here. Alternatively, the general n-dimensional Kramers–Moyal expansion can be developed by using several different approaches (see Section 4.1 of Risken).[36] The derivation outlined below does not require the use of derivatives of delta functions; in this respect it is simpler than the method of derivation used in Section 6.2. The approach outlined below does require the use of an auxiliary scalar-valued function $R(\vec{X})$ which does not appear in the final results. Function $R(\vec{X})$ is arbitrary except for two requirements. First, all of the function derivatives must be continuous. Second, the function, and all of its derivatives, must vanish as $\vec{X}$ approaches infinity.

In the development given here of the n-dimensional Kramers–Moyal expansion, it is necessary to consider the change in the transition density $f(\vec{X}, t \mid \vec{X}_o, t_0)$ which occurs over the time interval $t = t_1$ to $t = t_1 + \Delta t$; this change must be expressed in terms of the moments of the random vector increment $\Delta \vec{X} t_1 = \vec{X}(t_1 + \Delta t) - \vec{X}(t_1)$. To accomplish this, use the n-dimensional Chapman–Kolmogorov equation (in Equation 6.2-12 replace the scalar random variables by vector-valued random variables) and write

$$f(\vec{X}, t_1 + \Delta t \mid \vec{X}_o, t_o) - f(\vec{X}, t_1 \mid \vec{X}_o, t_o)$$
$$= \underset{\substack{n \\ \text{times}}}{\int} f(\vec{X}, t_1 + \Delta t \mid \vec{X}_1, t_1) f(\vec{X}_1, t_1 \mid \vec{X}_o, t_o) d\vec{X}_1 - f(\vec{X}, t_1 \mid \vec{X}_o, t_o), \quad (6.3\text{-}2)$$

where the integration is performed over the n-dimensional sample space associated with random vector $\vec{X}(t)$. Now, multiply both sides of Equation 6.3-2 by the scalar-valued function $R(\vec{X})$, and integrate this result over the n-dimensional sample space to obtain

$$\int\limits_{\substack{n\\\text{times}}} R(\vec{X})\left[f(\vec{X}, t_1 + \Delta t \mid \vec{X}_o, t_o) - f(\vec{X}, t_1 \mid \vec{X}_o, t_o)\right] d\vec{X}$$

$$= \int\limits_{\substack{n\\\text{times}}} R(\vec{X}) \int\limits_{\substack{n\\\text{times}}} f(\vec{X}, t_1 + \Delta t \mid \vec{X}_1, t_1) f(\vec{X}_1, t_1 \mid \vec{X}_o, t_o) d\vec{X}_1 d\vec{X} \quad (6.3\text{-}3)$$

$$- \int\limits_{\substack{n\\\text{times}}} R(\vec{X}) f(\vec{X}, t_1 \mid \vec{X}_o, t_o) d\vec{X}.$$

Next, expand $R(\vec{X})$ in a Taylor's series around the point $\vec{X}_1$ and obtain

$$R(\vec{X}) = R(\vec{X}_1) + \sum_{i=1}^{n}(x_i - x_{1i})\frac{\partial}{\partial x_i}R(\vec{X})\bigg|_{\vec{X}=\vec{X}_1}$$

$$+ \frac{1}{2!}\sum_{i=1}^{n}\sum_{j=1}^{n}(x_i - x_{1i})(x_j - x_{1j})\frac{\partial^2}{\partial x_i \partial x_j}R(\vec{X})\bigg|_{\vec{X}=\vec{X}_1} \quad (6.3\text{-}4)$$

$$+ \frac{1}{3!}\sum_{i=1}^{n}\sum_{j=1}^{n}\sum_{k=1}^{n}(x_i - x_{1i})(x_j - x_{1j})(x_k - x_{1k})\frac{\partial^3}{\partial x_i \partial x_j \partial x_k}R(\vec{X})\bigg|_{\vec{X}=\vec{X}_1}$$

$$+ \ldots,$$

where $\vec{X} = [x_1\ x_2\ \ldots\ x_n]^T$ and $\vec{X}_1 = [x_{11}\ x_{12}\ \ldots\ x_{1n}]^T$. Substitute this expansion for $R(\vec{X})$ into the first integral on the right-hand side of Equation 6.3-3; this 2n-fold integral can be written as

STOCHASTIC METHODS FOR THE NONLINEAR PLL MODEL 219

$$\int_{\substack{n \\ \text{times}}} R(\vec{X}) \int_{\substack{n \\ \text{times}}} f(\vec{X}, t_1 + \Delta t \mid \vec{X}_1, t_1) f(\vec{X}_1, t_1 \mid \vec{X}_0, t_0) d\vec{X}_1 d\vec{X}$$

$$= \int_{\substack{n \\ \text{times}}} R(\vec{X}_1) f(\vec{X}_1, t_1 \mid \vec{X}_0, t_0) d\vec{X}_1$$

$$+ \int_{\substack{n \\ \text{times}}} [\int_{\substack{n \\ \text{times}}} \sum_{i=1}^{n} (x_i - x_{1i}) f(\vec{X}, t_1 + \Delta t \mid \vec{X}_1, t_1) d\vec{X}]$$

$$\cdot \frac{\partial}{\partial x_i} R(\vec{X}) \Big|_{\vec{X} = \vec{X}_1} f(\vec{X}_1, t_1 \mid \vec{X}_0, t_0) d\vec{X}_1 \quad (6.3\text{-}5)$$

$$+ \int_{\substack{n \\ \text{times}}} [\int_{\substack{n \\ \text{times}}} \frac{1}{2!} \sum_{i=1}^{n} \sum_{j=1}^{n} (x_j - x_{1j})(x_i - x_{1i}) f(\vec{X}, t_1 + \Delta t \mid \vec{X}_1, t_1) d\vec{X}]$$

$$\cdot \frac{\partial^2}{\partial x_i \partial x_j} R(\vec{X}) \Big|_{\vec{X} = \vec{X}_1} f(\vec{X}_1, t_1 \mid \vec{X}_0, t_0) d\vec{X}_1$$

$$+ \cdots ,$$

Note that the first integral on the right-hand side of Equation 6.3-5 is n-fold since

$$\int_{\substack{n \\ \text{times}}} f(\vec{X}, t_1 + \Delta t \mid \vec{X}_1, t_1) d\vec{X} = 1. \quad (6.3\text{-}6)$$

Substitute Equation 6.3-5 into the right-hand side of Equation 6.3-3 and obtain

$$\int_{n \text{ times}} R(\vec{X})\left[f(\vec{X}, t_1+\Delta t | \vec{X}_0, t_0) - f(\vec{X}, t_1 | \vec{X}_0, t_0)\right] d\vec{X}$$

$$= \int_{n \text{ times}} \sum_{i=1}^{n} M_i^{(1)}(\vec{X}_1, t_1, \Delta t) f(\vec{X}_1, t_1 | \vec{X}_0, t_0) \frac{\partial}{\partial x_i} R(\vec{X})\bigg|_{\vec{X}=\vec{X}_1} d\vec{X}_1$$

$$+ \int_{n \text{ times}} \frac{1}{2!} \sum_{i=1}^{n} \sum_{j=1}^{n} M_{ij}^{(2)}(\vec{X}_1, t_1, \Delta t) f(\vec{X}_1, t_1 | \vec{X}_0, t_0)$$

$$\cdot \frac{\partial^2}{\partial x_i \partial x_j} R(\vec{X})\bigg|_{\vec{X}=\vec{X}_1} d\vec{X}_1 \quad (6.3\text{-}7)$$

$$+ \int_{n \text{ times}} \frac{1}{3!} \sum_{i=1}^{n} \sum_{j=1}^{n} \sum_{k=1}^{n} M_{ijk}^{(3)}(\vec{X}_1, t_1, \Delta t) f(\vec{X}_1, t_1 | \vec{X}_0, t_0)$$

$$\cdot \frac{\partial^3}{\partial x_i \partial x_j \partial x_k} R(\vec{X})\bigg|_{\vec{X}=\vec{X}_1} d\vec{X}_1$$

$$+ \cdots ,$$

where

$$M_i^{(1)}(\vec{X}_1, t_1, \Delta t) \equiv \int_{n \text{ times}} (x_i - x_{1i}) f(\vec{X}, t_1+\Delta t | \vec{X}_1, t_1) d\vec{X}$$

$$M_{ij}^{(2)}(\vec{X}_1, t_1, \Delta t) \equiv \int_{n \text{ times}} (x_i - x_{1i})(x_j - x_{1j}) f(\vec{X}, t_1+\Delta t | \vec{X}_1, t_1) d\vec{X}$$

$$M_{ijk}^{(3)}(\vec{X}_1, t_1, \Delta t) \quad (6.3\text{-}8)$$

$$\equiv \int_{n \text{ times}} (x_i - x_{1i})(x_j - x_{1j})(x_k - x_{1k}) f(\vec{X}, t_1+\Delta t | \vec{X}_1, t_1) d\vec{X}$$

$$\vdots$$

are conditional moments of the components belonging to the increment $\Delta \vec{X}_{t_1} = \vec{X}(t_1 + \Delta t) - \vec{X}(t_1)$. Finally, to the integrals on the right-hand side of Equation 6.3-7, apply integration by parts a sufficient number

STOCHASTIC METHODS FOR THE NONLINEAR PLL MODEL 221

of times in order to remove all derivatives of R from the integrands. For example, the first integral on the right-hand side of Equation 6.3-7 can be expressed as a sum of n integrals of the form

$$\underbrace{\int \cdots \int}_{n \text{ times}} M_i^{(1)}(\vec{X}_1, t_1, \Delta t)\, f(\vec{X}_1, t_1 | \vec{X}_o, t_o) \frac{\partial}{\partial x_i} R(\vec{X}) \bigg|_{\vec{X}=\vec{X}_1} d\vec{X}_1$$

$$= \underbrace{\int \cdots \int}_{n \text{ times}} M_i^{(1)}(\vec{X}, t_1, \Delta t)\, f(\vec{X}_1, t_1 | \vec{X}_o, t_o) \frac{\partial}{\partial x_i} R(\vec{X})\, d\vec{X} \qquad (6.3\text{-}9)$$

$$= -\underbrace{\int \cdots \int}_{n \text{ times}} R(\vec{X}) \frac{\partial}{\partial x_i} \Big[M_i^{(1)}(\vec{X}, t_1, \Delta t)\, f(\vec{X}, t_1 | \vec{X}_o, t_o) \Big] d\vec{X}$$

where i, $1 \le i \le n$, is the index in the sum. This result was obtained by using integration by parts combined with the fact that $R(\vec{X})$ and its derivatives vanish at the $\pm\infty$ limits of integration. Apply this integration by parts procedure a sufficient number of times, and express Equation 6.3-7 as

$$\underbrace{\int \cdots \int}_{n \text{ times}} R(\vec{X}) \Big[f(\vec{X}, t_1 + \Delta t | \vec{X}_0, t_0) - f(\vec{X}, t_1 | \vec{X}_0, t_0) \Big] d\vec{X}$$

$$= -\underbrace{\int \cdots \int}_{n \text{ times}} R(\vec{X}) \sum_{i=1}^{n} \frac{\partial}{\partial x_i} \Big[M_i^{(1)}(\vec{X}, t_1, \Delta t) f(\vec{X}, t_1 | \vec{X}_0, t_0) \Big] d\vec{X}$$

$$+ \underbrace{\int \cdots \int}_{n \text{ times}} \frac{1}{2!} R(\vec{X}) \sum_{i=1}^{n} \sum_{j=1}^{n} \frac{\partial^2}{\partial x_i \partial x_j} \Big[M_{ij}^{(2)}(\vec{X}, t_1, \Delta t) f(\vec{X}, t_1 | \vec{X}_0, t_0) \Big] d\vec{X} \qquad (6.3\text{-}10)$$

$$- \underbrace{\int \cdots \int}_{n \text{ times}} \frac{1}{3!} R(\vec{X}) \sum_{i=1}^{n} \sum_{j=1}^{n} \sum_{k=1}^{n} \frac{\partial^3}{\partial x_i \partial x_j \partial x_k} \Big[M_{ijk}^{(3)}(\vec{X}, t_1, \Delta t)$$

$$\cdot f(\vec{X}, t_1 | \vec{X}_0, t_0) \Big] d\vec{X}$$

$+ \cdots$.

Aside from the continuity and end conditions outlined above, function R is arbitrary. Hence, the integrands on both sides of Equation 6.3-10 must be equal so that

$$f(\vec{X}, t_1 + \Delta t \mid \vec{X}_0, t_0) - f(\vec{X}, t_1 \mid \vec{X}_0, t_0)$$

$$= -\sum_{i=1}^{n} \frac{\partial}{\partial x_i} \left[M_i^{(1)}(\vec{X}, t_1, \Delta t) f(\vec{X}, t_1 \mid \vec{X}_0, t_0) \right]$$

$$+ \frac{1}{2!} \sum_{i=1}^{n} \sum_{j=1}^{n} \frac{\partial^2}{\partial x_i \partial x_j} \left[M_{ij}^{(2)}(\vec{X}, t_1, \Delta t) f(\vec{X}, t_1 \mid \vec{X}_0, t_0) \right] \quad (6.3\text{-}11)$$

$$- \frac{1}{3!} \sum_{i=1}^{n} \sum_{j=1}^{n} \sum_{k=1}^{n} \frac{\partial^3}{\partial x_i \partial x_j \partial x_k} \left[M_{ijk}^{(3)}(\vec{X}, t_1, \Delta t) f(\vec{X}, t_1 \mid \vec{X}_0, t_0) \right]$$

$$+ \ldots .$$

This result expresses a change in the transition density $f(\vec{X}, t_1 \mid \vec{X}_0, t_0)$ in terms of the cross moments that are calculated by using components from the random vector increment $\Delta \vec{X}_{t_1} \equiv \vec{X}(t_1 + \Delta t) - \vec{X}(t_1)$.

The n-dimensional Kramers–Moyal expansion can be obtained from the previous result. In Equation 6.3-11, note that the subscript on the time variable t_1 can be dropped since it is not required, and it no longer serves to clarify the development. After applying this simple modification to Equation 6.3-11, divide both sides of the result by Δt, and allow Δt to approach zero to obtain

$$\frac{\partial}{\partial t} f(\vec{X}, t \mid \vec{X}_0, t_0)$$

$$= -\sum_{i=1}^{n} \frac{\partial}{\partial x_i} \left[K_i^{(1)}(\vec{X}, t) f(\vec{X}, t \mid \vec{X}_0, t_0) \right]$$

$$+ \frac{1}{2!} \sum_{i=1}^{n} \sum_{j=1}^{n} \frac{\partial^2}{\partial x_i \partial x_j} \left[K_{ij}^{(2)}(\vec{X}, t) f(\vec{X}, t \mid \vec{X}_0, t_0) \right] \quad (6.3\text{-}12)$$

$$- \frac{1}{3!} \sum_{i=1}^{n} \sum_{j=1}^{n} \sum_{k=1}^{n} \frac{\partial^3}{\partial x_i \partial x_j \partial x_k} \left[K_{ijk}^{(3)}(\vec{X}, t) f(\vec{X}, t \mid \vec{X}_0, t_0) \right]$$

$$+ \ldots ,$$

where

$$K_i^{(1)}(\vec{X},t) \equiv \lim_{\Delta t \to 0} \frac{M_i^{(1)}(\vec{X},t,\Delta t)}{\Delta t}$$

$$K_{ij}^{(2)}(\vec{X},t) \equiv \lim_{\Delta t \to 0} \frac{M_{ij}^{(2)}(\vec{X},t,\Delta t)}{\Delta t} \qquad (6.3\text{-}13)$$

$$K_{ijk}^{(3)}(\vec{X},t) \equiv \lim_{\Delta t \to 0} \frac{M_{ijk}^{(3)}(\vec{X},t,\Delta t)}{\Delta t}$$

$$\vdots$$

The scalars given by Equation 6.3-13 are the *intensity coefficients*; for the n-dimensional case, Equation 6.3-13 is the counterpart of Equation 6.2-25. Note that a superscript in parentheses is used to indicate the order of each intensity coefficient, and the subscripts show which components of $\vec{X}$ are used to compute the coefficient (no subscripts are required or used in the one-dimensional case discussed in Section 6.2). Equation 6.3-12 is the n-dimensional Kramers–Moyal expansion, and it is a generalization of Equation 6.2-24.

6.3.2 The n-Dimensional Fokker–Planck Equation

Third- and higher-order intensity coefficients vanish in applications where $\vec{X}$ is the state of an n-dimensional system driven by white Gaussian noise (Risken).[36] This means that incremental changes $\Delta\vec{X}_t \equiv [\vec{X}(t + \Delta t) - \vec{X}(t)]$ occur slowly enough so that their third- and higher-order moments vanish more rapidly than Δt. Under these conditions, Equation 6.3-12 reduces to the n-dimensional Fokker–Planck equation

$$\frac{\partial}{\partial t} f(\vec{X},t \,|\, \vec{X}_o, t_o) = L_{FP}\, f(\vec{X},t \,|\, \vec{X}_o, t_o), \qquad (6.3\text{-}14)$$

where

$$L_{FP} = -\sum_{i=1}^{n} \frac{\partial}{\partial x_i} K_i^{(1)}(\vec{X},t) + \frac{1}{2!}\sum_{i=1}^{n}\sum_{j=1}^{n} \frac{\partial^2}{\partial x_i \partial x_j} K_{ij}^{(2)}(\vec{X},t) \qquad (6.3\text{-}15)$$

denotes the n-dimensional *Fokker–Planck operator*. Note that Equation 6.3-14 is the n-dimensional counterpart of Equation 6.2-26.

The conditioning on the density $f(\vec{X}, t \mid \vec{X}_o, t_0)$ in Equation 6.3-14 can be removed. To accomplish this, note that for $t \geq t_0$

$$f(\vec{X}, t) = \underbrace{\int f(\vec{X}, t \mid \vec{X}_o, t_0) f(\vec{X}_o, t_0) d\vec{X}_o}_{n \text{ times}}, \qquad (6.3\text{-}16)$$

where the integration is n-fold. Hence, multiply both sides of Equation 6.3-14 by $f(\vec{X}_o, t_0)$, and integrate over the range of values taken on by $\vec{X}_o$ to obtain the Fokker–Planck equation

$$\frac{\partial}{\partial t} f(\vec{X}, t) = L_{FP} f(\vec{X}, t). \qquad (6.3\text{-}17)$$

The Fokker–Planck equation can be written as

$$\begin{aligned}\frac{\partial}{\partial t} f(\vec{X}, t) &= -\nabla \cdot \vec{\mathfrak{S}}(\vec{X}, t) \\ &= -\sum_{i=1}^{n} \frac{\partial}{\partial x_i} \mathfrak{S}_i(\vec{X}, t),\end{aligned} \qquad (6.3\text{-}18)$$

where

$$\vec{\mathfrak{S}}(\vec{X}, t) = \left[\mathfrak{S}_1(\vec{X}, t) \quad \mathfrak{S}_2(\vec{X}, t) \quad \ldots \quad \mathfrak{S}_n(\vec{X}, t)\right]^T \qquad (6.3\text{-}19)$$

is the probability current vector, and

$$\mathfrak{S}_i(\vec{X}, t) \equiv \left[K_i^{(1)}(\vec{X}, t) - \frac{1}{2} \sum_{j=1}^{n} \frac{\partial}{\partial x_j} K_{ij}^{(2)}(\vec{X}, t)\right] f(\vec{X}, t), \qquad (6.3\text{-}20)$$

$1 \leq i \leq n$, is the probability current component in the ith direction. Note the similarity of Equation 6.3-18 with the continuity equation Equation 6.2-34.

The probability current components $\mathfrak{S}_i$, $1 \leq i \leq n$, are composed of two main parts. The first part is the *drift current*:

$$\left[\mathfrak{S}_i\right]_{\text{drift}} = K_i^{(1)} f. \qquad (6.3\text{-}21)$$

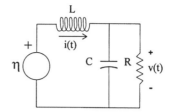

FIGURE 6.9
Simple RLC circuit driven by white Gaussian noise.

In the PLL applications considered in Chapters 7 and 8, $K_i^{(1)}$ results from a restoring force produced by system dynamics that are dependent on the input reference signal. This force tries to restore the PLL state to a stable equilibrium point. Through $K_i^{(1)}$, it is responsible for a probability current component that causes probability to accumulate around the equilibrium points. For this reason, the coefficients $K_i^{(1)}$, $1 \leq i \leq n$, are known as *drift coefficients*. The second part of Equation 6.3-20 is the *diffusion current*,

$$[\mathfrak{I}_i]_{\text{diffusion}} = -\frac{1}{2}\sum_{j=1}^{n} \frac{\partial}{\partial x_j} K_{ij}^{(2)}(\vec{X}, t) f(\vec{X}, t), \qquad (6.3\text{-}22)$$

where the coefficients $K_{ij}^{(2)}(\vec{X}, t)$, $1 \leq i, j \leq n$, are known as *diffusion coefficients*. The diffusion current in the ith direction is coupled to diffusions occurring in each of the n directions. The negative sign in Equation 6.3-22 implies that particles tend to diffuse in the direction of lower probability concentrations. When the diffusion coefficients are independent of displacement, it is easy to see that the diffusion current components are proportional to the gradient of the particle density. In the PLL applications considered in this book, the diffusion coefficients are independent of both $\vec{X}$ and t, and they are dependent on the additive noise on the reference.

6.3.3 A Simple Example

Figure 6.9 illustrates a simple second-order RLC circuit driven by white Gaussian noise. In this section, this circuit is described by a Fokker–Planck equation, and its steady-state behavior is examined. This example illustrates how to derive the Fokker–Planck equation that describes a simple system. Also, it illustrates some important well-known results for linear systems driven by white Gaussian noise. Namely, for such systems, the Fokker–Planck equation has drift coefficients that are linear in the state variables. Also, under steady-state conditions, the system state is described by a Gaussian random process.

The quantities i(t) and v(t) can serve as state variables that describe the RLC circuit depicted by Figure 6.9. As always, it is assumed that the initial conditions are independent of the forcing function $\eta(t)$. By generalizing the argument given in Section 6.2.1, it can be shown that the state vector $[\ i\ \ v\ ]^T$ is a Markov process. Finally, the state is described by a stochastic differential equation that is represented by

$$\frac{di}{dt} = \frac{1}{L}[\eta - v]$$
$$\frac{dv}{dt} = \frac{1}{C}[i - v/R].$$
(6.3-23)

Strictly speaking, Equation 6.3-23 only has meaning in that it implies process increment. These process increments can be written formally as

$$\Delta i_t \equiv i(t+\Delta t) - i(t) = -\frac{1}{L}\left[v(t)\Delta t - \int_t^{t+\Delta t}\eta\, dt\right]$$

$$\Delta v_t \equiv v(t+\Delta t) - v(t) = \frac{1}{C}[i(t) - v(t)/R]\Delta t,$$
(6.3-24)

and this representation is sufficient for the applications discussed here. In the first equation of Equation 6.3-24, the integral of η cannot be interpreted in the usual Rieman sense. That is, for small Δt, this integral *cannot* be approximated as $\eta(t)\Delta t$; this follows from the requirement that the state and process increments have finite variances. Instead, a *stochastic integral* should appear in the first equation of Equation 6.3-24, and it must be defined in a special manner (see Gardiner,[38] Jazwinskii,[35] or Arnold).[34] In the applications described in this book, only expectations of stochastic integrals are needed, and these are treated in the usual formal manner. That is, expectations are passed through integral signs to obtain

$$E\left[\int_t^{t+\Delta t}\eta\, dt\right] = \int_t^{t+\Delta t} E[\eta]\, dt = 0$$
(6.3-25)

and

$$E\left[\int_t^{t+\Delta t}\int_t^{t+\Delta t}\eta(t_1)\eta(t_2)\,dt_1dt_2\right] = \int_t^{t+\Delta t}\int_t^{t+\Delta t} E[\eta(t_1)\eta(t_2)]\,dt_1dt_2$$
$$= \frac{N_0}{2}\Delta t,$$
(6.3-26)

since the noise is zero mean and delta correlated.

The drift and diffusion coefficients can be calculated by computing Equations 6.3-8 and 6.3-13 for the increments given by Equation 6.3-24. The drift coefficients for the current i and voltage v are

$$K_i^{(1)} \equiv \lim_{\Delta t \to 0} \frac{E[\Delta i_t \mid i, v]}{\Delta t} \qquad (6.3\text{-}27)$$

$$= -v/L$$

and

$$K_v^{(1)} \equiv \lim_{\Delta t \to 0} \frac{E[\Delta v_t \mid i, v]}{\Delta t} \qquad (6.3\text{-}28)$$

$$= \frac{1}{C}[i - v/R],$$

respectively. In a similar manner, the diffusion coefficients can be computed as

$$K_{ii}^{(2)} \equiv \lim_{\Delta t \to 0} \frac{E[(\Delta i_t)^2 \mid i, v]}{\Delta t} \qquad (6.3\text{-}29)$$

$$= \frac{N_0}{2L^2}$$

$$K_{vv}^{(2)} \equiv \lim_{\Delta t \to 0} \frac{E[(\Delta v_t)^2 \mid i, v]}{\Delta t} \qquad (6.3\text{-}30)$$

$$= 0$$

$$K_{vi}^{(2)} \equiv \lim_{\Delta t \to 0} \frac{E[\Delta v_t \Delta i_t \mid i, v]}{\Delta t} \qquad (6.3\text{-}31)$$

$$= 0.$$

From Equations 6.3-27 and 6.3-28, note that the drift coefficients are linear in the state variables. This feature is a characteristic of the drift coefficients for linear systems.

The diffusion and drift coefficients can be used to write the Fokker–Planck equation for the circuit. From Equations 6.3-15 and 6.3-17, this equation can be written as

$$\frac{\partial}{\partial t}p(i, v, t) = L_{FP}\, p(i, v, t)$$

$$L_{FP} \equiv \frac{v}{L}\frac{\partial}{\partial i} - C^{-1}\frac{\partial}{\partial v}[i - v/R] + \frac{N_0}{4L^2}\frac{\partial^2}{\partial i^2}.$$

(6.3-32)

This equation describes the probabilistic nature of the circuit state. At $t = 0$, the current and voltage are known constants i_0 and v_0, respectively. Also, the probability density satisfies natural boundary conditions.

It is well known that a linear system forced by a Gaussian process has a Gaussian steady-state response. Hence, as $t \to \infty$, the state $[i(t)\ v(t)]^T$ approaches a Gaussian random process. In the limit, the steady-state solution of Equation 6.3-32 can be expressed as

$$p(\vec{X}) = \frac{1}{2\pi|\Lambda|^{1/2}}\exp[\vec{X}^T \Lambda^{-1} \vec{X}],$$

(6.3-33)

where $\vec{X} \equiv [i\ v]^T$ is the state vector. In Equation 6.3-33, the covariance matrix Λ can be expressed as

$$\Lambda \equiv \begin{bmatrix} \sigma_i^2 & \gamma\sigma_i\sigma_v \\ \gamma\sigma_i\sigma_v & \sigma_v^2 \end{bmatrix},$$

(6.3-34)

where the variance of i (of v) is denoted as σ_i^2 (as σ_v^2), and the correlation coefficient is denoted as γ.

The quantities in Equation 6.3-34 can be solved for. First, substitute Equation 6.3-33 and Equation 6.3-34 into

$$L_{FP}\, p(\vec{X}) = 0,$$

(6.3-35)

and equate to zero the various powers of i and v. This laborious algebraic procedure produces a system of equations that yield

$$\gamma = \frac{1}{\sqrt{1 + R^2 C/L}}$$

$$\sigma_i^2 = \frac{N_0}{4L^2}\left(\frac{RC}{1-\gamma^2}\right)$$

$$\sigma_v^2 = (\gamma R)^2 \sigma_i^2$$

(6.3-36)

as solutions.

Chapter 7

NOISE IN THE NONLINEAR PLL MODEL

Analyzing the effects of noise on a PLL becomes more difficult and challenging when the signal-to-noise ratio in the loop parameter ρ decreases to a point where the linear theory is no longer valid, and methods must be employed that utilize the nonlinear PLL model. However, the additional effort that is required to perform such a nonlinear analysis can be rewarded by interesting and useful results. For example, an analysis of the nonlinear PLL model can lead to results, such as an analysis of noise-induced cycle slips, that have no counterparts in the linear PLL theory. Also, only by analyzing the nonlinear problem can the accuracy of the linear theory be determined; it is possible to rely on linear methods and obtain results that contain large errors (without knowing that this has happened).

In this chapter, the nonlinear stochastic theory from Chapter 6 is applied. Also, material provided here lays the foundations for the powerful numerical techniques described in Chapter 8. Section 7.1 provides a qualitative description and commonly used models of the phase error process. In combination with techniques from Chapter 6, these results are applied to the first-order PLL in Section 7.2. This application of the theory to the first-order PLL produces results that can be computed easily; in fact, many closed-form results are given for the first-order case. Unfortunately, such straightforward results are not available for second- and higher-order PLLs. This becomes apparent in Section 7.3 where a general second-order PLL is considered as an example.

7.1 QUALITATIVE NATURE OF AND MODELS FOR THE PHASE ERROR

In the absence of noise, assume that the nth-order PLL phase locks at the equilibrium point $(\phi, \vec{Y}) = (\varphi_o, \vec{Y}_o)$. Vector $\vec{Y}$ is n-1 dimensional, and it contains the nonphase state variables $y_2, y_3, ..., y_n$. For each integer k, the PLL has a stable lock point at $(\phi, \vec{Y}) = (\varphi_o + 2\pi k, \vec{Y}_o)$. In

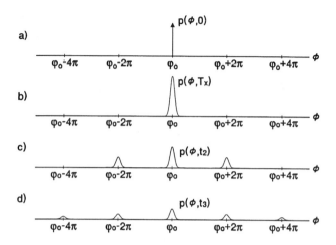

FIGURE 7.1
Marginal probability density function p(φ,t) illustrates the cycle slipping phenomenon.

this model, the phase variable is not restricted, and there are an infinite number of stable lock points (also known as stable states). For this reason, the model is referred to as the *unrestricted phase error model*.

Introduce noise into the discussion by considering the PLL model outlined in Section 2.5. This nth-order PLL model is supplied with a constant-frequency sinusoid that is embedded in Gaussian noise with a bandwidth that is large compared to the bandwidth of the PLL. Under these conditions, the state of the PLL is a random process, and it is described by a time-varying density function $p(\phi, \vec{Y}, t)$, $t \geq 0$, where ϕ is the unrestricted phase error variable. The qualitative nature of p is of interest in this section.

Suppose that at t = 0 the state starts at the stable equilibrium point $(\varphi_o, \vec{Y}_o)$. This implies that the density function used to describe the state must satisfy

$$p(\phi, \vec{Y}, 0) = \delta(\phi - \varphi_o) \prod_{k=2}^{n} \delta(y_k - y_{ok}), \quad (7.1\text{-}1)$$

where y_k and y_{ok} are the components of $\vec{Y}$ and $\vec{Y}_o$, respectively. This initial condition is suggested by Figure 7.1a, an illustration of the marginal density p(φ,0) for the phase variable.

Noise on the PLL reference causes the state to jitter around $(\varphi_o, \vec{Y}_o)$. As time increases, the density function $p(\phi, \vec{Y}, t)$ spreads out from Equation 7.1-1 to form the original, or primary, mode centered at $(\varphi_o, \vec{Y}_o)$; this primary mode is suggested by the marginal density depicted by Figure 7.1b. For the moderate noise case where cycle slips are infrequent, rare events (these conditions are discussed in Section 7.1.4), the general shape of this primary mode is determined by some T_x seconds; for $t > T_x$, the primary mode retains its general shape, but it decreases in scale. The quantity T_x is called the *relaxation time constant*; typically, it is within an order of magnitude of $1/B_L$.

Let R_0 denote a domain of the state space where the primary mode is significant. Centered on $(\varphi_o, \vec{Y}_o)$, the above-mentioned primary mode is approximately zero outside of domain R_0. Otherwise, the exact dimensions of R_0 are not important in this qualitative discussion. Now, replicate R_0 at each stable state of the PLL; for each integer k, define a domain R_k that surrounds stable state $(\varphi_k, \vec{Y}_o)$, $\varphi_k = \varphi_o - 2\pi k$. Basically, consider R_k as a simple translation of R_0. This procedure forms an infinite system of domains, and the domains are assumed to be pairwise disjoint (no domain overlaps any other domain). Such a system of domains can be constructed for practical PLLs that operate under the moderate noise conditions discussed in Section 7.1.4.

After an amount of time that, on the average, is large compared to T_x, the random noise will cause the PLL state to leave domain R_0. A single *cycle slip* is said to occur if the state enters domains R_{-1} or R_1. A slip consisting of a burst of cycles might occur; in this case the state leaves R_0, skips completely over intermediate domains, and enters directly a domain R_k, $|k| \geq 2$.

As time increases, the PLL continues to slip cycles in a random manner. Random slips consisting of bursts of cycles may occur as well. However, the average time between cycle slips is large compared to T_x (the phase error exhibits "quasi-stationary" behavior over time intervals on the order of T_x). As cycle slip activity occurs, the density function p takes on a multi-modal appearance as is suggested by Figures 7.1c and 7.1d (however, p may not be symmetric). The density function p develops secondary modes around an increasing number of equilibrium points. Finally, the density function disperses to an extent that

$$p(\phi, \vec{Y}, t) \to 0 \qquad (7.1\text{-}2)$$

as $t \to \infty$. In other terms, the phase error variance increases with time, and it approaches infinity as $t \to \infty$. Clearly, the unconstrained phase error ϕ has a steady-state behavior (as $t \to \infty$) that is of little interest.

For moderate to large values of in-loop SNR, cycle slips are extremely rare events. However, the cycle slip rate N (average number of slips per second) grows with decreasing in-loop SNR. As in-loop SNR decreases, a point is reached where the cycle slip rate is sufficient to render the PLL useless. The value of in-loop SNR at which this happens can be defined as the *loop threshold*. Obviously, this definition of threshold is subjective, but it (or something similar) is referenced frequently in the PLL literature. However, substantially different definitions of loop threshold have been given, and no one definition is prevalent (see Meyr and Ascheid).[8]

In many applications, a need exists for quantitative data on the phase error over a long timescale. As discussed in Sections 7.1.1 through 7.1.3, new phase error variables can be defined, and these new processes retain the pertinent information of the original unconstrained phase error process. Also, the new processes are described by density functions that, in the steady state, produce meaningful, nonzero results.

7.1.1 Modulo-2π Phase Error Model

As discussed above, the unrestricted phase error ϕ is described by a density function with Property 7.1-2. Clearly, knowledge of this steady-state behavior is of little practical value in applications. The basic problem is that there are an infinite number of domains R_k that can be reached by the PLL state; the state occupies any single domain with a probability that approaches zero as time approaches infinity.

Fortunately, the periodic nature of the PLL model can be exploited, and a new phase error variable can be defined that is described by a nonzero steady-state density function. To accomplish this, the PLL phase variable is redefined in a modulo-$2\pi m$ sense, and the domains are limited in number to m. By keeping integer m finite, appreciable probability accumulates in the domains as time approaches infinity, and the new phase error process is described by a useful, nonzero, steady-state probability density function.

As a first example of this approach, the unrestricted phase error variable ϕ can be taken modulo-2π to define the *modulo-2π phase error variable* ϕ_1, where $-\pi \leq \phi_1 < \pi$. Figure 7.2 provides an example of such a relationship. At t = 0, the PLL state vector starts at the equilibrium point $(\varphi_1, \vec{Y}_o)$, and the variable ϕ_1 starts at φ_1 on the figure (the coordinate directions associated with $\vec{Y}$ are not shown). As time increases, the variable ϕ_1 is shown to decrease until it reaches $-\pi$. Then, on the figure, it makes a discontinuous jump to π before decreasing towards the equilibrium point. As discussed in the remainder of this section, the steady-state density function that describes state $(\phi_1, \vec{Y})$ is concentrated in region R_1.

An infinite cylinder can be used to obtain ϕ_1 from ϕ as is illustrated by Figure 7.3. First, calibrate the cylinder circumference in values of ϕ_1, $-\pi \leq \phi_1 < \pi$. On the figure, angle is defined relative to the line $0 - 0'$, and positive angle is measured in a counterclockwise sense when looking into the open end of the cylinder (the t = 0 end). Then map the unrestricted phase error $\phi(t)$ onto the cylinder. At every instant of time t,

NOISE IN THE NONLINEAR PLL MODEL

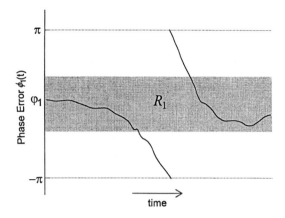

FIGURE 7.2
An example of the modulo-2π phase error function.

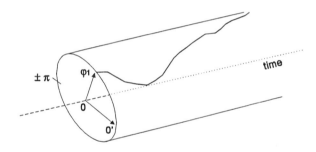

FIGURE 7.3
The unrestricted phase error mapped onto the surface of a cylinder.

$\phi(t)$ lies on the cylinder, and it is easy to obtain the corresponding value of ϕ_1 from the cylinder calibration. This method of determining ϕ_1 reinforces the idea that, for any integer k, the quantity $\phi(t) + 2\pi k$ maps to the same value of ϕ_1. In particular, the method illustrates the requirement that the points $\phi = \pm \pi$ map to the same value $\phi_1 = -\pi$ in the modulo-2π model.

The unrestricted phase error process ϕ can be expressed in terms of ϕ_1. The relationship is

$$\phi(t) = \phi_1(t) - 2\pi M(t), \qquad (7.1\text{-}3)$$

where M(t) denotes an integer-valued random counting process that changes value every time a cycle slip occurs. As time increases, the variance σ_ϕ^2 of ϕ grows without bounds. In fact, for large t, variance σ_ϕ^2 has the asymptotic approximation

$$\sigma_\phi^2 \approx \sigma_{\phi_1}^2 + (2\pi)^2 \, Dt, \qquad (7.1\text{-}4)$$

where $\sigma_{\phi_1}^2$ is the steady-state variance of ϕ_1, and D is a diffusion coefficient that is on the order of the average cycle slip rate N (see Chapter 9 of Meyr and Ascheid).[8] Thus, constant $\sigma_{\phi_1}^2$ characterizes the ability of the PLL to track with a small phase error, while diffusion constant D characterizes the PLL threshold behavior.

Let $p_1(\phi_1, \vec{Y}, t)$ denote the density function for the modulo-2π phase error. As discussed above, the phase variable ϕ_1 is restricted to the interval $[-\pi, \pi)$, and $\vec{Y}$ describes the remaining state variables y_2, $y_3, \ldots y_n$. Consider as part of the definition of p_1 a requirement on initial conditions. At $t = 0$, assume that the PLL state starts at the equilibrium point $\phi = \varphi_1$, $\vec{Y} \equiv [y_{02} \ldots y_{0n}]^T$. Then, density p_1 must satisfy the conditions

$$p_1(\phi_1, \vec{Y}, 0) = \delta(\phi_1 - \varphi_1) \prod_{i=2}^{n} \delta(y_i - y_{0i}), \quad -\pi \le \phi_1 < \pi,$$

$$\int_{-\pi}^{\pi} \underbrace{\int p_1(\phi_1, \vec{Y}, t) d\vec{Y}}_{n-1 \text{ times}} d\phi_1 = 1, \; t \ge 0. \qquad (7.1\text{-}5)$$

The density function $p_1(\phi_1, \vec{Y}, t)$ is a solution of a Fokker–Planck equation. As discussed in Section 6.3.2, this equation has the form

$$\frac{\partial}{\partial t} p_1(\phi_1, \vec{Y}, t) = -\nabla \cdot \vec{\Im}(\phi_1, \vec{Y}, t), \qquad (7.1\text{-}6)$$

where $\vec{\Im}$ is a vector-valued, n-dimensional probability current with components

$$\Im_i(\phi_1, \vec{Y}, t)$$

$$= \left[K_i^{(1)}(\phi_1, \vec{Y}) - \frac{1}{2} \frac{\partial}{\partial \phi_1} K_{i1}^{(2)} - \frac{1}{2} \sum_{j=2}^{n} \frac{\partial}{\partial y_j} K_{ij}^{(2)} \right] p_1(\phi_1, \vec{Y}, t), \qquad (7.1\text{-}7)$$

for $1 \le i \le n$. For the PLL model outlined in Section 2.5, the drift coefficients $K_i^{(1)}(\phi_1, \vec{Y})$, $1 \le i \le n$, are independent of time, and the diffusion coefficients $K_{ij}^{(2)}$, $1 \le i, j \le n$, are constants.

To find $p_1(\phi_1, \vec{Y}, t)$, Equation 7.1-6 should be solved subject to certain initial, boundary, and normalization conditions. The initial and normalization conditions are given by Equation 7.1-5. In the phase variable, periodic boundary conditions must be used; natural boundary conditions should be used in the nonphase variables. This means that both the density and the probability current should be 2π-periodic functions of phase. This condition can be written as

$$p_1(-\pi, \vec{Y}, t) = p_1(\pi, \vec{Y}, t)$$
$$\vec{\Im}(-\pi, \vec{Y}, t) = \vec{\Im}(\pi, \vec{Y}, t). \tag{7.1-8}$$

In the nonphase variables, the density and probability current should satisfy the natural boundary conditions

$$\lim_{y_i \to \pm\infty} p_1(\phi_1, \vec{Y}, t) = 0$$
$$\lim_{y_i \to \pm\infty} \vec{\Im}(\phi_1, \vec{Y}, t) = 0 \tag{7.1-9}$$

for each of the components y_i, $2 \le i \le n$, that make up vector $\vec{Y}$.

These boundary conditions ensure that p_1 is 2π-periodic in the phase variable. Also, they ensure that probability is conserved; that is, the area under p_1 remains unity for all time. This last claim is seen easily in the one-dimensional case where ϕ_1 is the only variable. For example, integrate a one-dimensional version of Equation 7.1-6 over the interval $-\pi \le \phi_1 < \pi$ to obtain

$$\frac{\partial}{\partial t}\int_{-\pi}^{\pi} p_1(\phi_1, t)\, d\phi_1 = -\int_{-\pi}^{\pi} \frac{\partial}{\partial \phi_1}\Im(\phi_1, t)\, d\phi_1$$
$$= -[\Im(\pi, t) - \Im(-\pi, t)]. \tag{7.1-10}$$

However, this result must be zero due to equal probability currents entering the boundaries. Hence, the area under p_1 is independent of time; by scaling, it can be taken as unity.

As $t \to \infty$, the nonzero limit of $p_1(\phi_1, \vec{Y}, t)$ is assumed to exist, and it describes the steady-state probabilistic behavior of the PLL state when phase error is taken modulo-2π. In what follows, this limiting density function is denoted by $p_1(\phi_1, \vec{Y})$. In a typical application described by the model outlined in Section 2.5, the density $p_1(\phi_1, \vec{Y})$ is a continuous function that is sharply peaked around the stable phase lock point.

Under so-called *moderate noise conditions* (see Ryter,[64,65] Ryter and Meyr,[66] and Ryter and Jordan),[67] the density $p_1(\phi_1, \vec{Y})$ is approximately zero outside of a small domain R_1 centered at the only equilibrium point $(\varphi_1, \vec{Y}_o)$. In its ϕ_1 coordinate, the domain R_1 spans only a small section of $[-\pi, \pi)$. On Figure 7.2, R_1 is illustrated by a shaded region.

The modulo-2π phase error model is not useful when it comes to analyzing the cycle slipping phenomenon in PLLs. The reason for this is that it is based on only one stable equilibrium point and only one domain R_1. A phase error model must contain two or more stable states before it can be used in an analysis of cycle slips (i.e., noise must be able to drive the system from one stable state to another). Hence, the modulo-2π phase error model can be used to analyze relaxation behavior (i.e., how $p_1(\phi_1, \vec{Y}, t)$ evolves from the initial $p_1(\phi_1, \vec{Y}, 0)$ to the steady-state $p_1(\phi_1, \vec{Y})$), but it cannot be used to analyze the cycle slipping phenomenon.

7.1.2 Bistable Cyclic Model

Similar in nature to the modulo-2π model, the *bistable phase error model* incorporates two stable equilibrium points, and phase is interpreted modulo-4π. In what follows, the quantity $\phi_2(t)$, $-2\pi \leq \phi_2 < 2\pi$, denotes the bistable phase error variable. The first stable equilibrium point of the model is denoted as $(\varphi_1, \vec{Y}_o)$, and it is surrounded by a domain R_1 that is similar in nature to the domain described above for the modulo-2π model. The second stable equilibrium point is displaced from the first by 2π radians in the phase variable, and it is surrounded by a domain R_2 that is similar in nature to R_1. Noise can drive the system state from one domain to the other. Unlike the modulo-2π phase error model, the bistable model is useful in analyzing the cycle slipping phenomenon.

Figure 7.4 depicts an example phase error function adapted to the bistable model. At t = 0, the PLL state starts at the stable equilibrium point $(\phi_2, \vec{Y}) = (\varphi_1, \vec{Y}_o)$. On the figure, φ_1 is illustrated as a positive quantity, and the coordinate directions associated with $\vec{Y}$ are not shown. Domain R_1 contains the point $(\varphi_1, \vec{Y}_o)$, and it is illustrated as a shaded region on Figure 7.4. As time increases, noise causes a cycle slip to occur by driving state $(\phi_2, \vec{Y})$ to domain R_2 centered at $(\varphi_1 - 2\pi, \vec{Y}_o)$. On Figure 7.4, this is illustrated by plotting ϕ_2 in R_2 for a brief period of time. Note that R_2 appears to be split into two sections; this observation is explained by the modulo-4π nature of ϕ_2. A second cycle slip occurs a short time after the first, and $(\phi_2, \vec{Y})$ returns back to domain R_1. Note that the state returns back to the domain from which it left after two successive cycle slips.

NOISE IN THE NONLINEAR PLL MODEL

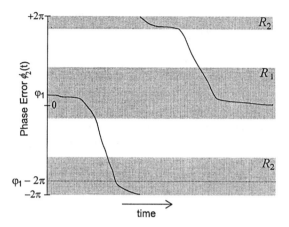

FIGURE 7.4
Phase error plot for the bistable cyclic model.

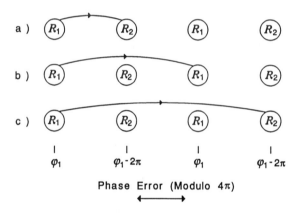

FIGURE 7.5
(a) A simple cyclic slip. (b) A slip of two cycles. (c) A slip of three cycles.

In an abstract manner, Figure 7.5 depicts the cycle slipping phenomenon as defined for the bistable model. On the figure, the two domains are depicted as being 2π radians apart. Also, the horizontal axis is calibrated in the modulo-4π value of the phase error. Figure 7.5a depicts a single cycle slip from R_1 to R_2. A cycle slip consisting of a burst of two cycles is illustrated by Figure 7.5b. Note that the trajectory leaves and returns to R_1, and it completely skips over domain R_2. Thus, a burst of two cycles is not seen as a transition between the domains of the bistable model. More generally, a burst consisting of an even number of cycles is not interpreted as a transition between the model states. Finally, Figure 7.5c depicts a slip involving a burst of three cycles. In this case, the trajectory leaves R_1, and it skips over a domain twice before entering domain R_2. Thus, a burst of three cycles is seen as a single transition between the domains of the bistable model. More generally, a burst consisting of an odd number of cycles is seen as a single transition between the model domains.

The transition rate between the domains of the bistable model can be approximated by using the methods outlined in Chapter 8. Due to the limitations of the bistable model, these methods provide incomplete results when it comes to a cycle slip containing a burst of cycles. They completely ignore a slip containing a burst of an even number of cycles. On the other hand, a slip containing a burst of an odd number of cycles is counted as a single cycle slip.

Basically, the bistable model is obtained by adding a second stable state to the modulo-2π phase error model. This additional state gives the bistable model some capability in the analysis of cycle slips. However, as outlined above, the bistable model is still limited in analyzing bursts of cycle slips. As discussed next, this limitation can be relaxed by including even more stable states in the phase error model.

7.1.3 Generalizations: the Multistable/m-Attractor Cyclic Model

The models discussed in the previous two sections can be incorporated into one generalized model containing m, m ≥ 1, stable equilibrium states. Also, each stable state is contained in a domain, and the domains R_k, $1 \le k \le m$, are well separated under the moderate noise conditions discussed in Section 7.1.4. The generalized model is called the *multistable cyclic model* or the *m-attractor cyclic model*. In this model, the m-attractor phase error variable is denoted as ϕ_m, and its modulo-$2\pi m$ nature implies that $-m\pi \le \phi_m < m\pi$. The case m = 2 (m = 1) corresponds to the bistable model (modulo-2π model).

For m ≥ 2, the multistable cyclic model is useful in the analysis of the cycle slip phenomenon. However, it is limited when the cycle slip consists of a burst of cycles. The model can distinguish between bursts consisting of up to m – 1 cycles. However, a burst consisting of m cycles is missed by using the m-attractor model. More generally, a burst consisting of m or more cycles is interpreted incorrectly as a burst of fewer than m cycles.

Let $p_m(\phi_m, \vec{Y}, t)$ denote a density function for the m-attractor model. The phase variable ϕ_m is restricted to [$-\pi m$, πm), and $\vec{Y}$ contains the remaining state variables. To facilitate the developments in Chapter 8, the density is assumed to satisfy the initial condition

$$p_m(\phi_m, \vec{Y}, 0) = \delta(\phi_m - \varphi_i) \prod_{k=2}^{n} \delta(y_k - y_{ok}), \quad -\pi m \le \phi_m < \pi m. \quad (7.1\text{-}11)$$

In this equation, initial condition (φ_i, y_{o2}, ..., y_{on}) denotes one of the m stable states described by the model (density p_m is defined up to a user-specified initial state). Also, the density must satisfy the normalization condition

$$\int_{-\pi m}^{\pi m} \underbrace{\int p_m(\phi_m, \vec{Y}, t) d\vec{Y}}_{n-1 \text{ times}} d\phi_m = 1 \qquad (7.1\text{-}12)$$

for $t \geq 0$.

In a manner similar to that used for p_1 in Section 7.1.1, density $p_m(\phi_m, \vec{Y}, t)$ can be found by solving a Fokker–Planck equation of the form given by Equations 7.1-6 and 7.1-7 (replace p_1 and ϕ_1 by p_m and ϕ_m, respectively, in these equations). This equation must be subjected to the initial and normalization conditions given by Equations 7.1-11 and 7.1-12, respectively. Furthermore, boundary conditions must be used that are periodic in the phase variable ϕ_m so that

$$p_m(-\pi m, \vec{Y}, t) = p_m(+\pi m, \vec{Y}, t)$$
$$\vec{\Im}(-\pi m, \vec{Y}, t) = \vec{\Im}(+\pi m, \vec{Y}, t) \qquad (7.1\text{-}13)$$

for $t \geq 0$. Natural boundary conditions should be used for the nonphase variables; this implies that

$$\lim_{y_i \to \pm\infty} p_m(\phi_m, \vec{Y}, t) = 0$$
$$\lim_{y_i \to \pm\infty} \vec{\Im}(\phi_m, \vec{Y}, t) = 0 \qquad (7.1\text{-}14)$$

for each component of $\vec{Y}$.

For the m-attractor phase error model, the steady-state density $p_m(\phi_m, \vec{Y})$ is assumed to exist. This density satisfies

$$L_{FP}\left[p_m(\phi_m, \vec{Y})\right] = 0, \qquad (7.1\text{-}15)$$

where L_{FP} denotes the Fokker–Planck operator that describes the PLL model under consideration. This operator must employ the phase error variable ϕ_m. For any $m > 1$, the steady-state density can be related to $p_1(\phi_1, \vec{Y})$, the steady-state density for the modulo-2π model, by

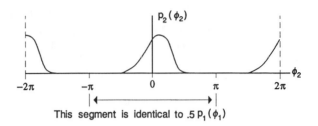

FIGURE 7.6
Steady-state density function p_2 for the bistable model. The points $\phi_2 = 2\pi$ and $\phi_2 = -2\pi$ are the same for this model.

$$p_m(\phi_m, \bar{Y}) = \frac{1}{m} p_1([\phi_m]_{2\pi}, \bar{Y}), \quad -\pi m \leq \phi_m < \pi m, \quad (7.1\text{-}16)$$

where $[\phi_m]_{2\pi}$ denotes the modulo-2π value of ϕ_m, where $-\pi \leq [\phi_m]_{2\pi} < \pi$.

To understand why this is true, consider Figure 7.6 which illustrates the bistable case m = 2 for the first-order PLL. When viewing this figure, keep in mind that the point $\phi = 2\pi$ maps to $\phi_2 = -2\pi$ in the 2-attractor model (alternatively, wrap Figure 7.6 around a cylinder of radius 4π). Let R_1 denote the domain of attraction that is near the origin; similarly, R_2 denotes the domain near -2π. Suppose that the PLL state starts in domain R_1 at t = 0. After a sufficiently long period of time, the state will enter domain R_2. As time approaches infinity, the state moves in a random manner back and forth between the two domains, and it is equally likely to find the state in either domain. In the limit, the PLL state shows no preference for either domain, and both domains have an equal probability of being occupied. The shape and scale of steady-state $p_2(\phi_2)$ must be identical on the two domains. On the restricted range $-\pi \leq \phi_2 < \pi$, take $p_2(\phi_2)$ as one half of $p_1(\phi_2)$, and extend this result to $-2\pi \leq \phi_2 < 2\pi$ by use of Equation 7.1-16. The density p_2 produced in this manner is a steady-state solution of the Fokker–Planck equation, and it satisfies the periodic boundary Conditions 7.1-13 with m = 2. The extension of this logic to the m > 2 case and higher-order PLLs is straightforward.

7.1.4 Qualitative Properties of the Multistable Model for the Moderate Noise Case

In PLL applications of practical interest, the in-loop SNR must be sufficiently high so that noise-induced cycle slips occur on a timescale that is large compared to $1/B_L$. That is, for the PLL to be useful, the average time between cycle slips must be large compared to the reciprocal of the loop noise-equivalent bandwidth. Qualitatively, a cycle slip must be a relatively "rare event" in practical systems. As discussed in Section 7.1.1, the so-called *moderate noise case* is said to apply under these circumstances (see Ryter,[64,65] Ryter and Meyr,[66] Ryter and Jordan,[67] and Meyr and Ascheid)[8].

The moderate noise case can be described qualitatively in terms of the steady-state density $p_m(\phi_m, \vec{Y})$ for the m-attractor model. As discussed by Ryter,[64] the noise is considered to be moderate when $p_m(\phi_m, \vec{Y})$ is concentrated onto well-separated domains R_k, $1 \leq k \leq m$, in the state space. That is, each of the m stable states can be enclosed by a domain, and domains from different states do not overlap. Furthermore, the density $p_m(\phi_m, \vec{Y})$ is approximately zero outside of the collection of domains. As an example, consider the bistable case depicted by Figure 7.6; on this figure, it is easy to define two disjoint domains that contain nearly all of the probability.

For the moderate noise case, and under steady-state conditions, the PLL state vector exhibits two general types of behavior that are of interest here. The first is motion of the state *within* a domain surrounding a stable equilibrium point. To an external observer, this state vector behavior manifests itself as a random phase jitter on the VCO output. This phenomenon occurs on a relatively fast timescale (referred to here as the *fast timescale*); within a time period of $1/B_L$ seconds, the state vector experiences noticeable motion within a domain. The second type of state vector behavior involves transitions *between* distinct domains of the m-attractor model. To an observer, this behavior is seen as cycle slips whereby the VCO gains or loses one or more cycles relative to the reference. As discussed at the beginning of this section, this second phenomenon occurs on a timescale that is large compared to $1/B_L$. The process spends relatively little time outside of a domain. Also, the average amount of time that the state occupies any domain is large (hence, the average time between domain transitions is large). On this *slow timescale*, the state vector can be modeled as a Markov jump process (this model development is discussed in Section 8.4.1), and the *transition rates* (i.e., number of jumps per unit time) between the domains is of interest. The numerical calculation of these transition rates is discussed in Chapter 8.

7.1.5 Phase Error Model Containing Absorbing Boundaries

A phase error model that is useful in the discussion of the cycle slipping phenomenon incorporates absorbing boundaries on the hyperplanes $\phi = \varphi_o - 2\pi$ and $\phi = \varphi_o + 2\pi$ in the n-dimensional state space. Such a phase error model is illustrated by Figure 7.7 for the case of a first-order PLL. At t = 0, the PLL model is started at the equilibrium point associated with $\phi = \varphi_o$. Noise in the model causes the state to

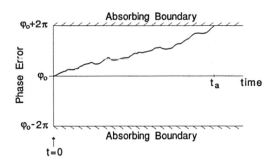

FIGURE 7.7
Phase error model with absorbing boundaries at $\varphi_o \pm 2\pi$.

jitter around the equilibrium point. Finally, at some time t_a, noise will drive the state of the model into one of the absorbing hyperplanes, and the phase error process terminates. Note that an absorbing hyperplane can be intersected only once, and only one hyperplane is intersected. This is illustrated on Figure 7.7 by the graph that terminates on the absorbing hyperplane passing through $\varphi_o + 2\pi$.

In what follows, this phase error model is used in two different ways. First, time t_a is a random variable, and its expected value is an important parameter in PLL analysis. In Section 7.2, the expected value of t_a is computed for a first-order PLL. Second, with nonzero probabilities, the particle can be absorbed by either of the two hyperplanes, and the ratio of the absorption probabilities is of interest. In Section 7.3, this ratio is discussed qualitatively for a second-order PLL.

7.2 NOISE IN THE FIRST-ORDER PLL

A first-order PLL supplied with a constant frequency reference sinusoid embedded in wideband Gaussian noise is considered in this section. The theory outlined in Section 6.2.7 is applied, and the steady-state density function is obtained for the modulo-2π phase error. Also, the frequency of slipping cycles is calculated by using the theory outlined in Sections 6.2.8 through 6.2.11.

As it turns out, only simple cycle slips are possible in the first-order PLL considered in this section (see Figure 7.5). Unlike the case for some second- and higher-order PLL, a cycle slip consisting of a burst of cycles cannot happen in the first-order PLL studied here. There is a simple reason for this. The single state variable of the first-order PLL has continuous sample functions, and it cannot bypass completely, or jump over, a domain of attraction.

Where it is possible to do so, the exact results of the nonlinear theory are compared to the approximate results obtained from the

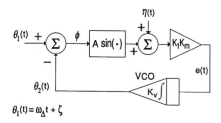

FIGURE 7.8
A nonlinear model for a first-order PLL with a noisy reference.

linear model of the loop. This comparison provides some insight into the limits of the linear PLL theory and illustrates the value of nonlinear PLL analysis.

7.2.1 Modeling the First-Order PLL with a Noisy Reference

The PLL sinusoidal reference signal has an amplitude of A volts RMS, and its frequency exceeds by ω_Δ the quiescent frequency of the VCO (i.e., the loop detuning parameter is ω_Δ). This reference signal is embedded in wideband Gaussian noise that has a double-sided spectral density of $N_0/2$ watts/Hz (see Figure 2.7). Furthermore, the input noise has a bandwidth 2B that is large compared to the PLL bandwidth. As a result it appears white and Gaussian to the loop.

The model for this first-order PLL is obtained from Section 2.5, and it is described here by Figure 7.8. On the figure, K_m and K_1 denote the phase comparator and the RMS value of the VCO output, respectively. The quantity η is modeled as white Gaussian noise, and it has a double-sided spectral density of $N_0/2$ watts/Hz. Finally, angles θ_1 and θ_2 represent the relative phase of the reference and VCO, respectively (see Equations 2.1-9 and 2.1-10). Note that θ_1 can be given as $\omega_\Delta t + \zeta$, where ζ is an arbitrary constant, since the reference frequency exceeds the quiescent frequency of the VCO by ω_Δ radians per second.

The equation that describes this PLL follows easily from the results outlined in Section 2.5. Since there is no loop filter, use $f(t) = \delta(t)$ in Equation 2.5-16 and obtain

$$\frac{d\phi}{dt} = \omega_\Delta - G\sin\phi - K_1 K_m K_v \eta, \tag{7.2-1}$$

since $\theta_1 = \omega_\Delta t + \zeta$. Equation 7.2-1 is a formal representation of the first-order stochastic differential equation used to describe the phase error process in the first-order PLL under consideration.

Physical interpretations of Equation 7.2-1 are of interest. A commonly provided interpretation is that this equation describes the motion of a particle in a force field that is both nonuniform (i.e., the force is a nonlinear function of displacement) and random. Under phase-locked conditions, the particle becomes trapped in the vicinity of one of the stable lock points given by Equation 5.1-3. The term $\omega_\Delta - G\sin\phi$ is analogous to a *restoring force* that tries to hold the particle at the stable lock point. On the other hand, the term $K_1 K_m K_v \eta$ is analogous to a random force that causes the particle to move in an erratic manner about the lock point.

The movement of the particle in the force field induces a change in the potential of the particle. The amount of potential change can be found by integrating the force over the distance traveled by the particle. Up to some constant, the force acting on the particle induces an *average potential* given by

$$U(\phi) = -\int^{\phi}(\omega_\Delta - G\sin x)dx \tag{7.2-2}$$
$$= -(\omega_\Delta \phi + G\cos\phi).$$

Figure 6.5 illustrates a typical potential function of the type given by Equation 7.2-2. The figure shows several local minima, or potential wells, that are spaced 2π radians apart. These wells occur at the PLL stable lock points. To continue the analogy under discussion, the particle tends to be trapped in a potential well. Inside of its well, the particle moves in a random manner. Also, movement between adjacent wells is possible.

The explicit use of G can be eliminated from Equation 7.2-1 if normalized time $\tau \equiv Gt$ is used. Since $G \equiv AK_1 K_m K_v$, Equation 7.2-1 becomes

$$\frac{d\phi}{d\tau} = \omega_\Delta' - \sin\phi - A^{-1}\eta \tag{7.2-3}$$
$$\omega_\Delta' \equiv \omega_\Delta/G$$

after time normalization. Note that in Equation 7.2-3, the dependence of ϕ and η on normalized time τ has been suppressed in order to simplify the notation.

It is important not to overlook the effects of time normalization on the statistical properties of the noise in Equation 7.2-3. Recall that in Equation 7.2-1, $\eta(t)$ is zero mean, and it has the autocorrelation

$$R_\eta(t) = E[\eta(t_1)\eta(t_1+t)]$$
$$= \frac{N_o}{2}\delta(t), \tag{7.2-4}$$

NOISE IN THE NONLINEAR PLL MODEL

where t is the real-time variable. Now, $\eta(\tau)$ appears in the normalized Equation 7.2-3, but this convenience is a universally employed abuse of notation. The differential equation Equation 7.2-3 could be written more precisely if it utilized (for example)

$$m(\tau) \equiv \eta(\tau/G) \tag{7.2-5}$$

as the noise disturbance term. By using Equation 7.2-4, the autocorrelation of Equation 7.2-5 is computed as

$$\begin{aligned} R_m(\tau_1) &= E\big[m(\tau+\tau_1)m(\tau)\big] \\ &= E\left[\eta\left(\frac{\tau+\tau_1}{G}\right)\eta\left(\frac{\tau}{G}\right)\right] \\ &= R_\eta(\tau_1/G) \\ &= \frac{N_o}{2}\delta(\tau_1/G). \end{aligned} \tag{7.2-6}$$

Hence, normalization of the time variable produces differential equation Equation 7.2-3 that contains noise term η with autocorrelation

$$R_\eta(\tau/G) = \frac{N_o}{2}\delta(\tau/G). \tag{7.2-7}$$

In what follows, Equation 7.2-7 is used as the autocorrelation of the noise term in the general time-normalized PLL model.

7.2.2 Fokker–Planck Equation for the Time-Normalized, First-Order PLL Model

As outlined in Section 6.2.4, the Fokker–Planck equation for the time-normalized, first-order PLL model can be expressed as

$$\frac{\partial}{\partial \tau}p(\phi,\tau) = -\frac{\partial}{\partial \phi}\big[K^{(1)}(\phi)p(\phi,\tau)\big] + \frac{1}{2}\frac{\partial^2}{\partial \phi^2}\big[K^{(2)}(\phi)p(\phi,\tau)\big], \tag{7.2-8}$$

where $K^{(1)}$ and $K^{(2)}$ are the drift and diffusion coefficients, respectively (subscripts on the intensity coefficients are not necessary in the one-dimensional case). As shown in this section, diffusion coefficient $K^{(2)}$ is independent of ϕ for the first-order PLL under consideration. Along with this equation, initial and boundary conditions are required for the determination of p(ϕ;τ).

As can be seen from Equation 6.2-25, coefficients $K^{(1)}$ and $K^{(2)}$ involve conditional moments of the process increment

$$\Delta\phi \equiv \phi(\tau+\Delta\tau)-\phi(\tau)$$
$$= [\omega_\Delta' - \sin\phi(\tau)]\Delta\tau - A^{-1}\int_\tau^{\tau+\Delta\tau}\eta(\xi)d\xi. \quad (7.2\text{-}9)$$

This increment was obtained by integrating both sides of Equation 7.2-3 over a $\Delta\tau$ interval. Substitute Equation 7.2-9 into Equation 6.2-25 and obtain

$$K^{(1)}(\phi) = \lim_{\Delta\tau\to 0}\frac{E[\Delta\phi\mid\phi]}{\Delta\tau}$$
$$= \omega_\Delta' - \sin\phi - \lim_{\Delta\tau\to 0}\frac{A^{-1}}{\Delta\tau}\int_\tau^{\tau+\Delta\tau}E[\eta(\xi)]\,d\xi. \quad (7.2\text{-}10)$$

However, the noise is zero mean, so that $E[\eta] = 0$. With Equation 7.2-10, this assumption leads to

$$K^{(1)}(\phi) = \omega_\Delta' - \sin\phi. \quad (7.2\text{-}11)$$

In a similar manner, the diffusion coefficient can be calculated as

$$K^{(2)}(\phi) = \lim_{\Delta\tau\to 0}\frac{E[(\Delta\phi)^2\mid\phi]}{\Delta\tau}$$
$$= \lim_{\Delta\tau\to 0}\Bigg[[\omega_\Delta' - \sin\phi]^2\Delta\tau - 2A^{-1}[\omega_\Delta' - \sin\phi]\int_\tau^{\tau+\Delta\tau}E[\eta(\xi)]d\xi$$
$$+ \frac{A^{-2}}{\Delta\tau}\int_\tau^{\tau+\Delta\tau}\int_\tau^{\tau+\Delta\tau}E[\eta(\xi_1)\eta(\xi_2)]d\xi_1 d\xi_2\Bigg] \quad (7.2\text{-}12)$$
$$= \lim_{\Delta\tau\to 0}\frac{A^{-2}}{\Delta\tau}\int_\tau^{\tau+\Delta\tau}\int_\tau^{\tau+\Delta\tau}E[\eta(\xi_1)\eta(\xi_2)]d\xi_1 d\xi_2.$$

Now, from Equation 7.2-7 note that

NOISE IN THE NONLINEAR PLL MODEL

$$\int_{\tau}^{\tau+\Delta\tau}\int_{\tau}^{\tau+\Delta\tau} E[\eta(\xi_1)\eta(\xi_2)]d\xi_1 d\xi_2$$

$$= \int_{\tau}^{\tau+\Delta\tau}\int_{\tau}^{\tau+\Delta\tau} \frac{N_o}{2}\delta\left(\frac{\xi_1-\xi_2}{G}\right)d\xi_1 d\xi_2 \quad (7.2\text{-}13)$$

$$= G\frac{N_o}{2}\Delta\tau.$$

Finally, substitute Equation 7.2-13 into Equation 7.2-12 and obtain

$$K^{(2)}(\phi) = G\frac{N_o}{2A^2}. \quad (7.2\text{-}14)$$

Diffusion coefficient $K^{(2)}$ can be expressed in terms of ρ, the signal-to-noise ratio within the loop. Combine Equations 2.5-20 and 2.5-21 to obtain

$$\rho = \frac{1}{\sigma_\phi^2} = \frac{A^2}{N_o B_L}. \quad (7.2\text{-}15)$$

As was derived in Section 3.1.3, the loop noise bandwidth for a first-order PLL is $B_L = G/4$. Now, use this value of B_L in Equation 7.2-15 and obtain

$$\rho = 2\left[G\frac{N_o}{2A^2}\right]^{-1}. \quad (7.2\text{-}16)$$

Finally, substitute Equation 7.2-16 into Equation 7.2-14 to obtain

$$K^{(2)}(\phi) = \frac{2}{\rho} \quad (7.2\text{-}17)$$

for the diffusion coefficient.

The Fokker–Planck equation for the time-normalized, first-order PLL model follows easily from the derivations of $K^{(1)}$ and $K^{(2)}$ given above. Substitute Equations 7.2-11 and 7.2-17 into Equation 6.2-26 and obtain

$$\frac{\partial}{\partial\tau}p(\phi,\tau) = L_{FP}\,p(\phi,\tau)$$

$$L_{FP} \equiv -\frac{\partial}{\partial\phi}(\omega_\Delta' - \sin\phi) + \frac{1}{\rho}\frac{\partial^2}{\partial\phi^2} \quad (7.2\text{-}18)$$

for the desired Fokker–Planck equation. For the remainder of Section 7.2, this equation is utilized in a noise analysis of the first-order PLL model.

The one-dimensional potential function for Equation 7.2-18 follows easily. Substitute Equations 7.2-11 and 7.2-17 into Equation 6.2-37 and obtain

$$U(\phi) = -\rho \int^{\phi} (\omega_\Delta' - \sin\xi) d\xi$$
$$= -\rho(\omega_\Delta' \phi + \cos\phi). \tag{7.2-19}$$

For $|\omega_\Delta'| < 1$, the potential function has periodically spaced local minima (known as *wells*) at $\phi_k = \sin^{-1}\omega_\Delta' + 2\pi k$ (see Figure 6.5). It is no coincidence that the local minima of U correspond to the PLL stable lock points. During practical operation of a properly designed first-order PLL, random process ϕ stays near a potential minimum for extended periods of time. Only rarely does the loop response due to noise move ϕ from one potential well to another.

7.2.3 Steady-State Density for the Modulo-2π Phase Error

A steady-state, 2π-periodic solution to Equation 7.2-18 is computed in this section. The solution represents the steady-state density function for the modulo-2π phase error in the first-order PLL. In terms of the notation established in Section 7.1.1, it is denoted as $p_1(\phi_1)$, where ϕ_1 is the modulo-2π phase error variable. Density p_1 can be represented in closed form for the case $\omega_\Delta' = 0$; on the other hand, when $\omega_\Delta' \neq 0$, it must be found by using numerical methods or infinite series.

The steady-state, 2π-periodic, density function $p_1(\phi_1)$ satisfies an equation obtained by setting to zero the left-hand side of Equation 7.2-18. That is, $p_1(\phi_1)$ satisfies the differential equation

$$0 = -\frac{d}{d\phi_1}\left[(\omega_\Delta' - \sin\phi_1)p_1 - \frac{1}{\rho}\frac{d}{d\phi_1}p_1\right]. \tag{7.2-20}$$

This equation can be integrated once to obtain

$$\mathfrak{I}_{ss} = (\omega_\Delta' - \sin\phi_1)p_1 - \frac{1}{\rho}\frac{d}{d\phi_1}p_1, \tag{7.2-21}$$

where constant $\Im_{ss}$ denotes the steady-state probability current. Constant $\Im_{ss}$ must be selected so that Equation 7.2-21 has the 2π-periodic solution $p_1(\phi_1)$ that satisfies the normalization condition

$$\int_{-\pi}^{\pi} p_1(\phi_1) d\phi_1 = 1. \tag{7.2-22}$$

The determination of $\Im_{ss}$ and p_1 is facilitated by transforming Equation 7.2-21 into a time-invariant equation driven by a periodic function. Represent p_1 as the product

$$p_1(\phi_1) = u_1(\phi_1) \exp(\rho \cos \phi_1), \tag{7.2-23}$$

and substitute this representation into Equation 7.2-21 to obtain

$$\frac{d}{d\phi_1} u_1 = \beta_\Delta u_1 - \rho \Im_{ss} \exp(-\rho \cos \phi_1), \tag{7.2-24}$$

where

$$\beta_\Delta \equiv \rho \, \omega_\Delta'. \tag{7.2-25}$$

A 2π-periodic solution of Equation 7.2-24 must be found for use in Equation 7.2-23.

This can be accomplished easily when loop detuning parameter ω_Δ' is zero (so that $\beta_\Delta = 0$). In this case, $\Im_{ss}$ must be zero, and u_1 must be constant as can be seen from inspection of Equation 7.2-24 with $\beta_\Delta = 0$. The Constraint 7.2-22 leads to

$$u_1 = \left[\int_{-\pi}^{\pi} \exp(\rho \cos \phi_1) d\phi_1 \right]^{-1}$$

$$= \frac{1}{2\pi I_o(\rho)}. \tag{7.2-26}$$

Now, use this result in Equation 7.2-23 to obtain

$$p_1(\phi_1) = \frac{\exp(\rho \cos \phi_1)}{2\pi I_o(\rho)}, \quad -\pi \le \phi_1 < \pi, \tag{7.2-27}$$

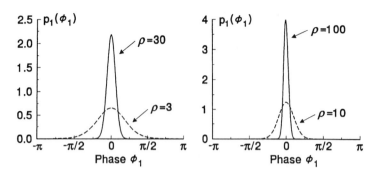

FIGURE 7.9
Steady-state density functions for the modulo-2π phase error when $\omega_\Delta' = 0$.

for the density of the modulo-2π, steady-state phase error in a first-order PLL with $\omega_\Delta' = 0$.

Figure 7.9 depicts plots of Equation 7.2-27 for several values of ρ. Note that Equation 7.2-27 and the plots are symmetrical around $\phi_1 = 0$. This symmetry implies that $E[\phi_1] = 0$ for the case $\omega_\Delta' = 0$ under consideration. Also, note from Figure 7.9 that the density function $p_1(\phi_1)$ becomes more sharply peaked as ρ increases. This implies that the modulo-2π phase error has a variance that is inversely related to the signal-to-noise ratio ρ.

For large values of ρ, the asymptotic behavior of Equation 7.2-27 is of interest. Modified Bessel function $I_0(\rho)$ has the well-known asymptotic approximation

$$I_0(\rho) \approx \frac{\exp(\rho)}{\sqrt{2\pi\rho}} \qquad (7.2\text{-}28)$$

for large ρ. Use this approximation in Equation 7.2-27 to obtain

$$\begin{aligned}
p_1(\phi_1) &\approx \frac{\exp\left(\rho[\cos\phi_1 - 1]\right)}{\sqrt{2\pi(1/\rho)}}, \\
&= \frac{\exp\left(-\rho[\phi_1^2/2 - \phi_1^4/4! \ \cdots]\right)}{\sqrt{2\pi}\ \sigma_\phi} \qquad (7.2\text{-}29) \\
&= \frac{\exp\left(-(\phi_1^2/2\sigma_\phi^2)\right)\left[1 - 2\phi_1^2/4! + 2\phi_1^4/6! \ \cdots\right]}{\sqrt{2\pi}\ \sigma_\phi}
\end{aligned}$$

for large ρ. Note that $\sigma_\phi^2 \equiv 1/\rho$ has been used here. As discussed in Chapter 2, σ_ϕ^2 represents the variance of the phase error in the first-order PLL linear model. Relative to $p_1(0)$, $p_1(\phi_1)$ is small for all values of ϕ_1 in $[-\pi, \pi)$ that lie outside of a small neighborhood of $\phi_1 = 0$; equivalently, when ρ is large, $p_1(\phi_1)$ is only significant for small values of ϕ_1. For these small values of ϕ_1, fourth- and higher-order terms in Equation 7.2-29 become insignificant. This observation leads to the approximation

$$p_1(\phi_1) \approx \frac{\exp(-\phi_1^2/2\sigma_\phi^2)}{\sqrt{2\pi}\,\sigma_\phi}, \quad -\pi \le \phi_1 < \pi. \tag{7.2-30}$$

Hence, phase error ϕ_1 is approximately Gaussian distributed for large values of signal-to-noise ratio ρ. Recall that this claim was made in Section 2.5 when noise in the PLL linear model was considered.

Density function $p_1(\phi_1)$ cannot be expressed in closed form when detuning $\omega_\Delta' \neq 0$. For this case, the well known Jacobi–Anger expansion (Fourier series)

$$\exp(\pm\rho\cos\phi_1) = I_0(\rho) + 2\sum_{n=1}^{\infty}(\pm1)^n\,I_n(\rho)\cos n\phi_1 \tag{7.2-31}$$

can be used in Equation 7.2-24. In Equation 7.2-31, the modified Bessel functions I_n, $n = 0, 1, 2, \ldots$, have the real-valued argument ρ. By using superposition, the 2π-periodic solution of Equation 7.2-24 can be expressed as

$$u_1(\phi_1) = \mathfrak{I}_{ss}\frac{I_0(\rho)}{\omega_\Delta'}\left[1 + 2\sum_{n=1}^{\infty}(-1)^n\left[\frac{I_n(\rho)}{I_0(\rho)}\right]\Psi_n(\phi_1)\right], \tag{7.2-32}$$

where

$$\Psi_n(\phi_1) \equiv \frac{\cos n\phi_1 - (n/\beta_\Delta)\sin n\phi_1}{1 + (n/\beta_\Delta)^2}. \tag{7.2-33}$$

Now, substitute Equation 7.2-32 into Equation 7.2-23 to obtain

$$p_1(\phi_1) = \mathfrak{I}_{ss}\frac{I_0}{\omega_\Delta'}\exp(\rho\cos\phi_1)\left[1 + 2\sum_{n=1}^{\infty}(-1)^n(I_n/I_0)\Psi_n(\phi_1)\right] \tag{7.2-34}$$

as an expression for p_1 in terms of the unknown current $\Im_{ss}$. In Equation 7.2-34, the dependence of the Bessel functions on parameter ρ is suppressed in order to simplify the notation.

Constant $\Im_{ss}$ must be selected so that Equation 7.2-34 satisfies the normalization condition Equation 7.2-22. To find $\Im_{ss}$, substitute Equation 7.2-31 for the exponential in Equation 7.2-34, and use the orthogonality of the trigonometric functions to obtain

$$\int_{-\pi}^{\pi} p_1(\phi_1) d\phi_1 = \Im_{ss} \frac{I_o}{\omega_\Delta'} \left[2\pi I_o + 2 \sum_{n=1}^{\infty} (-1)^n (I_n/I_o) \left[\frac{2\pi I_n}{1+(n/\beta_\Delta)^2} \right] \right]. \quad (7.2\text{-}35)$$

When this result is equated to unity, the steady-state probability current

$$\Im_{ss} = \frac{\omega_\Delta'}{2\pi I_o^2} \left[1 + 2 \sum_{n=1}^{\infty} (-1)^n \left[\frac{(I_n/I_o)^2}{1+(n/\beta_\Delta)^2} \right] \right]^{-1} \quad (7.2\text{-}36)$$

is obtained. Equations 7.2-34 and 7.2-36 provide a practical method of approximating p_1. For small values of ρ, the Bessel function series converge quickly, and a good approximation can be obtained with a moderate number of terms.

Alternatively, for the case $\omega_\Delta' \neq 0$, numerical techniques can be applied directly to Equation 7.2-21. For $\omega_\Delta' = \sin(\pi/4)$ and several values of ρ, numerical approximations of p_1 are depicted on Figure 7.10. As expected, density p_1 remains sharply peaked for large values of ρ, and the peak is centered at the steady-state value of static phase error that would result if the reference signal was noise free. However, for a fixed value of ρ, the density function "spreads out" (its variance increases), its peak value declines, and it becomes more asymmetric as

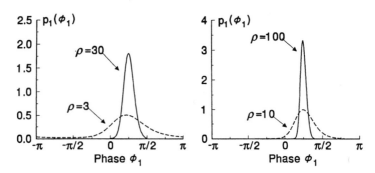

FIGURE 7.10
Steady-state density functions for the modulo-2π phase error when $\omega_\Delta' = \sin(\pi/4)$.

NOISE IN THE NONLINEAR PLL MODEL

ω_Δ' increases. This can be seen by comparing the widths and heights of the plots depicted on Figures 7.9 and 7.10.

7.2.4 Steady-State Distribution for the Modulo-2π Phase Error

The modulo-2π phase error has a steady-state, distribution-type function that is of interest in this section. This function is defined by

$$P_1(\phi_1) = \int_{-\phi_1}^{\phi_1} p_1(\phi)\,d\phi, \qquad (7.2\text{-}37)$$

and its importance results from the fact that it indicates the percentage of time during which modulo-2π phase error has an absolute magnitude that is less than ϕ_1. Note that Equation 7.2-37 differs from the definition of a distribution function that is given in elementary probability theory.

Function $P_1(\phi_1)$ can be expressed as an infinite series, or it can be computed numerically. When $\omega_\Delta' = 0$, the series is relatively simple, and a partial sum can yield an accurate approximation for $P_1(\phi_1)$. Substitute Equations 7.2-31 and 7.2-27 into Equation 7.2-37 and obtain

$$\begin{aligned}
P_1(\phi_1) &= \int_{-\phi_1}^{\phi_1} \frac{\exp(\rho\cos\phi)}{2\pi I_0(\rho)}\,d\phi \\
&= \frac{1}{2\pi I_0(\rho)} \int_{-\phi_1}^{\phi_1} \left[I_0(\rho) + 2\sum_{n=1}^{\infty} I_n(\rho)\cos n\phi \right] d\phi \qquad (7.2\text{-}38) \\
&= \frac{\phi_1}{\pi} + \frac{2}{\pi I_0(\rho)} \sum_{n=1}^{\infty} \frac{1}{n} I_n(\rho)\sin n\phi_1
\end{aligned}$$

for the case $\omega_\Delta' = 0$. This series converges quickly, and its first ten terms were used to produce the results depicted by Figure 7.11.

7.2.5 Mean of the Steady-State, Modulo-2π Phase Error

For $\omega_\Delta' \neq 0$, the mean of the modulo-2π, steady-state phase error is nonzero. By starting with Equation 7.2-34, it can be represented as a double sum of Bessel functions (see p. 391 of Lindsey).[4] Alternatively, the product $\phi_1 p_1(\phi_1)$, where p_1 is represented by Equation 7.2-34, can be integrated numerically to calculate the mean. This numerical method was used to produce the results discussed below.

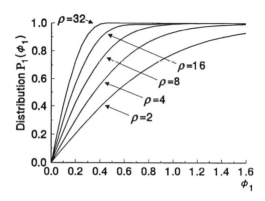

FIGURE 7.11
Steady-state distribution functions for the case $\omega_\Delta' = 0$.

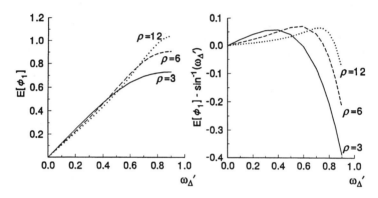

FIGURE 7.12
$E[\phi_1]$ and $E\phi_1] - \sin^{-1}(\omega_\Delta')$ as functions of ω_Δ'.

Figures 7.12 (a) and (b) depict plots of $E[\phi_1]$ and $E[\phi_1] - \sin^{-1}(\omega_\Delta')$, respectively, vs. ω_Δ' for several values of ρ. Note that $\sin^{-1}(\omega_\Delta')$ is the closed-loop phase error in the absence of noise on the reference. Observe that $E[\phi_1]$ increases with ω_Δ', but the rate of increase becomes small as ω_Δ' approaches unity. Also, note that $E[\phi_1] - \sin^{-1}(\omega_\Delta')$ has a magnitude that increases as ω_Δ' approaches unity. These observations suggest that, as far as $E[\phi_1]$ is concerned, the influence of the signal diminishes relative to the influence of the noise as ω_Δ' approaches unity.

7.2.6 Variance of the Steady-State, Modulo-2π Phase Error

The variance σ_ϕ^2 of the steady-state, modulo-2π phase error is an important parameter in assessing the influence of noise on a PLL. In what follows, it is expressed in series form for the first-order, nonlinear

PLL model under consideration. These results are compared with the phase error variance obtained from the linear model. This comparison helps quantify the limitations of the linear theory.

For the case $\omega_\Delta' = 0$, variable ϕ_1 has zero mean, and variance σ_ϕ^2 is given by

$$\sigma_\phi^2 = \int_{-\pi}^{\pi} \phi_1^2 \, p_1(\phi_1) \, d\phi_1. \qquad (7.2\text{-}39)$$

Now, use Equations 7.2-27 and 7.2-31 to obtain

$$\sigma_\phi^2 = \int_{-\pi}^{\pi} \phi_1^2 \, p_1(\phi_1) \, d\phi_1$$

$$= \frac{1}{2\pi I_0(\rho)} \int_{-\pi}^{\pi} \phi_1^2 \left[I_0(\rho) + 2\sum_{n=1}^{\infty} I_n(\rho) \cos n\phi_1 \right] d\phi_1 \qquad (7.2\text{-}40)$$

$$= \frac{\pi^2}{3} + \frac{4}{I_0(\rho)} \sum_{n=1}^{\infty} (-1)^n \frac{1}{n^2} I_n(\rho).$$

This series converges very rapidly, and excellent results can be obtained by truncating it to ten terms. This method was used to obtain the plot of σ_ϕ^2 that is depicted by the solid-line graph on Figure 7.13 (and it is given the label "exact"). The dashed-line graph on the figure is a plot of the variance obtained from the linear theory of the PLL ($\sigma_\phi^2 = 1/\rho$ for the linear theory; see Equation 2.5-20). The linear theory produces results that are in error by approximately 17% for $\rho = 4$; the error that results from use of the linear theory decreases rapidly with larger values of ρ.

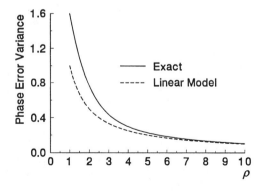

FIGURE 7.13
Steady-state phase variance for the case $\omega_\Delta' = 0$.

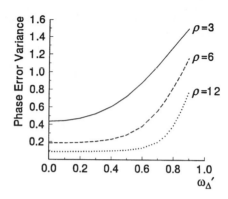

FIGURE 7.14
Phase error variance as a function of gain-normalized detuning.

Calculation of the variance is more difficult when $\omega_\Delta' \neq 0$. By starting with Equation 7.2-34, it is possible to represent the variance as a double sum of Bessel functions (see p. 392 of Lindsey).[4] Alternatively, the first and second moments of the phase error can be computed by using Equation 7.2-34 in a numerical integration, and these moments can be used to determine the variance. This numerical method can produce accurate results for moderate values of ρ where it is possible to calculate accurately and easily a sufficient number of Bessel functions.

This numerical method was used to produce the results depicted by Figure 7.14. As expected, the variance is inversely related to ρ for all values of ω_Δ'. Also, for fixed values of ρ, the variance is an increasing function of ω_Δ'. This observation can be explained intuitively by considering Figure 7.8. The effective gain of the phase detector is proportional to the slope of its output characteristic function at the operating point $E[\phi_1]$. From Figure 7.8, note that the effective phase detector gain is proportional to the cosine of the average phase error. Hence, an increase in ω_Δ' and $E[\phi_1]$ causes a decrease in effective phase detector gain and effective signal level within the loop. However, the noise η is introduced after the phase detector, and its influence on the phase error is not diminished significantly (relative to that of the signal) with decreasing effective phase detector gain. Hence, an increase in ω_Δ' and $E[\phi_1]$ produces an effective signal-to-noise ratio decrease within the loop, and this results in an increase in the phase error variance.

7.2.7 Noise-Induced Cycle Slips

As discussed in Section 7.1, noise-induced cycle slips can occur in a first-order PLL. Unlike what can occur in higher-order PLLs, only simple cycle slips occur in the first-order loop under consideration. A burst of cycle slips, where one or more domains of attraction are skipped (see Figure 7.5), cannot occur. Sample functions of the phase error are continuous, and they cannot skip over a domain of attraction. This implies that the mean time to first slip is equal to the reciprocal of the average number of cycle slips per second in the first-order PLL discussed here.

In Section 7.1.5, a phase error model containing absorbing boundaries was discussed. This model serves as the basis for the analysis of the cycle slipping phenomenon in the first-order PLL. Consider Figure 7.7, which shows the system starting at the equilibrium point φ_o at time $t = 0$ ($\varphi_o = \sin^{-1}(\omega_\Delta')$). The noise-driven phase error process can wander around the equilibrium point for a period of time. However, as suggested by the figure, a time t_a will come when (for the first time) the phase error advances or retards 2π radians relative to the equilibrium point φ_o, and a cycle slip occurs. As discussed in Sections 6.2.7 through 6.2.10, time t_a is known as the time to first passage, it is a random variable, and its expected value $E[t_a]$ denotes the average amount of time between cycle slips. Furthermore, the reciprocal of $E[t_a]$ is the *frequency of slipping cycles* in a first-order PLL, an important parameter in PLL design. This section is devoted to a quantitative analysis of cycle slipping in the first-order PLL.

Let N_+ (N_-) denote the average cycle slip rate in the positive (negative) direction under steady-state conditions. That is, every second the loop slips an average of N_+ (N_-) cycles to the right (left), and the phase error increases (decreases) an average of $2\pi N_+$ ($2\pi N_-$) radians. Then $N_+ + N_-$ represents the average slip rate regardless of direction, and the phase error increases an average net value of $N_+ - N_-$ complete cycles every second. Note that real time t (*not* slow-time τ) is used in the definitions of N_+ and N_-.

This last quantity can be related to the average frequency error $E[d\phi_1/dt]$ in the PLL modulo-2π model. On the average, the loop phase error gains $(2\pi)^{-1}E[d\phi_1/dt]$ complete cycles every second so that

$$\frac{1}{2\pi}E\left[\frac{d\phi_1}{dt}\right] = \frac{G}{2\pi}E\left[\frac{d\phi_1}{d\tau}\right] = (N_+ - N_-) \qquad (7.2\text{-}41)$$

since $\tau \equiv Gt$. The average frequency error $E[d\phi_1/dt]$ (also called the *residual frequency detuning*) is a measure of the frequency tracking ability of a PLL with a noisy reference signal.

The average frequency error can be computed easily. Equation 7.2-3 and $p_1(\phi_1)$ can be used to write

$$E\left[\frac{d\phi_1}{dt}\right] = GE\left[\frac{d\phi_1}{d\tau}\right] = G\int_{-\pi}^{\pi}[\omega_\Delta' - \sin\phi_1]p_1(\phi_1)d\phi_1, \quad (7.2\text{-}42)$$

since the noise term η has an average value of zero. Now, substitute Equation 7.2-21 into this last result and obtain

$$E\left[\frac{d\phi_1}{dt}\right] = G\int_{-\pi}^{\pi}\left[\mathfrak{I}_{ss} + \frac{1}{\rho}\frac{d}{d\phi_1}p_1(\phi_1)\right]d\phi_1$$

$$= 2\pi G\mathfrak{I}_{ss} \quad (7.2\text{-}43)$$

since p_1 is 2π-periodic. Finally, combine Equations 7.2-41 and 7.2-43 to write

$$N_+ - N_- = G\mathfrak{I}_{ss}. \quad (7.2\text{-}44)$$

A numerical value for the net slip rate $N_+ - N_-$ can be computed by using Equation 7.2-36 to compute $\mathfrak{I}_{ss}$ for use in Equation 7.2-44. Note that for $\omega_\Delta' > 0$ ($\omega_\Delta' < 0$), the net slip rate is positive (negative), and the trend is for the sinusoidal reference to gain (lose) cycles relative to the VCO.

In addition to the net slip rate ($N_+ - N_-$), the ratio N_+/N_- can be obtained from the results developed in Section 6.2.11. For the first-order PLL under consideration, this ratio of slip rates in the forward and reverse directions is equivalent to the ratio of probability currents flowing into the boundaries $b_2 = \varphi_0 + 2\pi$ and $b_1 = \varphi_0 - 2\pi$. From Equation 6.2-95, this is given by

$$\frac{N_+}{N_-} = -\frac{\mathbb{C}_0 - 1}{\mathbb{C}_0}, \quad (7.2\text{-}45)$$

where

$$\mathbb{C}_0 = \frac{\int_{\varphi_0}^{\varphi_0+2\pi} \exp U(\alpha)d\alpha}{\int_{\varphi_0-2\pi}^{\varphi_0+2\pi} \exp U(\alpha)d\alpha} \quad (7.2\text{-}46)$$

is obtained from Equations 6.2-82 and 6.2-85. Now, $\mathbb{C}_0$ can be written as

NOISE IN THE NONLINEAR PLL MODEL

$$\mathbb{C}_o = \frac{\int_{\varphi_0}^{\varphi_0+2\pi} \exp U(\alpha)\,d\alpha}{\int_{\varphi_0-2\pi}^{\varphi_0} \exp U(\alpha)\,d\alpha + \int_{\varphi_0}^{\varphi_0+2\pi} \exp U(\alpha)\,d\alpha}$$

$$= \left[1 + \frac{\int_{\varphi_0-2\pi}^{\varphi_0} \exp U(\alpha)\,d\alpha}{\int_{\varphi_0}^{\varphi_0+2\pi} \exp U(\alpha)\,d\alpha}\right]^{-1}.$$

(7.2-47)

A change in variable and the simple nature of potential Function 7.2-19 can be used to write

$$\int_{\varphi_0-2\pi}^{\varphi_0} \exp U(\alpha)\,d\alpha = \int_{\varphi_0}^{\varphi_0+2\pi} \exp U(\alpha-2\pi)\,d\alpha$$

$$= \exp\left[2\pi\rho\omega_\Delta'\right] \int_{\varphi_0}^{\varphi_0+2\pi} \exp U(\alpha)\,d\alpha.$$

(7.2-48)

This last equation can be used in Equation 7.2-47 to write

$$\mathbb{C}_o = \frac{1}{1+\exp\left[2\pi\rho\omega_\Delta'\right]}.$$

(7.2-49)

Finally, this simplified form of $\mathbb{C}_o$ can be substituted into Equation 7.2-45 to obtain

$$\frac{N_+}{N_-} = \exp\left[2\pi\rho\omega_\Delta'\right]$$

(7.2-50)

as the ratio of average cycle slip rates. This result states that the slip ratio depends exponentially on $2\pi\rho\,\omega_\Delta' = U(\phi) - U(\phi + 2\pi)$, the difference in potential between adjacent wells (see Equation 7.2-19).

The average slip rates N_+ and N_- are of interest, and they can be obtained easily from the developments in this section. Equations 7.2-44 and 7.2-50 can be solved for

$$N_+ = +\frac{G\mathfrak{I}_{ss}}{1-\exp\left[-2\pi\rho\omega_\Delta'\right]}$$

(7.2-51)

$$N_- = -\frac{G\mathfrak{J}_{ss}}{1-\exp[+2\pi\rho\omega_\Delta']}. \qquad (7.2\text{-}52)$$

The sum of these produces

$$N_+ + N_- = G\mathfrak{J}_{ss}\coth(\pi\rho\omega_\Delta') \qquad (7.2\text{-}53)$$

as the average slip rate independent of direction. Numerical values for these slip rates can be obtained by using Equation 7.2-36 to compute the steady-state probability current $\mathfrak{J}_{ss}$. Finally, the reciprocal of $N_+ + N_-$ is the mean time to first slip in the first-order PLL under consideration.

For $\omega_\Delta' = 0$, the slip rate in the forward direction is equal to the slip rate in the reverse direction. This case can be analyzed by taking the limit of Equation 7.2-51 (or Equation 7.2-52) as ω_Δ' approaches zero. First, note that Equation 7.2-36 implies

$$\lim_{\omega_\Delta' \to 0}\left[\frac{\mathfrak{J}_{ss}}{\omega_\Delta'}\right] = \frac{1}{2\pi I_0(\rho)^2}. \qquad (7.2\text{-}54)$$

Now, this result can be used with Equation 7.2-51 to produce

$$\lim_{\omega_\Delta' \to 0} N_+ = \lim_{\omega_\Delta' \to 0} N_- = \frac{G}{(2\pi I_0(\rho))^2 \rho}. \qquad (7.2\text{-}55)$$

For the case $\omega_\Delta' = 0$, the average slip rate independent of direction is twice Equation 7.2-55. Finally, for large ρ, approximation 7.2-28 can be used with this last result to obtain

$$\lim_{\omega_\Delta' \to 0} N_+ = \lim_{\omega_\Delta' \to 0} N_- \approx \frac{Ge^{-2\rho}}{2\pi}. \qquad (7.2\text{-}56)$$

Figure 7.15 depicts plots of gain-normalized cycle slip rates (i.e., rates based on the slow-time variable τ) in a first-order PLL. This figure was constructed by using Equations 7.2-51, 7.2-52, and 7.2-55; for these formulas, probability current $\mathfrak{J}_{ss}$ was approximated by using Equation 7.2-36. For $\omega_\Delta' = 0.4$, the figure depicts plots of N_+/G and N_-/G for a range of ρ. Also, the figure depicts a plot of the one-sided slip rate $N_+/G = N_-/G$ for $\omega_\Delta' = 0$ and $0.5 \leq \rho \leq 5$. As can be seen from Figure 7.15 and Equation 7.2-50, a small value of detuning ω_Δ' causes a strongly-favored slip rate direction.

NOISE IN THE NONLINEAR PLL MODEL

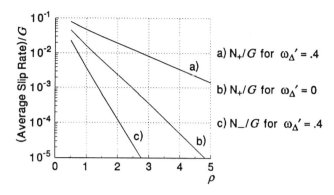

FIGURE 7.15
Average one-direction slip rate for $0.5 \leq \rho \leq 5$.

7.3 NOISE IN SECOND-ORDER PLL

Second-order PLLs supplied with a constant-frequency reference sinusoid embedded in wideband Gaussian noise are considered in this section. The model used is general; by setting parameter values appropriately, it covers the perfect and imperfect integrator cases considered in Sections 3.2 and 3.3, respectively. Also, the model describes the second-order PLL that results when a simple single-pole RC lowpass filter is used within the loop. The perfect integrator case is given the most attention, and an approximation is developed for the steady-state, modulo-2π phase error density function. Similar approximation techniques can be applied to the imperfect integrator case and to higher-order PLLs.

7.3.1 Nonlinear Model for a Second-Order PLL with a Noisy Reference

The PLLs under consideration have a loop filter given by

$$F(s) = a_0 + \frac{a_1}{s+b}. \qquad (7.3\text{-}1)$$

Use $b = 0$, $a_0 = 1$, and $a_1 = \alpha > 0$ in Equation 7.3-1 to obtain the perfect integrator case first introduced in Section 3.2 (α is the integrator gain for this case). The imperfect integrator case introduced in Section 3.3 results when $a_0 = 1$, $a_1 > 0$, and $b > 0$. Finally, set $a_0 = 0$ and $a_1 = b > 0$

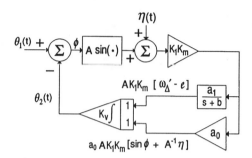

FIGURE 7.16
A nonlinear model for a second-order PLL with a noisy reference. The input is $\theta_1(t) = \omega_\Delta t$.

to realize a simple loop filter that can be synthesized from one resistor and a single capacitor ($a_1 = b = 1/RC$ in this case).

The nonlinear model discussed in Section 2.5 can be applied to this PLL. The relevant integral–differential equation is obtained by substituting the loop filter impulse response

$$f(t) = a_0\, \delta(t) + a_1 e^{-bt} U(t) \tag{7.3-2}$$

into Equation 2.5-16. Since it is assumed that $\theta_1 = \omega_\Delta t$, this substitution results in

$$\frac{d}{dt}\phi(t) = Ge(t) - Ga_0\left[\sin\phi(t) + A^{-1}\eta(t)\right]$$

$$e(t) \equiv \omega_\Delta' - a_1 \int_0^t \left[\sin\phi(\xi) + A^{-1}\eta(\xi)\right] e^{-b(t-\xi)} d\xi, \tag{7.3-3}$$

where $\omega_\Delta' \equiv \omega_\Delta/G$ is the gain-normalized detuning parameter, and η is white Gaussian noise with a double-sided spectral density of $N_0/2$ watts/Hz. Note that Equation 7.3-3 provides a mathematical description of the diagram depicted by Figure 7.16 (this figure follows from Figure 2.9).

This last equation can be reformulated as a system of first-order ordinary differential equations that define a vector Markov process. To accomplish this, simply differentiate the second equation of Equation 7.3-3 and obtain

$$\frac{d}{dt}\phi = Ge - a_0 G\left[\sin\phi + A^{-1}\eta\right]$$

$$\frac{d}{dt}e = -a_1\left[-b\int_0^t \left[\sin\phi(\xi) + A^{-1}\eta(\xi)\right]e^{-b(t-\xi)}\,d\xi + \sin\phi + A^{-1}\eta\right]$$

$$= -a_1\left[+b\left\{\frac{e - \omega_\Delta'}{a_1}\right\} + \sin\phi + A^{-1}\eta\right] \qquad (7.3\text{-}4)$$

$$= -b\left[e - \omega_\Delta'\right] - a_1\left[\sin\phi + A^{-1}\eta\right].$$

By using $\tau \equiv Gt$, the explicit dependence on G can be removed; this step yields

$$\frac{d}{d\tau}\phi = e - a_0\left[\sin\phi + A^{-1}\eta\right]$$

$$\frac{d}{d\tau}e = -b'\left[e - \omega_\Delta'\right] - a_1'\left[\sin\phi + A^{-1}\eta\right], \qquad (7.3\text{-}5)$$

where $b' \equiv b/G$ and $a_1' \equiv a_1/G$. In Equation 7.3-5, note that parameters a_0 and A are not gain normalized. This system of equations defines the vector Markov process $[\phi\ e]^T$.

System Equation 7.3-5 describes all of the commonly used second-order PLLs. This system with $a_0 \equiv 1$, $b' \equiv 0$, and $a_1' \equiv \alpha'$ (the integrator gain) describes the perfect integrator case. Note that Equation 7.3-5 does not depend on ω_Δ' when $b' = 0$ is used for the perfect integrator case. Set $a_0 \equiv 1$, $b' > 0$, and $a_1' > 0$ in Equation 7.3-5 to obtain a set of equations that describes the imperfect integrator case. Finally, set $a_0 \equiv 0$, and $b' = a_1' > 0$ to model the PLL containing an RC lowpass filter.

7.3.2 Fokker–Planck Equation for the Second-Order PLL

Subject to some given initial conditions, Equation 7.3-5 is satisfied by a vector Markov process denoted here as $[\phi(\tau)\ e(\tau)]^T$. Let $p(\phi, e, \tau)$ denote the joint density function for this vector process. It satisfies a Fokker–Planck equation that is developed in this section.

The drift and diffusion coefficients must be developed for the Fokker–Planck equation. They follow from the process increments

$$\Delta\phi(\tau) = [e(\tau) - a_0 \sin\phi(\tau)]\Delta\tau - a_0 A^{-1} \int_{\tau}^{\tau+\Delta\tau} \eta(\xi)d\xi$$

$$\Delta e(\tau) = -b'[e(\tau) - \omega_{\Delta}']\Delta\tau - a_1'[\sin\phi(\tau)]\Delta\tau - a_1' A^{-1} \int_{\tau}^{\tau+\Delta\tau} \eta(\xi)d\xi$$

(7.3-6)

that are obtained by integrating Equation 7.3-5 over the normalized time increment $[\tau, \tau + \Delta\tau]$. Now, use Equations 6.3-8 and 6.3-13 to write the drift coefficients

$$K_1^{(1)}(\phi, e) = \lim_{\Delta\tau \to 0} \frac{E[\Delta\phi | \phi, e]}{\Delta\tau}$$

$$= e - a_0 \sin\phi$$

(7.3-7)

$$K_2^{(1)}(\phi, e) = \lim_{\Delta\tau \to 0} \frac{E[\Delta e | \phi, e]}{\Delta\tau}$$

$$= -b'[e - \omega_{\Delta}'] - a_1' \sin\phi.$$

(7.3-8)

The simplicity of these drift coefficients results from the fact that noise η has zero mean. With the aid of Equation 7.2-7, the first diffusion coefficient can be computed as

$$K_{11}^{(2)}(\phi, e) = \lim_{\Delta\tau \to 0} \frac{E[(\Delta\phi)^2 | \phi, e]}{\Delta\tau}$$

$$= a_0^2 A^{-2} \lim_{\Delta\tau \to 0} \frac{1}{\Delta\tau} \int_{\tau}^{\tau+\Delta\tau} \int_{\tau}^{\tau+\Delta\tau} E[\eta(\xi_1)\eta(\xi_2)]d\xi_1 d\xi_2$$

$$= a_0^2 A^{-2} \lim_{\Delta\tau \to 0} \frac{1}{\Delta\tau} \int_{\tau}^{\tau+\Delta\tau} \int_{\tau}^{\tau+\Delta\tau} \frac{N_o}{2}\delta\left[\frac{\xi_1 - \xi_2}{G}\right]d\xi_1 d\xi_2 \quad (7.3\text{-}9)$$

$$= a_0^2 A^{-2} \lim_{\Delta\tau \to 0} \frac{1}{\Delta\tau}\left[G\frac{N_o}{2}\Delta\tau\right]$$

$$= a_0^2 \frac{GN_o}{2A^2}.$$

The remaining diffusion coefficients

NOISE IN THE NONLINEAR PLL MODEL

$$K_{12}^{(2)}(\phi,e) = K_{21}^{(2)}(\phi,e) = \lim_{\Delta\tau\to 0} \frac{E[(\Delta\phi)(\Delta e)\mid\phi,e]}{\Delta\tau}$$

$$= a_0 a_1' \frac{GN_o}{2A^2}$$

(7.3-10)

$$K_{22}^{(2)}(\phi,e) = \lim_{\Delta\tau\to 0} \frac{E[(\Delta e)^2 \mid \phi,e]}{\Delta\tau}$$

$$= (a_1')^2 \frac{GN_o}{2A^2}$$

(7.3-11)

follow in a manner similar to that used to obtain $K_{11}^{(2)}$.

These drift and diffusion coefficients can be used to write a Fokker–Planck equation for the second-order PLL. First, use Equation 6.3-20 and the coefficients developed above to write the probability currents

$$\mathfrak{I}_1(\phi,e,\tau) = \left[e - a_0 \sin\phi - a_0 \rho_2^{-1}\left(a_0 \frac{\partial}{\partial\phi} + a_1' \frac{\partial}{\partial e}\right)\right] p(\phi,e,\tau) \quad (7.3\text{-}12)$$

$$\mathfrak{I}_2(\phi,e,\tau)$$
$$= \left[-b'(e - \omega_\Delta') - a_1' \sin\phi - a_1' \rho_2^{-1}\left(a_0 \frac{\partial}{\partial\phi} + a_1' \frac{\partial}{\partial e}\right)\right] p(\phi,e,\tau),$$

(7.3-13)

where

$$\rho_2 \equiv \frac{4A^2}{GN_o}.$$

(7.3-14)

With the aid of Equation 6.3-18, these probability currents can be used to write the Fokker–Planck equation

$$\frac{\partial}{\partial\phi}\mathfrak{I}_1(\phi,e,\tau) + \frac{\partial}{\partial e}\mathfrak{I}_2(\phi,e,\tau) + \frac{\partial}{\partial\tau} p(\phi,e,\tau) = 0 \quad (7.3\text{-}15)$$

for the second-order PLL described by Equation 7.3-5.

7.3.3 A Simple Example: a PLL with an RC Lowpass Loop Filter

In this section, the Fokker–Planck theory is illustrated by a simple example. A PLL is considered that contains a single-pole, lowpass loop filter. This PLL is described by Equation 7.3-5 with $a_0 = 0$ and $a_1 = b > 0$. While not often used in practice, this PLL provides a simple example of the theory; it is one of the few cases of a second-order phase-locked loop for which closed-form, steady-state results can be obtained. Also, in Chapter 8, it provides an appropriate starting example for useful numerical methods.

For this example, the steady-state Fokker–Planck equation is obtained from Equations 7.3-12 through 7.3-15. In the development, use the modulo-2π phase error model, and replace p and ϕ by p_1 and ϕ_1, respectively. Also, use $a_0 = 0$ and $a_1' = b' > 0$ to write

$$\frac{\partial}{\partial \phi_1} e\, p_1(\phi_1, e) - b' \frac{\partial}{\partial e}\left[(e - \omega_\Delta') + \sin\phi_1 + \sigma_e^2 \frac{\partial}{\partial e}\right] p_1(\phi_1, e) = 0, \quad (7.3\text{-}16)$$

where

$$\sigma_e^2 \equiv b' \rho_2^{-1}, \quad (7.3\text{-}17)$$

and ρ_2 is given by Equation 7.3-14. For the RC loop filter case under consideration, ρ_2 is the signal-to-noise ratio in the loop parameter ρ, first introduced in Section 2.5.4 (this is not generally true for second-order PLLs). Also, it is easily shown that Equation 7.3-17 describes the variance of e in the linear model of the PLL under consideration.

Equation 7.3-16 has a closed-form solution when detuning $\omega_\Delta' \equiv 0$. This relatively simple solution is given by

$$p_1(\phi_1, e) = \frac{1}{(2\pi)^{3/2} \sigma_e I_0(\rho_2)} \exp\left[-\frac{e^2}{2\sigma_e^2} + \rho_2 \cos\phi_1\right]. \quad (7.3\text{-}18)$$

To verify this result, simply substitute Equation 7.3-18 into Equation 7.3-16 with $\omega_\Delta' = 0$. Note that this joint density satisfies the boundary conditions given by steady-state versions of Equations 7.1-8 and 7.1-9, and it satisfies the normalization condition given in Equation 7.1-5. Unfortunately, the solution of Equation 7.3-16 cannot be expressed in closed form if $\omega_\Delta' \neq 0$. However, the numerical methods outlined in Chapter 8 produce good results for this case.

Equation 7.3-18 has several interesting features. It can be factored into a product $p_1(\phi_1)p(e)$ of densities; therefore, ϕ_1 and e are statistically independent when $\omega_\Delta' = 0$. The factor $p_1(\phi_1)$ is identical to the Density 7.2-27 that describes the first-order PLL with $\omega_\Delta' = 0$. Also, density $p(e)$ is Gaussian with zero mean and variance σ_e^2 given by Equation 7.3-17.

7.3.4 Equation of Flow in the ϕ_1 Direction

In many applications of second-order PLLs, the probabilistic nature of the nonstationary, modulo-2π phase error ϕ_1 is of interest. For these applications, the Fokker–Planck equation is obtained from Equation 7.3-15 by replacing ϕ and p by ϕ_1 and p_1, respectively. Furthermore, the boundary conditions outlined in Section 7.1.1 apply to this case. In terms of the variables ϕ_1 and e, these boundary conditions are expressed as

$$p_1(-\pi,e,\tau) = p_1(\pi,e,\tau)$$

$$\Im_k(-\pi,e,\tau) = \Im_k(\pi,e,\tau), \quad k=1,2$$

$$\lim_{e \to \pm\infty} p_1(\phi_1,e,\tau) = 0 \qquad (7.3\text{-}19)$$

$$\lim_{e \to \pm\infty} \Im_k(\phi_1,e,\tau) = 0, \quad k=1,2.$$

Finally, for $\tau \geq 0$, the joint density function must satisfy the normalization condition

$$\int_{-\pi}^{\pi} \int_{-\infty}^{\infty} p_1(\phi_1,e,\tau) \, de \, d\phi_1 = 1. \qquad (7.3\text{-}20)$$

Convergence of this last integral requires that p_1 approach zero faster than $1/e$ as e approaches infinity.

The variable e can be eliminated as a dependent variable, and a one-dimensional equation can be obtained that describes the probability flow in the ϕ_1 direction. To obtain this simplified equation, replace ϕ and p in Equation 7.3-15 by ϕ_1 and p_1, respectively, and then integrate this modified equation to obtain

$$\frac{\partial}{\partial \phi_1} \int_{-\infty}^{\infty} \Im_1(\phi_1,e,\tau) \, de + \Im_2(\phi_1,e,\tau) \bigg|_{e=-\infty}^{e=\infty} + \frac{\partial}{\partial \tau} p_1(\phi_1,\tau) = 0, \qquad (7.3\text{-}21)$$

where $p_1(\phi_1,\tau)$ denotes the marginal density for the modulo-2π phase error ϕ_1. Now, the middle term in Equation 7.3-21 vanishes as a result of Boundary Conditions 7.3-19. Hence, this last equation can be written as

$$\frac{\partial}{\partial \phi_1}\mathfrak{I}_\phi(\phi_1,\tau)+\frac{\partial}{\partial \tau}p_1(\phi_1,\tau)=0, \qquad (7.3\text{-}22)$$

where

$$\mathfrak{I}_\phi(\phi_1,\tau) \equiv \int_{-\infty}^{\infty} \mathfrak{I}_1(\phi_1,e,\tau)\,de. \qquad (7.3\text{-}23)$$

Partial differential Equation 7.3-22 describes the flow of probability current in the ϕ_1 direction. In fact, the quantity $\mathfrak{I}_\phi(\zeta,\tau)$ describes the net amount of probability current that, in a unit time period, crosses the hyperplane $\phi_1 = \zeta$; here, the probability current is assumed to flow in a positive direction. In what follows, $\mathfrak{I}_\phi(\phi_1,\tau)$ is referred to as the net probability current in the ϕ_1 direction.

For the class of PLLs under consideration, the probability current $\mathfrak{I}_\phi(\phi_1,\tau)$ can be written in a standard canonical form. To accomplish this, substitute Equation 7.3-12 into Equation 7.3-23 and use the Boundary Conditions 7.3-19 to write

$$\mathfrak{I}_\phi(\phi_1,\tau) = \int_{-\infty}^{\infty}\left[e - a_0 \sin\phi_1 - a_0 \rho_2^{-1}\left(a_0\frac{\partial}{\partial \phi_1}+a_1'\frac{\partial}{\partial e}\right)\right]p_1(\phi_1,e,\tau)\,de$$

$$\qquad (7.3\text{-}24)$$

$$= \int_{-\infty}^{\infty} e p_1(\phi_1,e,\tau)\,de - a_0\left(\sin\phi_1 + a_0\rho_2^{-1}\frac{\partial}{\partial \phi_1}\right)p_1(\phi_1,\tau).$$

Now, use Bayes' rule for density functions to obtain

$$\int_{-\infty}^{\infty} e p_1(\phi_1,e,\tau)\,de = \int_{-\infty}^{\infty} e p_1(e,\tau \mid \phi_1,\tau)\,p_1(\phi_1,\tau)\,de$$

$$\qquad (7.3\text{-}25)$$

$$= E[e(\tau) \mid \phi_1(\tau)]\,p_1(\phi_1,\tau).$$

In Equation 7.3-25, $E[e(\tau) \mid \phi_1(\tau)]$ denotes the conditional expectation of e at time τ given a value for ϕ_1 at time τ; it should be thought of as a function of the variable τ and the given value of ϕ_1 at time τ. With the use of Equation 7.3-25, Equation 7.3-24 can be written as

NOISE IN THE NONLINEAR PLL MODEL

$$\mathfrak{S}_\phi(\phi_1,\tau) = \left[K_\phi^{(1)}(\phi_1,\tau) - \frac{1}{2}K_{11}^{(2)}\frac{\partial}{\partial \phi_1} \right] p_1(\phi_1,\tau), \quad (7.3\text{-}26)$$

where $K_{11}^{(2)}$ is given by Equation 7.3-9 and

$$K_\phi^{(1)}(\phi_1,\tau) \equiv \mathrm{E}\!\left[e(\tau)\,|\,\phi_1\right] - a_0 \sin\phi_1 \quad (7.3\text{-}27)$$

(use the argument ϕ_1 given in $K_\phi^{(1)}(\phi_1,\tau)$ for the value of $\phi_1(\tau)$ in $\mathrm{E}[e(\tau)\,|\,\phi_1(\tau)]$) represents a new drift coefficient that depends on the conditional expectation $\mathrm{E}[e(\tau)\,|\,\phi_1(\tau)]$. In terms of p_1, Equation 7.3-22 describes the net flow of probability current in the ϕ_1 direction. The combination of Equations 7.3-22 and 7.3-26 yields a partial differential equation that describes the evolution of the time-dependent marginal density $p_1(\phi_1,\tau)$; it is known as the *reduced Fokker–Planck equation* for the example under consideration.

Alternate representations for the probability current $\mathfrak{S}_\phi(\phi_1,\tau)$ are helpful for simplifying the analysis that follows. For the special case $a_0 = 0$ (i.e., the RC loop filter case), the diffusion coefficient $K_{11}^{(2)}$ is zero (see Equation 7.3-9), and the probability current $\mathfrak{S}_\phi(\phi_1,\tau)$ has a particularly simple form. When $a_0 \neq 0$, $\mathfrak{S}_\phi$ can be written in the canonical form

$$\mathfrak{S}_\phi(\phi_1,\tau) = -\frac{1}{2}K_{11}^{(2)} \exp\!\left[-U(\phi_1,\tau)\right] \frac{\partial}{\partial \phi_1}\!\left[p_1(\phi_1,\tau) \exp\!\left[U(\phi_1,\tau)\right]\right], \quad (7.3\text{-}28)$$

where the potential function is given by

$$U(\phi_1,\tau) \equiv -\int^{\phi_1}\!\left[\frac{1}{2}K_{11}^{(2)}\right]^{-1} K_\phi^{(1)}(\phi_1,\tau)\,d\phi. \quad (7.3\text{-}29)$$

7.3.5 Conditional Expectation Method for Approximating $p_1(\phi_1)$

In many applications, the steady-state marginal density $p_1(\phi_1)$ is of interest. Fortunately, the results just developed can be used to write a canonical form for p_1. In the steady state, Equations 7.3-22 and 7.3-28 imply that

$$\begin{aligned}\mathfrak{S}_\phi(\phi_1) &= -\frac{1}{2}K_{11}^{(2)} \exp\!\left[-U(\phi_1)\right]\frac{\partial}{\partial \phi_1}\!\left[p_1(\phi_1)\exp\!\left[U(\phi_1)\right]\right] \\ &= \mathfrak{S}_{ss},\end{aligned} \quad (7.3\text{-}30)$$

where constant $\mathfrak{I}_{ss}$ represents the steady-state net probability current in the ϕ_1 direction. In Equation 7.3-30, the quantity $U(\phi_1)$ is the steady-state potential. It is given by a steady-state version of Equation 7.3-29 that employs $K_\phi^{(1)}(\phi_1) \equiv E_s[e \mid \phi_1] - a_0 \sin\phi_1$, where

$$E_s\left[e \mid \phi_1\right] \equiv \lim_{\tau \to \infty} E\left[e(\tau) \mid \phi_1(\tau)\right] \qquad (7.3\text{-}31)$$

is the conditional expectation under steady-state conditions. Finally, Equation 7.3-30 can be integrated to obtain the canonical form

$$p_1(\phi_1) = C_0 \exp\left[-U(\phi_1)\right]\left[1 + D_0 \int_{-\pi}^{\phi_1} \exp\left[U(\phi)\right] d\phi\right], \qquad (7.3\text{-}32)$$

where C_0 and D_0 are constants that must be determined.

These constants are obtained by invoking boundary and normalization conditions. From the boundary condition $p_1(-\pi) = p_1(\pi)$ comes the result

$$D_0 = \frac{\exp\left[-U(-\pi)\right] - \exp\left[-U(\pi)\right]}{\exp\left[-U(\pi)\right]\int_{-\pi}^{\pi} \exp\left[U(\phi)\right]d\phi}. \qquad (7.3\text{-}33)$$

The constant C_0 must be selected to normalize p_1 so that it is a proper density function.

Equation 7.3-32 is a general result for the class of second-order PLLs considered in this section. In the PLL literature, its development and use are referred to as the *conditional expectation method*. The results described in this section can be generalized easily to higher-order PLLs. For the general nth-order PLL, a canonical form similar to Equation 7.3-32 can be written for the closed-loop phase error (see Chapter 11 of Lindsey).[4]

Equation 7.3-32 provides a canonical form for the steady-state marginal density $p_1(\phi_1)$ in a second-order PLL. This formula depends on the steady-state potential function given by Equations 7.3-29 and 7.3-27 with $E[e(\tau) \mid \phi_1(\tau)]$ replaced by the steady-state expectation $E_s[e \mid \phi]$. Hence, in order to use Equation 7.3-32, the conditional expectation given by Equation 7.3-31 must be known or approximated. One such approximation is discussed in the next section.

7.3.6 Approximating the Conditional Expectation

This section discusses an approximation to the conditional expectation given by Equation 7.3-31. As given here, the development of the approximation relies on formal algebraic manipulation; however, the nonrigorous approach that is used produces simple and useful approximations for $E_s[e \mid \phi_1]$ and $p_1(\phi_1)$. This fact is shown in Section 7.3.7, where the theory is applied to the second-order PLL that contains a perfect integrator in its loop filter, and a simple approximation is determined for $p_1(\phi_1)$. In Chapter 8, this approximation is shown to compare well with numerical results computed for various values of in-loop signal-to-noise ratio.

The conditional expectation $E[e(\tau) \mid \phi_1(\tau)]$ that is needed on the right-hand side of Equation 7.3-31 can be approximated by using the second equation of Equation 7.3-5 to obtain

$$\frac{d}{d\tau_1}\left[\exp(b'\tau_1)e(\tau_1)\right]$$
$$= \left[b'\omega_\Delta' - a_1'\sin\phi_1(\tau_1) - a_1'A^{-1}\eta(\tau_1)\right]\exp(b'\tau_1), \quad (7.3\text{-}34)$$

where independent variable τ_1 is used to temporarily replace τ. On the quantity $\phi_1(\tau)$, condition the expectation of Equation 7.3-34 and write

$$E\left[\frac{d}{d\tau_1}\left[\exp(b'\tau_1)e(\tau_1)\right] \mid \phi_1(\tau)\right]$$
$$= E\left[b'\omega_\Delta' - a_1'\sin\phi_1(\tau_1) - a_1'A^{-1}\eta(\tau_1) \mid \phi_1(\tau)\right]\exp(b'\tau_1) \quad (7.3\text{-}35)$$

for $\tau \leq \tau_1$. On the left-hand side of this last equation, interchange the order of differentiation and expectation to obtain

$$\frac{d}{d\tau_1}E\left[e(\tau_1) \mid \phi_1(\tau)\right]\exp(b'\tau_1)$$
$$= E\left[b'\omega_\Delta' - a_1'\sin\phi_1(\tau_1) - a_1'A^{-1}\eta(\tau_1) \mid \phi_1(\tau)\right]\exp(b'\tau_1). \quad (7.3\text{-}36)$$

Integrate this result over the interval $\tau \leq \tau_1 < \infty$ to get

$$\int_\tau^\infty \frac{d}{d\tau_1}E\left[e(\tau_1) \mid \phi_1(\tau)\right]\exp(b'\tau_1)d\tau_1$$
$$= \int_\tau^\infty E\left[b'\omega_\Delta' - a_1'\sin\phi_1(\tau_1) \mid \phi_1(\tau)\right]\exp(b'\tau_1)d\tau_1 \quad (7.3\text{-}37)$$
$$+ \int_\tau^\infty E\left[-a_1'A^{-1}\eta(\tau_1) \mid \phi_1(\tau)\right]\exp(b'\tau_1)d\tau_1.$$

The fact that η is white noise implies that $\eta(\tau_1)$, $\tau < \tau_1 < \infty$, is independent of $\phi_1(\tau)$. Furthermore, the noise has zero mean; both of these facts imply that, on the right-hand side of Equation 7.3-37, the second integral is zero. Because of this, Equation 7.3-37 can be used to produce

$$\lim_{\tau_1 \to \infty} E\big[e(\tau_1) \,|\, \phi_1(\tau)\big]\exp(b'\tau_1) - E\big[e(\tau) \,|\, \phi_1(\tau)\big]\exp(b'\tau)$$

$$= \int_\tau^\infty E\big[b'\omega_\Delta' - a_1'\sin\phi_1(\tau_1) \,|\, \phi_1(\tau)\big]\exp(b'\tau_1)\,d\tau_1. \qquad (7.3\text{-}38)$$

The desired conditional expectation $E[e(\tau) \,|\, \phi_1(\tau)]$ appears on the left-hand side of the last equation. To obtain a usable expression for it, substitute for τ_1 the quantity $\tau_1 = \tau + v$; also, assume that τ is fixed, and v approaches infinity, as τ_1 goes to infinity. In Equation 7.3-38, this change of variable produces

$$E\big[e(\tau) \,|\, \phi_1(\tau)\big]\exp(b'\tau)$$

$$= -\int_0^\infty E\big[b'\omega_\Delta' - a_1'\sin\phi_1(\tau+v) \,|\, \phi_1(\tau)\big]\exp[b'(\tau+v)]\,dv \qquad (7.3\text{-}39)$$

$$+ \lim_{v \to \infty} E\big[e(\tau+v) \,|\, \phi_1(\tau)\big]\exp[b'(\tau+v)].$$

Both sides of this last equation can be multiplied by $\exp[-b'\tau]$ to obtain

$$E\big[e(\tau) \,|\, \phi_1(\tau)\big]$$

$$= -\int_0^\infty E\big[b'\omega_\Delta' - a_1'\sin\phi_1(\tau+v) \,|\, \phi_1(\tau)\big]\exp(b'v)\,dv \qquad (7.3\text{-}40)$$

$$+ \lim_{v \to \infty} E\big[e(\tau+v) \,|\, \phi_1(\tau)\big]\exp(b'v).$$

Now, to the right-hand side of this last equation, add and subtract the quantity

$$\int_0^\infty E\big[b'\omega_\Delta' - a_1'\sin\phi_1(\tau+v)\big]\exp[b'v]\,dv \qquad (7.3\text{-}41)$$

to obtain

$$E\big[e(\tau)\,|\,\phi_1(\tau)\big]$$

$$= a_1' \int_0^\infty \left(E\big[\sin\phi_1(\tau+v)\,|\,\phi_1(\tau)\big] - E\big[\sin\phi_1(\tau+v)\big]\right)\exp(b'v)\,dv$$

$$- \int_0^\infty E\big[b'\omega_\Delta' - a_1'\sin\phi_1(\tau+v)\big]\exp(b'v)\,dv \qquad (7.3\text{-}42)$$

$$+ \lim_{v\to\infty} E\big[e(\tau+v)\,|\,\phi_1(\tau)\big]\exp(b'v)$$

after some rearranging.

In general, the expectations in Equation 7.3-42 depend on τ. However, under steady-state conditions (large τ), the expectations become independent of τ. As was introduced by Equation 7.3-31, $E_s[e(\tau)\,|\,\phi_1(\tau)]$ denotes the steady state limit of $E[e(\tau)\,|\,\phi_1(\tau)]$; this quantity depends on ϕ_1, but it does not depend on τ (even though τ appears in $E_s[e(\tau)\,|\,\phi_1(\tau)]$). Likewise, under steady-state conditions, $E[\sin\phi_1(\tau + v)]$ becomes the *constant* (i.e., DC) component in $\sin\phi_1(\tau)$, denoted here as

$$E_s\big[\sin\phi_1(\tau)\big] \equiv \lim_{\tau\to\infty} E\big[\sin\phi_1(\tau)\big] \qquad (7.3\text{-}43)$$

$$\equiv \overline{\sin\phi_1(\tau)}.$$

In what follows, the overbar notation introduced in Equation 7.3-43 is used interchangeably with the operator E_s notation. Also, to simplify and shorten equations, the variable τ can be suppressed; steady-state expectations $E_s[e(\tau)\,|\,\phi_1(\tau)]$ and $E_s[e\,|\,\phi_1]$ denote the same τ-independent quantities (expectations $E[e(\tau)\,|\,\phi_1(\tau)]$ and $E[e\,|\,\phi_1]$ denote the same τ-dependent quantities). Finally, expectation $E_s[e(\tau + v)\,|\,\phi_1(\tau)]$ denotes the steady-state limit of $E[e(\tau + v)\,|\,\phi_1(\tau)]$; this quantity depends on ϕ_1 and v, but it does not depend on τ.

In terms of the notation just introduced, Equation 7.3-42 becomes

$$E_s\big[e(\tau)\,|\,\phi_1(\tau)\big]$$

$$= a_1' \int_0^\infty \left[E_s\big[\sin\phi_1(\tau+v)\,|\,\phi_1(\tau)\big] - \overline{\sin\phi_1(\tau+v)}\right]\exp(b'v)\,dv$$

$$- \big[b'\omega_\Delta' - a_1'\overline{\sin\phi_1(\tau+v)}\big]\frac{1}{b'}\Big[\lim_{v\to\infty}\exp(b'v) - 1\Big] \qquad (7.3\text{-}44)$$

$$+ \lim_{v\to\infty} E_s\big[e(\tau+v)\,|\,\phi_1(\tau)\big]\exp(b'v)$$

in the steady state. But this last equation can be written as

$$E_s\left[e(\tau)\,|\,\phi_1(\tau)\right]$$

$$= a_1'\int_0^\infty E_s\left[\sin\phi_1(\tau+v) - \overline{\sin\phi_1(\tau+v)}\,|\,\phi_1(\tau)\right]\exp(b'v)dv \quad (7.3\text{-}45)$$

$$+ \omega_\Delta' - \frac{a_1'}{b'}\overline{\sin\phi_1(\tau+v)} - Q$$

where,

$$Q \equiv \lim_{v\to\infty}\left[\omega_\Delta' - \frac{a_1'}{b'}\overline{\sin\phi_1(\tau+v)} - E_s\left[e(\tau+v)\,|\,\phi_1(\tau)\right]\right]\exp(b'v). \quad (7.3\text{-}46)$$

Unfortunately, Q has the indeterminate form of zero times infinity. To see this, analyze the DC component on both sides of the second equation in Equation 7.3-5; from such an analysis, draw the conclusion that

$$\lim_{v\to\infty}\left[\omega_\Delta' - \frac{a_1'}{b'}\overline{\sin\phi_1(\tau+v)} - E_s\left[e(\tau+v)\,|\,\phi_1(\tau)\right]\right]$$

$$= \omega_\Delta' - \frac{a_1'}{b'}\overline{\sin\phi_1} - \bar{e} \quad (7.3\text{-}47)$$

$$= 0,$$

where $\overline{\sin\phi_1(\tau)}$ and $\overline{e(\tau)}$ denote the DC components of $\sin\phi_1(\tau)$ and $e(\tau)$, respectively. In the context of the general nonlinear problem, little more can be said about Q. However, it is important to note that Q is zero when the problem is linearized (i.e., use $\sin\phi_1 \approx \phi_1$) and the assumption $a_1' > 0$ is made. This fact can be shown by applying linear methods to Equation 7.3-5.

In the remainder of this section, a relatively simple approximation for $E_s[e\,|\,\phi_1]$ is developed by using Equation 7.3-45 with Q = 0. In what follows, $E_s[e\,|\,\phi_1]$ is approximated as the sum of a constant and a sinusoidal term. In applications, this simple approximation degrades with decreases in in-loop signal-to-noise ratio, but the numerical results given in Chapter 8 show that it can provide usable results. Furthermore, the approximation leads to a theory that describes easy-to-observe nonlinear phenomena in the PLL. That is, it is used in Section 7.3.10 to explain asymmetries in both $p_1(\phi_1)$ and the phenomenon of noise-induced cycle slips.

NOISE IN THE NONLINEAR PLL MODEL

The integral on the right-hand side of Equation 7.3-45 must be approximated. More specifically, an approximation must be obtained for the conditional expectation

$$Y(\phi_1, v) \equiv E_s\left[\sin\phi_1(\tau+v) - \overline{\sin\phi_1(\tau+v)} \,\big|\, \phi_1(\tau)\right] \qquad (7.3\text{-}48)$$

that appears in the integrand of Equation 7.3-45. As defined by Equation 7.3-48, the quantity Y depends on the variables ϕ_1 and v, but it does not depend on τ (and $\overline{\sin\phi_1(\tau+v)}$ is a constant).

The approximations for $p_1(\phi_1)$ and $E_s[e \mid \phi_1]$ that are developed in this section must be 2π-periodic in ϕ_1. This requirement is achieved by approximating Y by a periodic function of ϕ_1. It is reasonable to approximate Y in this manner. After all, phase $\phi_1(\tau + v)$ is correlated with $\phi_1(\tau)$ (at least for small v), and a ramp in $\phi_1(\tau)$ should be expected to cause oscillatory changes in $E_s[\sin\phi_1(\tau + v) \mid \phi_1(\tau)]$ for a fixed, small v. Now, a simple periodic approximation for Y is

$$Y(\phi_1, v) \approx \rho_x(v) X(\phi_1), \qquad (7.3\text{-}49)$$

where $\rho_x(v)$ is a function to be determined, and $X(\phi_1)$ is defined as

$$X(\phi_1) \equiv \sin\phi_1 - \overline{\sin\phi_1(\tau)}. \qquad (7.3\text{-}50)$$

In Equation 7.3-50, the quantity $\overline{\sin\phi_1(\tau)}$ is the DC component in $\sin\phi_1(\tau)$; also, it is expressible as the ensemble average

$$\overline{\sin\phi_1(\tau)} = \int_{-\pi}^{\pi}\left[\sin\phi_1\right]p_1(\phi_1)d\phi_1. \qquad (7.3\text{-}51)$$

The function $\rho_x(v)$ in Equation 7.3-49 must be selected to complete the approximation of Y. A reasonable way to accomplish this is by using a mean-square error criteria. Select $\rho_x(v)$ to minimize the mean-square error between Y and its Approximation 7.3-49; that is, minimize

$$\text{MSE} \equiv E_{\phi_1}\left[\left\{Y(\phi_1, v) - \rho_x(v) X(\phi_1)\right\}^2\right], \qquad (7.3\text{-}52)$$

where the ensemble average is computed by using the steady-state density $p_1(\phi_1)$.

This formulation of the approximation problem has several advantages. The most important advantage is that the optimum ρ_x can be

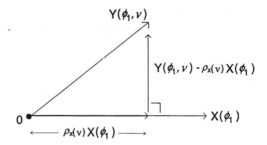

FIGURE 7.17
Minimum MSE occurs when the error is orthogonal to X.

determined by a simple application of the Projection Theorem (also known as the Orthogonality Principle). By abstractly depicting Y and X as vectors, Figure 7.17 illustrates the use of this powerful theorem in the present context. The figure shows that the error $Y - \rho_x X$ is minimized when it is orthogonal to X. Hence, the quantity ρ_x should be selected to force

$$E_{\phi_1}\left[\left[Y(\phi_1,v) - \rho_x(v)X(\phi_1)\right]X(\phi_1)\right] = 0, \qquad (7.3\text{-}53)$$

a condition stating that $(Y - \rho_x X)$ is orthogonal to X. In Equation 7.3-53, the ensemble average is computed by using the steady-state density $p_1(\phi_1)$.

The expectations in Equation 7.3-53 can be evaluated by using standard probability theory and the autocorrelation function for X. These expectations can be written as (see the paragraph that contains Equation 7.2-6)

$$E_{\phi_1}\left[Y(\phi_1,v)X(\phi_1)\right] = E_{\phi_1}\left[E_s\left[X(\phi_1(\tau+v))\mid \phi_1(\tau)\right]X(\phi_1)\right]$$

$$= E_s\left[X(\phi_1(\tau+v))X(\phi_1(\tau))\right] \qquad (7.3\text{-}54)$$

$$= R_x(v/G)$$

and

$$E_{\phi_1}\left[X(\phi_1)X(\phi_1)\right] = E_s\left[X(\phi_1(\tau))X(\phi_1(\tau))\right]$$

$$= R_x(0/G) \qquad (7.3\text{-}55)$$

$$= \sigma_x^2,$$

where

$$R_x(t) \equiv E_s\left[X(\phi_1(t_1))X(\phi_1(t_1+t))\right] \quad (7.3\text{-}56)$$

is the autocorrelation of the real-time process $X(\phi_1(t))$.

These results can be combined to form simple approximations for Y and $E_s[e \mid \phi_1]$. First, with the aid of Equations 7.3-54 and 7.3-55, Equation 7.3-53 can be solved for

$$\rho_x(v) = \frac{R_x(v/G)}{\sigma_x^2}. \quad (7.3\text{-}57)$$

Note that the desired quantity ρ_x is nothing more than the correlation function (i.e., normalized autocorrelation function) of X. Now, this last result and Equation 7.3-50 can be substituted into Equation 7.3-49 to obtain

$$Y(\phi_1, v) \approx \frac{R_x(v/G)}{\sigma_x^2}\left[\sin\phi_1 - \overline{\sin\phi_1(\tau)}\right]. \quad (7.3\text{-}58)$$

Finally, this equation and Equation 7.3-45 can be combined to form

$$E_s[e \mid \phi_1] \approx a_1'\left[\sin\phi_1 - \overline{\sin\phi_1(\tau)}\right]\int_0^\infty \frac{R_x(v/G)}{\sigma_x^2}\exp(b'v)dv$$

$$+ \omega_\Delta' - \frac{a_1'}{b'}\overline{\sin\phi_1(\tau)}, \quad (7.3\text{-}59)$$

where the assumption is made that Q = 0 (see the discussion that follows Equation 7.3-46).

In many practical applications, the output of the phase detector has a correlation time that is small compared to 1/b (equivalently, this output has a spectral bandwidth which is large compared to b). In these applications, the exponential in Equation 7.3-59 changes very little over the range of v where R_x is relevant. Under this condition, Equation 7.3-59 can be simplified to produce

$$E_s[e \mid \phi_1] \approx a_1\frac{\sin\phi_1 - \overline{\sin\phi_1(\tau)}}{2\sigma_x^2}S_x(0) + \omega_\Delta' - \frac{a_1'}{b'}\overline{\sin\phi_1(\tau)}, \quad (7.3\text{-}60)$$

where

$$S_x(\omega) \equiv \int_{-\infty}^{\infty} R_x(v/G) \exp[-j\omega(v/G)] d(v/G) \qquad (7.3\text{-}61)$$

is the power spectrum of $X(\phi_1(t))$.

Equations 7.3-59 and 7.3-60 provide a relatively simple approximation to the constant and fundamental components of $E_s[e \mid \phi_1]$. That is, the conditional expectation can be expressed as

$$E_s[e \mid \phi_1] \approx E_{dc} + E_f \sin\phi_1, \qquad (7.3\text{-}62)$$

where

$$\begin{aligned} E_{dc} &\equiv -a_1' \left[\frac{1}{b'} + \int_0^\infty \frac{R_x(v/G)}{\sigma_x^2} \exp(b'v) dv \right] \overline{\sin\phi_1(\tau)} + \omega_\Delta' \\ &\approx -a_1 \left[\frac{1}{b} + \frac{S_x(0)}{2\sigma_x^2} \right] \overline{\sin\phi_1(\tau)} + \omega_\Delta' \end{aligned} \qquad (7.3\text{-}63)$$

is the constant (i.e., DC) component, and

$$\begin{aligned} E_f &\equiv a_1' \int_0^\infty \frac{R_x(v/G)}{\sigma_x^2} \exp(b'v) dv \\ &\approx a_1 \frac{S_x(0)}{2\sigma_x^2} \end{aligned} \qquad (7.3\text{-}64)$$

is the fundamental component.

These results lead to a simple approximation of the potential function for the second-order PLL under consideration. Simply use Equations 7.3-9, 7.3-14, and 7.3-62 with steady-state versions of Equations 7.3-27 and 7.3-29 to obtain

$$U(\phi_1) \approx -a_0^{-2} \rho_2 \left[E_{dc} \phi_1 - (E_f - a_0) \cos\phi_1 \right]. \qquad (7.3\text{-}65)$$

Hence, U can be approximated by a corrugated plane, an important potential function in the classical theory of Brownian motion. As shown in Section 7.3.8, the ramp term in U causes asymmetries in both $p_1(\phi_1)$ and the direction in which the PLL tends to slip cycles.

7.3.7 Approximating $p_1(\phi_1)$ for the Perfect Integrator Case

In this section the conditional expectation method is applied to the second-order PLL with a perfect integrator in its loop filter. For this case, the linear PLL theory developed in Chapter 3 is applied to the results outlined in the last section to develop an approximation for the steady-state conditional expectation $E_s[e \mid \phi_1]$. The results of this effort are used with Equations 7.3-32 and 7.3-65 to write a non-Gaussian approximation for the steady-state marginal density $p_1(\phi_1)$. In Chapter 8, this approximation for $p_1(\phi_1)$ is compared with numerical results obtained for the PLL under consideration.

The second-order PLL that contains a perfect integrator in its loop filter is described by Equation 7.3-5 with $a_0 = 1$, $a_1' = \alpha'$, and $b' = 0$ (recall that α' is the integrator gain normalized by G). The detuning constant ω_Δ' does not appear in Equation 7.3-5 with $b' = 0$; without loss of generality, it can be set to zero.

An approximation must be developed for the steady-state, 2π-periodic $E_s[e \mid \phi_1]$. The ideas used to develop Equation 7.3-62 can be applied here. First, note that $\overline{\sin \phi_1} = 0$ for the perfect integrator case (otherwise the integrator output ramps to infinity). Hence, the conditional expectation $E_s[e \mid \phi_1]$ must have a constant component E_{dc} that is zero, and the potential function U must contain no ramp term. The fundamental component E_f of the conditional expectation can be estimated by using the linear approximation $X = \sin \phi_1 \approx \phi_1$ for Equation 7.3-50. Then, Equation 2.5-17 results in

$$S_x(0) \approx S_\phi(0)$$
$$\approx \frac{N_o}{2A^2}. \tag{7.3-66}$$

Furthermore, Equations 2.5-21 and 3.2-12 can be combined to produce

$$\sigma_x^2 \approx \sigma_\phi^2$$
$$= [1 + \alpha']\frac{GN_o}{4A^2}. \tag{7.3-67}$$

Finally, combine these last two results with Equations 7.3-62 and 7.3-64 to write the simple approximation

$$E_s[e \mid \phi_1] \approx \frac{\alpha'}{1+\alpha'} \sin \phi_1. \tag{7.3-68}$$

For the perfect integrator case under consideration, the steady-state potential function is periodic in nature. An approximation of this function can be obtained by substituting E_f, given by Equations 7.3-64 and 7.3-66, and $L_{dc} = 0$ into Equation 7.3-65 to obtain

$$U(\phi_1) \approx -\frac{P_2}{1+\alpha'} \cos\phi_1. \qquad (7.3\text{-}69)$$

An approximation to $p_1(\phi_1)$ follows easily by substituting Equation 7.3-69 into Equation 7.3-32. The fact that p_1 must be 2π-periodic implies that constant $D_0 \equiv 0$. Constant C_0 is chosen to normalize p_1 and obtain

$$p_1(\phi_1) = \frac{\exp(\rho \cos\phi_1)}{2\pi I_0(\rho)} \qquad (7.3\text{-}70)$$

where

$$\rho \equiv \frac{P_2}{1+\alpha'} = \frac{A^2}{N_0 B_L}. \qquad (7.3\text{-}71)$$

Note that ρ represents the in-loop SNR parameter for the linear model of the second-order PLL containing a perfect integrator in its loop filter (see Section 2.5.4).

Equation 7.3-70 is a non-Gaussian approximation for the closed-loop phase error. This feature differentiates it from the results described in Section 2.5.4. As should be expected, the accuracy of this result degrades with decreasing values of the in-loop SNR parameter ρ. However, the numerical results provided in Chapter 8 suggest that Equation 7.3-70 is a good approximation for values of ρ which are nominal for many practical applications ($\rho \geq 6$ dB).

7.3.8 Noise-Induced Cycle Slips in Second-Order PLL

In second-order PLLs, the phenomenon of noise-induced cycle slips is of interest in many applications. In a manner similar to the first-order PLL, simple slips can occur where the PLL state transitions between adjacent domains of attraction. However, unlike the first-order loop considered in Section 7.2, bursts of cycle slips can occur in second- and higher-order PLL. When this happens, the loop state "skips over" domains of attraction as is illustrated by Figure 7.5.

NOISE IN THE NONLINEAR PLL MODEL

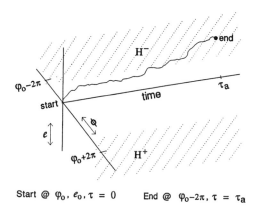

FIGURE 7.18
State space with absorbing boundaries.

Two related types of results on cycle slips are discussed in the PLL literature. The first is based on the well-known first-passage time problem. Consider a second-order PLL with state variables ϕ and e. At time $\tau = 0$, the PLL state is assumed to start from a stable equilibrium point $\phi = \varphi_0$, $e = e_0$. As is illustrated by Figure 7.18, the system noise causes the state to wander around this equilibrium point; after some time $\tau = \tau_a$, the state reaches either one of the hyperplanes

$$\begin{aligned} H^+ &\equiv \left[(\varphi_0 + 2\pi, e);\ -\infty < e < \infty\right] \\ H^- &\equiv \left[(\varphi_0 - 2\pi, e);\ -\infty < e < \infty\right] \end{aligned} \qquad (7.3\text{-}72)$$

for the first time. The random variable $\tau_a = t_a/G$ is the gain-normalized time to first passage, and a theory can be developed that provides expressions for its various moments.

The second type of result involves the average number of cycles slipped per unit of time under steady-state conditions. As discussed in Section 7.2.7, let N_+ (N_-) denote the average number of cycles slipped in the forward (backward) direction every second. Then, the quantity $N_+ - N_-$ represents the average net number of cycles advanced (i.e., in the direction of increasing phase) every second. In a manner that is independent of direction, the quantity $N_+ + N_-$ denotes the average number of cycle slips every second. In general, for second- and higher-order PLLs, this latter quantity is not equal to the reciprocal of $E[t_a]$.

The quantities $N_+ - N_-$ and $N_+ + N_-$ are important in many applications. For example, these parameters are important in tracking applications, where estimates of velocity and changes in range are made from Doppler measurements. The average net number of cycles advanced per second is a measure of the average error introduced into this type of Doppler measurement. The average of the number of slips independent of direction is a measure of noise-induced system degradation. By using the results given in Section 6.2, a theory involving $N_+ - N_-$ and $N_+ + N_-$ can be developed, and it is discussed in the remainder of this section.

7.3.9 Average Cycle Slip Rate in the Forward Direction

Under steady state conditions, the average net number of cycles slipped per second in the forward direction is related to the constant probability current that flows in the ϕ_1 direction. To obtain this relationship, first note that $N_+ - N_-$ is related to the residual frequency detuning by Equation 7.2-41. Here, it is important to note that the definition of residual detuning is based on the modulo-2π phase error model. Now, substitute the first of Equation 7.3-5 into Equation 7.2-41 to obtain

$$N_+ - N_- = \frac{G}{2\pi} E\left[\frac{d\phi_1}{d\tau}\right]$$

$$= \frac{G}{2\pi} \int_{-\pi}^{\pi} \int_{-\infty}^{\infty} \left[-a_0 \sin\phi_1 + e\right] p_1(\phi_1, e) \, de \, d\phi_1. \quad (7.3\text{-}73)$$

However, recall that the joint density $p_1(\phi_1,e)$ is 2π-periodic in ϕ_1, and it satisfies natural boundary conditions in the variable e. These facts imply that Equation 7.3-73 can be written as

$$N_+ - N_- = \frac{G}{2\pi} \int_{-\pi}^{\pi} \int_{-\infty}^{\infty} \left[(-a_0 \sin\phi_1 + e) \right.$$

$$\left. - a_0 \rho_2^{-1}(a_0 \frac{\partial}{\partial \phi_1} + a_1' \frac{\partial}{\partial e})\right] p_1(\phi_1, e) \, de \, d\phi_1 \quad (7.3\text{-}74)$$

$$= \frac{G}{2\pi} \int_{-\pi}^{\pi} \int_{-\infty}^{\infty} \mathfrak{I}_1(\phi_1, e) \, de \, d\phi_1$$

where $\mathfrak{I}_1$ denotes a steady-state version of the probability Current 7.3-12. Note that a steady-state version of Equation 7.3-23 can be used with Equation 7.3-74 to write

$$N_+ - N_- = \frac{G}{2\pi} \int_{-\pi}^{\pi} \mathfrak{I}_{\phi ss} \, d\phi_1, \quad (7.3\text{-}75)$$

NOISE IN THE NONLINEAR PLL MODEL

where $\Im_{\phi ss}$ is the steady-state probability current in the ϕ_1 direction. Now, Equation 7.3-22 implies that $\Im_{\phi ss}$ is constant; when combined with the previous equation, this fact leads to

$$N_+ - N_- = G\Im_{\phi ss} \tag{7.3-76}$$

for the average number of cycles slipped in a forward direction every second (time is not normalized in this result). Note that this last equation is similar in form to Equation 7.2-44 for the first-order PLL (where probability current $\Im_{ss}$ is in the ϕ_1 direction).

The average slip rate in the forward direction can be related to the amount of change that occurs in the steady-state potential $U(\phi_1)$ over a 2π interval of phase. To see this, substitute Equation 7.3-32 and the steady-state version of Equation 7.3-29 into Equation 7.3-30 to obtain

$$\Im_{\phi ss} = -\rho_2^{-1} a_0^2 C_0 D_0. \tag{7.3-77}$$

However, this last equation can be used with Equation 7.3-33 to write

$$\Im_{\phi ss} = -\rho_2^{-1} a_0^2 C_0 \frac{\exp[-\Delta U] - 1}{\int_{-\pi}^{\pi} \exp U(x) dx}$$

$$= \frac{2\rho_2^{-1} a_0^2 C_0}{\int_{-\pi}^{\pi} \exp U(x) dx} \exp\left[-\frac{\Delta U}{2}\right] \sinh\left[\frac{\Delta U}{2}\right], \tag{7.3-78}$$

where

$$\Delta U \equiv U(-\pi) - U(\pi). \tag{7.3-79}$$

Now, combine Equation 7.3-78 with Equation 7.3-76 to obtain

$$N_+ - N_- = G \frac{2\rho_2^{-1} a_0^2 C_0}{\int_{-\pi}^{\pi} \exp U(x) dx} \exp\left[-\frac{\Delta U}{2}\right] \sinh\left[\frac{\Delta U}{2}\right]. \tag{7.3-80}$$

This result implies that the average cycle slip rate in the forward direction is a function of ΔU, the change in potential that occurs over a 2π interval of phase.

The average slip rate in the forward direction is a function of the DC component in the 2π-periodic conditional expectation $E_s[e \mid \phi_1]$. To see this, use the steady-state potential given by Equation 7.3-65 to write

$$\Delta U \equiv 2\pi\, a_0^{-2}\, p_2 E_{dc}, \qquad (7.3\text{-}81)$$

where E_{dc} denotes the DC component in $E_s[e \mid \phi_1]$. Combine this result with Equation 7.3-80 to see that the average cycle slip rate in the forward direction is a function of E_{dc}. A nonzero value of E_{dc} causes asymmetry in the direction in which the PLL tends to slip cycles. Conversely, when $E_{dc} = 0$, the average cycle slip rate in the forward direction is zero; that is, relative to the PLL reference signal, the VCO is just as likely to gain a cycle as it is to fall behind by a cycle.

Finally, it is possible to relate the average cycle slip rate in the forward direction to the average value of $\sin\phi_1$ under steady-state conditions. First, note that Equation 7.3-73 implies

$$N_+ - N_- = \frac{G}{2\pi}\Big[E_s[e] - a_0 E_s\big[\sin\phi_1\big]\Big]. \qquad (7.3\text{-}82)$$

However, the second equation in Equation 7.3-5 can be used to write

$$E_s[e] = \omega_\Delta' - \frac{a_1'}{b'} E_s\big[\sin\phi_1\big]. \qquad (7.3\text{-}83)$$

Now, combine the previous two equations to obtain

$$N_+ - N_- = \frac{G}{2\pi}\left[\omega_\Delta' - \left(\frac{a_1'}{b'} + a_0\right) E_s\big[\sin\phi_1\big]\right]. \qquad (7.3\text{-}84)$$

This expresses the average cycle slip rate in the forward direction in terms of the average value $\overline{\sin\phi_1(\tau)} = E_s[\sin\phi_1(\tau)]$ of the phase detector output (after appropriate scaling).

In this section, the average cycle slip rate in the forward direction has been related to the PLL potential function and several other physical quantities which are relevant to PLL analysis. While largely qualitative in nature, the theory developed here is useful in numerically computing an approximation to $N_+ - N_-$. For example, the numerical algorithm developed in Sections 8.1 through 8.3 can be used to approximate $E_s[\sin\phi_1]$ for use in Equation 7.3-84; the quantity $N_+ - N_-$ can be computed in this manner.

NOISE IN THE NONLINEAR PLL MODEL

7.3.10 Ratio of Cycle Slip Rates for the Second-Order PLL

As was accomplished in Section 7.2.7 for the first-order PLL, the individual quantities N_+ and N_- for the second-order PLL can be represented in terms of the difference $N_+ - N_-$, discussed in the previous section, and the ratio N_+/N_-. In this section this ratio of cycle slip rates is discussed.

The quantity N_+/N_- can be expressed as the ratio of probabilities. At $\tau = 0$, assume that the PLL state starts at (φ_0, e_0), a stable equilibrium point in the absence of noise. As shown on Figure 7.18, noise causes the PLL state to wander around (φ_0, e_0) until $\tau = \tau_a$, when one of the absorbing hyperplanes is reached for the first time. Either hyperplane can be reached, and the probability of reaching a hyperplane approaches unity as time approaches infinity. In a manner similar to that used in the discussion of Figure 6.7, let $p_-(\infty)$ $(p_+(\infty))$ denote the probability that the hyperplane H^- (H^+) is reached. Then, the sum of $p_-(\infty)$ and $p_+(\infty)$ is unity, and the quantity $p_+(\infty)/p_-(\infty)$ is the desired ratio N_+/N_- of cycle slip rates.

Let $q(\phi, e, \tau \mid \varphi_0, e_0, 0)$ denote the solution of Fokker–Planck Equation 7.3-15 subject to the initial condition

$$q(\phi, e, \tau \mid \varphi_0, e_0, 0)\bigg|_{\tau=0} = \delta(\phi - \varphi_0)\, \delta(e - e_0) \qquad (7.3\text{-}85)$$

and the boundary conditions

$$q(\phi, e, \tau \mid \varphi_0, e_0, 0)\bigg|_{\phi = \varphi_0 - 2\pi} = q(\phi, e, \tau \mid \varphi_0, e_0, 0)\bigg|_{\phi = \varphi_0 + 2\pi} = 0$$

$$\lim_{e \to \pm\infty} q(\phi, e, \tau \mid \varphi_0, e_0, 0) = 0 \qquad (7.3\text{-}86)$$

$$\lim_{e \to \pm\infty} \mathfrak{I}_k(\phi, e, \tau) = 0,$$

for $k = 1, 2$. In what follows, the conditioning on φ_0 and e_0 at $\tau = 0$ is understood, but suppressed, and $q(\phi, e, \tau \mid \varphi_0, e_0, 0)$ is written as $q(\phi, e, \tau)$. As time increases, the total area under solution q decreases, and it approaches zero in the limit as time approaches infinity.

The reason for this is that probability is flowing into the hyperplanes H^- and H^+ as time increases. To describe this flow, note from Equation 7.3-12 that the ϕ component of the probability current vector is given by

$$\Im_1(\phi,e,\tau) = \left[e - a_0 \sin\phi - a_0 p_2^{-1}\left(a_0 \frac{\partial}{\partial \phi} + a_1' \frac{\partial}{\partial e}\right)\right] q(\phi,e,\tau). \quad (7.3\text{-}87)$$

Now, Equation 7.3-87 can be used to write

$$\Im_\phi(\phi,\tau) = \int_{-\infty}^{\infty} \Im_1(\phi,e,\tau)\, de, \quad (7.3\text{-}88)$$

where the quantity $\Im_\phi(\varphi_0 + 2\pi, \tau)$ describes the net amount of probability current that flows into the hyperplane H^+ per unit time. Also, the total probability that enters H^+ is given by

$$P_+(\infty) = \int_0^\infty \Im_\phi(\varphi_0 + 2\pi,\tau)\, d\tau = \int_0^\infty \int_{-\infty}^\infty \Im_1(\varphi_0 + 2\pi, e, \tau)\, de\, d\tau$$

$$= -a_0^2 p_2^{-1} \frac{d}{d\phi} Q(\phi) \bigg|_{\phi = \varphi_0 + 2\pi}$$

(7.3-89)

where

$$Q(\phi) = \int_0^\infty \int_{-\infty}^\infty q(\phi,e,\tau)\, de\, d\tau. \quad (7.3\text{-}90)$$

Note that the boundary Condition 7.3-86 was used to obtain Equation 7.3-89. Finally, in a manner identical to that used to obtain Equation 7.3-89, the total probability that crosses hyperplane H^- is

$$P_-(\infty) = -\int_0^\infty \int_{-\infty}^\infty \Im_1(\varphi_0 - 2\pi, e, \tau)\, de\, d\tau$$

$$= a_0^2 p_2^{-1} \frac{d}{d\phi} Q(\phi) \bigg|_{\phi = \varphi_0 - 2\pi}$$

(7.3-91)

The sign difference between Equations 7.3-89 and 7.3-91 results from the fact that probability current entering H^- must flow in a direction that is opposite to the probability current that flows into H^+.

The ratio of cycle slip rates can be expressed in terms of the results outlined above. With the aid of Equations 7.3-89 and 7.3-91, this ratio can be written as

$$\frac{N_+}{N_-} = \frac{p_+(\infty)}{p_-(\infty)}$$

$$= -\frac{\left.\frac{d}{d\phi}Q(\phi)\right|_{\phi=\varphi_0+2\pi}}{\left.\frac{d}{d\phi}Q(\phi)\right|_{\phi=\varphi_0-2\pi}}. \qquad (7.3\text{-}92)$$

Note the similarity of this result with that given by Equation 6.2-92 for the first-order case.

Equation 7.3-92 can be reformulated to appear like Equation 7.2-45, the ratio of slip rates for the first-order PLL. To accomplish this, use the initial and boundary conditions given by Equations 7.3-85 and 7.3-86, respectively, and integrate over $-\infty < e < \infty$, $0 < \tau < \infty$ a version of Equation 7.3-15 for $q(\phi, e, \tau)$. This procedure produces

$$-\delta(\phi-\varphi_0) = -\frac{d}{d\phi}\left[K_\phi^{(1)}(\phi) - \frac{1}{2}K_{11}^{(2)}\frac{d}{d\phi}\right]Q(\phi), \qquad (7.3\text{-}93)$$

where

$$K_\phi^{(1)}(\phi) \equiv \frac{\int_0^\infty\int_{-\infty}^\infty eq(\phi,e,\tau)de\,d\tau}{\int_0^\infty\int_{-\infty}^\infty q(\phi,e,\tau)de\,d\tau} - a_0\sin\phi, \qquad (7.3\text{-}94)$$

and the diffusion coefficient $K_{11}^{(2)}$ is given by Equation 7.3-9.

Equation 7.3-93 has a form that is identical to Equation 6.2-79, so it can be used with the theory outlined in Sections 6.2.10 and 6.2.11 to simplify the Ratio 7.3-92. This approach results in the ratio

$$\frac{N_+}{N_-} = \frac{p_+(\infty)}{p_-(\infty)}$$

$$= -\frac{C_0-1}{C_0}, \qquad (7.3\text{-}95)$$

where

$$C_0 = \frac{\int_{\varphi_0-2\pi}^{\varphi_0+2\pi} \mu(\alpha) U(\alpha - \varphi_0)\, d\alpha}{\int_{\varphi_0-2\pi}^{\varphi_0+2\pi} \mu(\alpha)\, d\alpha} \tag{7.3-96}$$

$$\mu(\phi) \equiv \exp\left[-\frac{2}{K_{11}^{(2)}} \int^{\phi} K_{\phi}^{(1)}(\alpha)\, d\alpha\right]. \tag{7.3-97}$$

Finally, individual expressions for N_+ and N_- can be obtained by using Equation 7.3-95 with any of the expressions for $N_+ - N_-$ developed in Section 7.3.9.

Note that Equation 7.3-95 has a form that is identical to Equation 7.2-45 (see also Equation 6.2-95), the ratio of cycle slip rates for a first-order system. This is not surprising since Equation 7.3-93 is identical in form to Equation 6.2-79. However, there is a significant difference between the two cases. For the first-order PLL, the general theory leads to Equations 7.2-51 and 7.2-52, closed-form formulas for the average cycle slip rates. For the second-order PLL, similar formulas for N_+ and N_- can be obtained with the use of Equation 7.3-95 only if the drift Coefficient 7.3-94 can be approximated.

Sections 7.3.9 and 7.3.10 present a qualitative analysis of the cycle slip phenomenon in second-order PLLs. Basically, by using the conditional expectation method, the two-dimensional problem is recast as one dimensional. Then, this new one-dimensional problem is analyzed by using the methods outlined in Section 7.2.7. In much the same manner, the "dimension-reducing" conditional expectation method can be applied to the general nth-order PLL model (see Chapter 11 of Lindsey).[4] Perhaps this generality is the most appealing feature of the theory.

Chapter 8

NUMERICAL METHODS FOR NOISE ANALYSIS IN THE NONLINEAR PLL MODEL

As the in-loop signal-to-noise ratio parameter ρ decreases in a PLL model, a point is reached where the classical linear theory (described in Section 2.5) no longer applies, and methods must be used that are based on the information presented in Chapters 6 and 7. In general, methods based on the nonlinear PLL model do not yield closed-form results, and they must be implemented numerically. Fortunately, numerical results can be obtained in many instances; in part, they are made practical by the powerful desktop computers and software that have become ubiquitous in the modern work environment. All of the numerical results described in this book were obtained by using an inexpensive personal computer.

Sections 8.1 through 8.3 describe and utilize an algorithm for the calculation of steady-state density functions. Emphasis is placed on the density function that describes the steady-state modulo-2π phase error in the closed loop. As discussed in Section 8.1, the method is based on a generalized Fourier series expansion of the density function. It is practical for many low-order PLLs. In Section 8.2, the method is applied to the second-order PLL that contains an RC loop filter (this PLL was described in Section 7.3.3). In Section 8.3, a similar application is made to the second-order PLL that contains a perfect integrator loop filter. For this second case, the computed results are compared with those obtained by the conditional expectation method described in Section 7.3.5. This useful comparison of different techniques helps to establish the limitations of the well-known conditional expectation method.

Sections 8.4 through 8.6 describe and utilize an algorithm for approximating the cycle slip rates in the nonlinear PLL model. As discussed in Section 8.4, the phenomenon of cycle slips can be formulated in terms of a finite-state Markov jump process model, where slips become transitions between the model states. By imposing a cyclic structure on the model and simplifying the notion of cycle slip, it is possible to relate the rates at which the simplified slips occur to some small (in magnitude) eigenvalues of the Fokker–Planck operator used to describe the statistical behavior of the PLL (in the literature, the technique is often referred to as the *eigenvalue method*). In Section 8.5, this theory is applied to the first-order PLL, and numerical results are compared with the exact, closed-form results developed in Section 7.2. This material on the first-order PLL serves to illustrate the strengths and limitations of the eigenvalue method. Finally, in Section 8.6, the theory is applied to a second-order system consisting of a first-order PLL model driven by a frequency-modulated reference embedded in additive white noise. The message used to frequency modulate the reference is generated by a first-order system driven by its own Gaussian noise source.

8.1 COMPUTING AN APPROXIMATION TO STEADY-STATE $P_1(\phi_1, \vec{Y})$

As discussed in Section 7.1, the quantity $p_1(\phi_1, \vec{Y}, t)$ denotes a density function of interest in PLL analysis. The quantity ϕ_1 represents the modulo-2π phase error. Vector $\vec{Y}$ describes the remaining state variables $y_2, y_3, \ldots, y_n$. The density function is assumed to satisfy the initial and normalization conditions given by Equation 7.1-5. Also, for large time, the PLL state is assumed to approach a stationary random process described by a density function denoted as $p_1(\phi_1, \vec{Y})$, the limit of $p_1(\phi_1, \vec{Y}, t)$ as time approaches infinity. In this section, a series expansion is given for this steady-state density. Also, a numerical algorithm is discussed for computing the terms in the series expansion.

In Sections 8.2 and 8.3, applications of this series-based method are illustrated. In the example discussed in Section 8.2, the method is applied to a second-order PLL containing an RC low-pass loop filter; this simple PLL is introduced in Section 7.3.3. In Section 8.3, a numerical analysis is given of the often-used, second-order PLL that contains a perfect integrator in its loop filter. In both examples, a version of the algorithm is used that generates approximations to the steady-state marginal density $p_1(\phi_1)$ (i.e., the density obtained from $p_1(\phi_1, y)$ by integrating out the nonphase variable y). For the case involving the perfect integrator loop filter, the computed $p_1(\phi_1)$ is compared with the approximation of this marginal density that is developed in Section 7.3.7 by using the conditional expectation method.

8.1.1 Expansion of the Joint Density in a Complete Orthonormal Set

Sets of complete orthonormal basis functions are needed to write the formal expansion of $p_1(\phi_1, \vec{Y})$. To match the nature of this joint density function, basis functions in one of these sets must be 2π-periodic, and the remaining $n - 1$ basis sets must contain functions that satisfy natural boundary conditions (review Section 6.2.6 on boundary conditions). In what follows, the set $\{z_k(\phi_1)\}$ (k is an index to the set) denotes a complete orthonormal basis for the vector space of 2π-periodic, square-integrable (on the interval $[0, 2\pi]$) functions. In a similar manner, for each i, $2 \le i \le n$, the set $\{h_k^{(i)}\}$ serves as a basis for functions (that satisfy natural boundary conditions) in the y_i dimension.

The necessary framework is complete to write a generalized Fourier series expansion of the joint density. In terms of the basis functions described above, the formal expansion of p_1 has the form

$$p_1(\phi_1, \vec{Y}) = \sum_{k_1}\left(\sum_{k_2} \cdots \sum_{k_n} A_{k_1 \cdots k_n}\, h_{k_2}^{(2)}(y_2) \cdots h_{k_n}^{(n)}(y_n)\right) z_{k_1}(\phi_1), \quad (8.1\text{-}1)$$

where the $A_{k_1 \cdots k_n}$ represent generalized Fourier coefficients. For the algorithm described in this section, a better representation for p_1 is

$$p_1(\phi_1, \vec{Y}) = \sum_{k_2} \cdots \sum_{k_n} h_{k_2}^{(2)}(y_2) \cdots h_{k_n}^{(n)}(y_n) p_{k_2 \cdots k_n}(\phi_1), \quad (8.1\text{-}2)$$

where the $p_{k_2 \cdots k_n}(\phi_1)$ are 2π-periodic functions that must be determined. Equations 8.1-1 and 8.1-2 represent generalized Fourier series expansions of the joint density p_1 into a complete orthonormal set of functions.

In practical applications involving Equation 8.1-2, the series must be truncated to a finite number of terms, and the unknown 2π-periodic weight functions must be computed (this task is discussed in more detail in the numerical examples given in Sections 8.2 and 8.3). As might be suspected, the computational requirements grow rapidly with system order n. However, for most second-order PLLs, practical algorithms exist for calculating the necessary functions, and density p_1 can be accurately approximated in this manner, at least over a limited range of in-loop signal-to-noise ratio ρ. Furthermore, for the second-order examples discussed in Sections 8.2 and 8.3, only modest computational resources are required to compute the 2π-periodic functions. In fact, the numerical results given in the examples were obtained by using an inexpensive personal computer.

An algorithm for calculating the 2π-periodic weight functions in Equation 8.1-2 can be discussed most conveniently by considering the case n = 2 that applies for second-order PLLs. At least conceptually, for n > 2, the computational procedure is a direct extension of the n = 2 case (the case n > 2 is discussed briefly at the end of this section). The expansion can use the weighted, orthonormal Hermite polynomials ψ_k, $0 \le k < \infty$, for the basis set in the one nonphase y-coordinate direction. In Appendix 8.1.1, these orthonormal polynomials are introduced and their important properties are given.

In terms of the weighted Hermite polynomials, the expansion for p_1 is given by

$$p_1(\phi_1, y) = \sum_{k=0}^{\infty} p_k(\phi_1) \psi_k(y; c), \qquad (8.1\text{-}3)$$

for the case n = 2. As shown in the next section, the p_k, $0 \le k < \infty$, are 2π-periodic functions that satisfy an infinite dimensional system of differential equations with 2π-periodic coefficients. From a theoretical standpoint, in the Hermite polynomials used in Equation 8.1-3, the value of the positive constant c is not critical. However, as discussed in the numerical examples covered in Sections 8.2 and 8.3, a heuristic approach is to set constant c equal to the standard deviation of y computed by using the linear model of the PLL.

8.1.2 A Coupled System of Differential Equations for the p_k

In Equation 8.1-3, the p_k, $0 \le k < \infty$, are unknown 2π-periodic functions. They satisfy an infinite dimensional system of second-order differential equations with periodic coefficients. To find this system, substitute Equation 8.1-3 into the steady-state Fokker–Planck equation for the second-order PLL under consideration (a steady-state version of Equation 7.3-15), and equate similar-ordered terms in the Hermite polynomials. For computational purposes, this infinite system must be truncated to finite order. In the remainder of this section, these steps are discussed in detail. Then, in Sections 8.2 and 8.3 they are put to practice.

For a second-order PLL model, the Fokker–Planck operator L_{FP} transforms Expansion 8.1-3 into a second expansion of weighted Hermite polynomials. As is evident from the results developed in Section 7.3.2, operator L_{FP} contains first- and second-order partial derivatives with respect to the nonphase variable y, and these derivatives operate on the ψ_k in the expansion. By using the results of Appendix 8.1.1 (specifically, see Equations 8.1.1-15 through 8.1.1-17), these partial derivatives of ψ_k can be expressed in terms of ψ_{k-2}, ψ_{k-1}, ψ_k, ψ_{k+1}, and ψ_{k+2}. Hence, the result $L_{FP}[p_1]$ can be expressed as a weighted sum of Hermite polynomials. Finally, in the expansion of p_1, the 2π-periodic p_k must be selected so that $L_{FP}[p_1] = 0$.

As outlined above, the algebraic details necessary to develop the infinite system of differential equations are straightforward. The result $L_{FP}[p_1] = 0$ can be expressed as

$$L_{FP}(p_1) = L_{FP}\left[\sum_{k=0}^{\infty} p_k(\phi_1)\psi_k(y;c)\right]$$

$$= \sum_{k=0}^{\infty} Q_k(\phi_1)\psi_k(y;c) \qquad (8.1\text{-}4)$$

$$= 0,$$

where the Q_k have the form

$$Q_k = L_k^{(-2)} p_{k-2} + L_k^{(-1)} p_{k-1} + L_k^{(0)} p_k + L_k^{(+1)} p_{k+1} + L_k^{(+2)} p_{k+2} \qquad (8.1\text{-}5)$$

(this expansion is developed in detail for the examples covered in Sections 8.2 and 8.3). For $k \geq 0$ and $-2 \leq j \leq 2$, the $L_k^{(j)}$ are linear differential operators in the ϕ_1 variable (note: $L_k^{(j)} = 0$ for $k + j < 0$), and they may contain trigonometric functions and partial derivatives (first and second order) with respect to ϕ_1.

Equation 8.1-4 implies that $Q_k = 0$ for $k \geq 0$. This leads to an infinite-dimensional system of ordinary differential equations that can be expressed in operator form as

$$k = 0: \quad L_0^{(0)} p_0 + L_0^{(+1)} p_1 + L_0^{(+2)} p_2 = 0$$

$$k = 1: \quad L_1^{(-1)} p_0 + L_1^{(0)} p_1 + L_1^{(+1)} p_2 + L_1^{(+2)} p_3 = 0$$

$$k = 2: \quad L_2^{(-2)} p_0 + L_2^{(-1)} p_1 + L_2^{(0)} p_2 + L_2^{(+1)} p_3 + L_2^{(+2)} p_4 = 0 \qquad (8.1\text{-}6)$$

$$\vdots \qquad \vdots \qquad \vdots$$

$$k: \quad L_k^{(-2)} p_{k-2} + L_k^{(-1)} p_{k-1} + L_k^{(0)} p_k + L_k^{(+1)} p_{k+1} + L_k^{(+2)} p_{k+2} = 0$$

$$\vdots \qquad \vdots \qquad \vdots$$

This linear system has 2π-periodic coefficients. Also, the system has a "band structure"; at most, each p_k is coupled to p_{k-2} through p_{k+2}.

For computational purposes, System 8.1-6 must be truncated to finite order. In applications, this system can be truncated to $k_0 + 1$ equations (in the computational algorithm, include equations numbered k, $0 \le k \le k_0$), where k_0 is selected so that the approximated density p_1 does not change significantly with increases in k_0. Fortunately, in many practical applications involving PLLs, the number of required equations is small, and the computational requirements are moderate. From the truncated system, retain only those terms involving p_k and $\dot{p}_k \equiv dp_k/d\phi_1$, $0 \le k \le k_0$. In the truncated system, each equation may be second-order (i.e., involve first- and second-order derivatives with respect to the independent variable ϕ_1); hence, the truncated system may be written as a first-order system of dimension no higher than $2k_0 + 2$.

8.1.3 Computing a 2π-Periodic Solution of the Coupled System

A 2π-periodic solution of the truncated system must be computed. Also, the scale of the solution must be determined so that the normalization Condition 7.1-5 is met. Of course, there are several ways to accomplish these tasks. One method involves expanding each p_k in an exponential Fourier series and solving for the Fourier coefficients (see Risken[36] for coverage of this approach). The numerical method discussed in the remainder of this section takes a different approach; it better utilizes the computational capabilities of modern computers and software. It employs standard numerical methods implemented by routines that are available from both commercial and public domain sources. In Sections 8.2 and 8.3, application of the numerical method is illustrated in detail.

When numerically integrating the truncated system to obtain its periodic solution, the problem is one of obtaining initial conditions. If the system is started at the proper initial conditions, then the desired periodic solution can be computed by a numerical integration over the interval $-\pi \le \phi_1 \le \pi$. Hence, the problem is reduced to one of computing initial conditions. The examples described in Sections 8.2 and 8.3 show how to compute the necessary initial conditions and periodic solutions. In general, practical problems may have known symmetries and other features that can be exploited; hence, the best approach to use when computing the initial conditions and periodic solutions may vary from problem to problem. However, some general ideas relevant to the numerical computations are given before concluding this section.

The periodic solution of the truncated system is assumed to be unique up to an arbitrary multiplicative scale factor. This scale factor should be chosen so that the normalization Condition 7.1-5 is met. Hence, when computing a periodic solution, the first initial condition $p_{0\varnothing} \equiv p_0(0)$ can be set arbitrarily (scaling is done after the periodic solution is computed). As a result, there are at most $2k_0 + 1$ unknown initial conditions

$$p_{k\varnothing} \equiv p_k(0)$$
$$\dot{p}_{k'\varnothing} \equiv \dot{p}_{k'}(0), \qquad (8.1\text{-}7)$$

$1 \le k \le k_0$, $0 \le k' \le k_0$, that must be solved for. Often, as shown in Section 8.3, symmetry and other problem-dependent properties can be used to reduce this number. A system of *algebraic equations* can be formulated that the unknown initial conditions satisfy. Finally, this algebraic system can be solved (and the desired initial conditions can be computed) by using an iterative root-finding procedure.

For $0 \le k \le k_0$, let $p_k^{(f)}$ and $\dot{p}_k^{(f)}$ denote the numerical solution that is computed in a forward direction (increasing ϕ_1) over the interval $[0, \pi]$. Of course, these solutions depend on the initial conditions; this functional dependence is denoted by including the initial conditions as arguments of the $p_k^{(f)}$ and $\dot{p}_k^{(f)}$. That is, it is possible to write

$$p_k^{(f)}\left(\phi_1; p_{0\varnothing}, \dot{p}_{0\varnothing}, p_{1\varnothing}, \dot{p}_{1\varnothing}, \cdots, p_{k_0\varnothing}, \dot{p}_{k_0\varnothing}\right)\Big|_{\phi_1=0} = p_{k\varnothing}$$
$$\dot{p}_k^{(f)}\left(\phi_1; p_{0\varnothing}, \dot{p}_{0\varnothing}, p_{1\varnothing}, \dot{p}_{1\varnothing}, \cdots, p_{k_0\varnothing}, \dot{p}_{k_0\varnothing}\right)\Big|_{\phi_1=0} = \dot{p}_{k\varnothing}. \qquad (8.1\text{-}8)$$

In a similar manner, let $p_k^{(b)}$ and $\dot{p}_k^{(b)}$ denote the numerical solution that is computed in a backward direction (decreasing ϕ_1) over the interval $[0, -\pi]$. In a manner similar to Equation 8.1-8, it is possible to write

$$p_k^{(b)}\left(\phi_1; p_{0\varnothing}, \dot{p}_{0\varnothing}, p_{1\varnothing}, \dot{p}_{1\varnothing}, \cdots, p_{k_0\varnothing}, \dot{p}_{k_0\varnothing}\right)\Big|_{\phi_1=0} = p_{k\varnothing}$$
$$\dot{p}_k^{(b)}\left(\phi_1; p_{0\varnothing}, \dot{p}_{0\varnothing}, p_{1\varnothing}, \dot{p}_{1\varnothing}, \cdots, p_{k_0\varnothing}, \dot{p}_{k_0\varnothing}\right)\Big|_{\phi_1=0} = \dot{p}_{k\varnothing}. \qquad (8.1\text{-}9)$$

Now, due to periodicity, require that the $p_k^{(f)}$ and $p_k^{(b)}$ match up at the ends of their integration intervals; similar constraints can be placed on all of the $\dot{p}_k^{(f)}$, $\dot{p}_k^{(b)}$. Since the truncated system has a periodic solution that is defined up to a multiplicative constant, the first initial condition in the list can be set to unity ($p_{0\emptyset} \equiv p_0(0) = 1$), and the remaining $2k_0 + 1$ initial conditions become the solution of the algebraic system

$$g_k\left(\dot{p}_{0\emptyset}, p_{1\emptyset}, \dot{p}_{1\emptyset}, \cdots, p_{k_0\emptyset}, \dot{p}_{k_0\emptyset}\right)$$

$$\equiv p_k^{(f)}\left(\pi; p_{0\emptyset} = 1, \dot{p}_{0\emptyset}, \cdots, p_{k_0\emptyset}, \dot{p}_{k_0\emptyset}\right)$$

$$- p_k^{(b)}\left(-\pi; p_{0\emptyset} = 1, \dot{p}_{0\emptyset}, \cdots, p_{k_0\emptyset}, \dot{p}_{k_0\emptyset}\right) = 0$$

$$f_{k'}\left(\dot{p}_{0\emptyset}, p_{1\emptyset}, \dot{p}_{1\emptyset}, \cdots, p_{k_0\emptyset}, \dot{p}_{k_0\emptyset}\right) \qquad (8.1\text{-}10)$$

$$\equiv \dot{p}_{k'}^{(f)}\left(\pi; p_{0\emptyset} = 1, \dot{p}_{0\emptyset}, \cdots, p_{k_0\emptyset}, \dot{p}_{k_0\emptyset}\right)$$

$$- \dot{p}_{k'}^{(b)}\left(-\pi; p_{0\emptyset} = 1, \dot{p}_{0\emptyset}, \cdots, p_{k_0\emptyset}, \dot{p}_{k_0\emptyset}\right) = 0,$$

where $1 \le k \le k_0$ and $0 \le k' \le k_0$.

Iterative root-finding routines can be used to solve this system of algebraic equations. For the examples discussed in Sections 8.2 and 8.3, the IMSL™ routines DNEQNF (for solving the algebraic equations) and DIVPRK (for numerical integration) were used (see IMSL™).[80] Also, the public-domain routine HYBRD1 (see MINPACK[81]) was used successfully to substitute for DNEQNF. Both of these root-finding routines use a modified Powell hybrid algorithm; they are relatively easy to use since they compute a finite-difference approximation to the Jacobian (a user-supplied Jacobian is not required). Alternate versions of these root-finding routines are available that require a user-supplied Jacobian (that can be computed by integrating an auxiliary system of differential equations); under some conditions, they have better convergence properties than their simpler-to-use counterparts that compute a finite-difference approximation to the Jacobian.

Initial estimates of the unknowns must be supplied to start the iterative root-finding algorithm. In general, selection of these initial estimates is less critical for small values of in-loop signal-to-noise ratio ρ (see the last paragraph in this section) and small values of integer k_0. If necessary, an auxiliary computation involving an exponential Fourier series expansion of the relevant p_k (such as described in Chapter 11 of Risken)[36] can be used to generate initial estimates for the iterative algorithm discussed in this section. However, the author has never had to rely on such a "starter algorithm". In most cases, small ρ and k_0 can be used to obtain convergence easily, and the computed results can be used to restart the iterative algorithm with larger values of ρ and k_0.

The algorithm outlined in this section was used to analyze several common second-order PLL models. Section 8.2 contains an application of the algorithm to the PLL model that contains a simple RC loop filter. In Section 8.3, similar results are given for the second-order PLL model that contains a loop filter based on a perfect integrator. In most cases involving second-order PLL models, these examples can be utilized as a starting point, and excellent results can be obtained with modest computational requirements.

However, the algorithm does have its limitations. A significant limitation involves the value of in-loop signal-to-noise ratio parameter ρ. In most cases, the system of differential equations that must be integrated becomes "stiff" for large values of ρ, and this makes numerical convergence (i.e., convergence of DNEQNF) more difficult to obtain. As shown by Equation 8.1-10, the algorithm computes the difference of forward and backward numerical integrations (as opposed to forcing boundary conditions on a single forward integration). In many PLL applications, this feature tends to reduce numerical convergence problems, and it increases the range of ρ over which the algorithm is practical.

8.1.4 The Case n > 2

At least in principle, the algorithm outlined above can be applied to third- and higher-order PLLs. However, complexity and computational requirements grow rapidly with PLL order n. In each nonphase coordinate direction i, $2 \leq i \leq n$, of the n-dimensional Expansion 8.1-2, a set of Hermite polynomials $\{\psi_k(\phi_1;c_i)\}$ can be used as the basis functions. This multidimensional expansion can be substituted into the steady-state Fokker–Planck equation, and terms involving like products of Hermite polynomials can be grouped. When these groups are equated to zero, an infinite system of differential equations with periodic coefficients is obtained. This system can be truncated to finite order by retaining only those terms that involve products of Hermite polynomials with indices that sum to an integer less than or equal to user-specified k_0 (i.e., $k_2 + k_3 + \ldots + k_n \leq k_0$). Finally, this truncated system must be solved for its periodic solution.

8.2 APPROXIMATING $P_1(\phi_1)$ FOR A PLL CONTAINING AN RC LOWPASS LOOP FILTER

In this section, the algorithm outlined in Section 8.1 is applied to a second-order PLL model based on an RC lowpass loop filter and a noisy reference signal. In Section 7.3.3, this simple PLL model is introduced, and its Fokker–Planck equation is developed. In the same section, this Fokker–Planck equation is shown to have a steady-state, closed-form solution for the special case $\omega_\Delta' = 0$. Unfortunately, for $\omega_\Delta' \ne 0$, the steady-state solution is not closed-formed, and it must be computed numerically. For $\omega_\Delta' = 0.25$, this calculation was performed and the results are described in this section.

The steady-state Fokker–Planck equation for the example under consideration is given by Equation 7.3-16, and it is rewritten here as

$$L_{FP}\left[p_1(\phi_1, e)\right] = 0, \qquad (8.2\text{-}1)$$

where L_{FP} denotes the Fokker–Planck operator given by

$$L_{FP} \equiv \frac{\partial}{\partial \phi_1} e - b' \frac{\partial}{\partial e}\left[(e - \omega_\Delta') + \sin\phi_1 + \sigma_e^2 \frac{\partial}{\partial e}\right]. \qquad (8.2\text{-}2)$$

The quantity σ_e^2 in Equation 8.2-2 is given by

$$\sigma_e^2 \equiv b' \frac{GN_0}{4A^2}$$

$$= b' \rho_2^{-1}, \qquad (8.2\text{-}3)$$

where $\rho_2 = 4A^2/GN_0$ is the signal-to-noise ratio in the loop parameter for the PLL containing an RC lowpass loop filter. Also, constant σ_e^2 represents the variance of e in the linear model of the PLL.

8.2.1 Transformation of the Fokker–Planck Operator

As given by Equation 8.2-2, the linear operator L_{FP} has a special structure which should be taken advantage of before computing a solution to Equation 8.2-1. It can be transformed into a new operator which will simplify the process of computing the desired numerical solution. First, note that Equation 8.2-2 can be written as

$$L_{FP} = L_1 - L_2, \qquad (8.2\text{-}4)$$

where

$$L_1 \equiv \frac{\partial}{\partial \phi_1} e - b' \frac{\partial}{\partial e}\left[-\omega_\Delta' + \sin\phi_1\right] \qquad (8.2\text{-}5)$$

$$L_2 \equiv b' \frac{\partial}{\partial e}\left[e + \sigma_e^2 \frac{\partial}{\partial e}\right]. \qquad (8.2\text{-}6)$$

Operator L_2 involves only the e variable, and it operates exclusively on the ψ_k terms in the series expansion of $p_1(\phi_1, e)$. The computational algorithm for solving Equation 8.2-1 is simplified by transforming L_2 into a new operator that has the ψ_k, $k \geq 0$ as eigenfunctions.

The desired transformation follows from the fact that

$$L_2\left[\exp\left[-e^2/2\sigma_e^2\right]\right] = 0. \qquad (8.2\text{-}7)$$

Because of this, operator L_2 can be brought into the desired simplified (Hermitian) form by a right-side operation with $\exp[-e^2/4\sigma_e^2]$ followed by a left-side operation with $\exp[+e^2/4\sigma_e^2]$. To see this, first note that the partial differential operator $\partial/\partial e$ transforms according to

$$\exp\left[e^2/4\sigma_e^2\right]\frac{\partial}{\partial e}\exp\left[-e^2/4\sigma_e^2\right] = -\frac{1}{\sigma_e}\left[\frac{1}{2}\frac{e}{\sigma_e} - \sigma_e\frac{\partial}{\partial e}\right]. \qquad (8.2\text{-}8)$$

Now, use this fact to transform L_2 into operator $\hat{L}_2$ given by

$$\begin{aligned}\hat{L}_2 &\equiv \exp\left[e^2/4\sigma_e^2\right]L_2\exp\left[-e^2/4\sigma_e^2\right] \\ &= -\frac{b'}{\sigma_e}\left[\frac{1}{2}\frac{e}{\sigma_e} - \sigma_e\frac{\partial}{\partial e}\right]\left[e - \sigma_e\left(\frac{1}{2}\frac{e}{\sigma_e} - \sigma_e\frac{\partial}{\partial e}\right)\right] \\ &= -b'\left[\frac{1}{2}\frac{e}{\sigma_e} - \sigma_e\frac{\partial}{\partial e}\right]\left[\frac{1}{2}\frac{e}{\sigma_e} + \sigma_e\frac{\partial}{\partial e}\right]. \end{aligned} \qquad (8.2\text{-}9)$$

Note that this last result can be written as

$$\hat{L}_2 = -b'\, b^+ b, \qquad (8.2\text{-}10)$$

where

$$b^+ \equiv \frac{1}{2}\frac{e}{\sigma_e} - \sigma_e\frac{\partial}{\partial e} \qquad (8.2\text{-}11)$$

$$b \equiv \frac{1}{2}\frac{e}{\sigma_e} + \sigma_e\frac{\partial}{\partial e} \qquad (8.2\text{-}12)$$

are linear operators. Operator b^+ is the formal *adjoint* of b. In Equations 8.1.1-11 and 8.1.1-12 of Appendix 8.1.1, the constant c_2 is replaced by σ_e to obtain Equations 8.2-11 and 8.2-12. From inspection of Equations 8.1.1-17 and 8.2-10, it is easy to see that transformed operator $\hat{L}_2$ has the weighted Hermite polynomials ψ_k, $k \geq 0$, as eigenfunctions, a very useful observation. These weighted polynomials are given by Equation 8.1.1-6 with c_2 replaced with σ_e.

The transformation must be applied to operator L_1 as well. Use Equation 8.2-8 in Equation 8.2-5 to obtain

$$\hat{L}_1 \equiv \exp[e^2/4\sigma_e^2] L_1 \exp[-e^2/4\sigma_e^2]$$
$$= \frac{\partial}{\partial \phi_1} e + \frac{b'}{\sigma_e}\left[\frac{1}{2}\frac{e}{\sigma_e} - \sigma_e \frac{\partial}{\partial e}\right]\left[-\omega_\Delta' + \sin\phi_1\right]. \qquad (8.2\text{-}13)$$

From Equations 8.1.1-13, 8.2-11, and 8.2-12, note that variable e can be expressed as $e = \sigma_e(b^+ + b)$. Use this representation for e, and use Equation 8.2-11 to write Equation 8.2-13 as

$$\hat{L}_1 = \sigma_e \frac{\partial}{\partial \phi_1}(b^+ + b) + \frac{b'}{\sigma_e}\left[-\omega_\Delta' + \sin\phi_1\right]b^+. \qquad (8.2\text{-}14)$$

The transformed Fokker–Planck operator follows from these developments. Simply substitute Equations 8.2-10 and 8.2-14 into a transformed version of Equation 8.2-4 to obtain

$$\hat{L}_{FP} = \exp[e^2/4\sigma_e^2] L_{FP} \exp[-e^2/4\sigma_e^2]$$
$$= \hat{L}_1 - \hat{L}_2 \qquad (8.2\text{-}15)$$
$$= \sigma_e \frac{\partial}{\partial \phi_1}(b^+ + b) + \frac{b'}{\sigma_e}\left[-\omega_\Delta' + \sin\phi_1\right]b^+ + b'\ b^+b.$$

For operating on a sum of weighted Hermite polynomials, the transformed operator $\hat{L}_{FP}$ is preferred over L_{FP}. As given by Equation 8.2-15, the operator $\hat{L}_{FP}$ is expressed in terms of b and b^+; when it is applied to a sum of weighted Hermite polynomials, the simple formulas given by Equations 8.1.1-15 through 8.1.1-17 can be used.

8.2.2 Form of Series Expansion

The form of Equation 8.2-15 suggests that solution $p_1(\phi_1, e)$ of Equation 8.2-1 should be expanded as

$$p_1(\phi_1, e) = \exp[-e^2/4\sigma_e^2] \sum_{k=0}^{\infty} p_k(\phi_1) \psi_k(e; \sigma_e). \qquad (8.2\text{-}16)$$

There are several reasons for choosing Equation 8.2-16 as the expansion of $p_1(\phi_1, e)$. First, from inspection of the first line of Equation 8.2-15, it is seen easily that the unknown coefficients p_k should be chosen to satisfy

$$\hat{L}_{FP}\left[\sum_{k=0}^{\infty} p_k(\phi_1) \psi_k(e; \sigma_e)\right] = 0. \qquad (8.2\text{-}17)$$

That is, $\hat{L}_{FP}$ operates directly on the sum of weighted polynomials. A second convenience that results from using the Expansion 8.2-16 is that the desired marginal density $p_1(\phi_1)$ is proportional to the first 2π-periodic coefficient $p_0(\phi_1)$. This result follows from using Equations 8.1.1-6 and 8.1.1-7 to obtain

$$p_1(\phi_1) = \int_{-\infty}^{\infty} p_1(\phi_1, e)\, de$$

$$= \sum_{k=0}^{\infty} p_k(\phi_1) \int_{-\infty}^{\infty} \exp[-e^2/4\sigma_e^2] \psi_k(e; \sigma_e)\, de \qquad (8.2\text{-}18)$$

$$= [2\pi \sigma_e^2]^{1/4} p_0(\phi_1).$$

This fact simplifies computation of the marginal density $p_1(\phi_1)$.

The 2π-periodic coefficients p_k satisfy a coupled system of linear differential equations with periodic coefficients. To obtain this coupled system, use Equations 8.1.1-15 through 8.1.1-17 to rewrite Equation 8.2-17 as

$$\hat{L}_{FP}\left[\sum_{k=0}^{\infty} p_k \psi_k\right] = \sum_{k=0}^{\infty} Q_k \psi_k \qquad (8.2\text{-}19)$$

$$= 0,$$

where

$$Q_k \equiv \sqrt{k}\left[\frac{b'}{\sigma_e}\left(-\omega_\Delta' + \sin\phi_1\right) + \sigma_e \frac{d}{d\phi_1}\right]p_{k-1}$$
$$+ b'k p_k + \sigma_e \sqrt{k+1}\frac{d}{d\phi_1}p_{k+1}.$$

(8.2-20)

In Equation 8.2-20, note that ordinary derivatives have been used instead of partial derivatives since the p_k are functions of ϕ_1 alone. In order to satisfy Equation 8.2-19, each of the terms Q_k, $k \geq 0$ must vanish. This requirement generates a system of differential equations with nearest-neighbor coupling between the equations. That is, Equation 8.2-20 can be used to write the implicit and infinite system

$$k=0: \quad \sigma_e \frac{d}{d\phi_1}p_1 = 0 \tag{8.2-21}$$

$$k=1: \quad \sqrt{1}\left[\frac{b'}{\sigma_e}\left(-\omega_\Delta' + \sin\phi_1\right) + \sigma_e \frac{d}{d\phi_1}\right]p_0 + b'\, p_1 + \sigma_e\sqrt{2}\frac{d}{d\phi_1}p_2 = 0$$

$$k=2: \quad \sqrt{2}\left[\frac{b'}{\sigma_e}\left(-\omega_\Delta' + \sin\phi_1\right) + \sigma_e \frac{d}{d\phi_1}\right]p_1 + 2b'\, p_2 + \sigma_e\sqrt{3}\frac{d}{d\phi_1}p_3 = 0$$

$$k=3: \quad \sqrt{3}\left[\frac{b'}{\sigma_e}\left(-\omega_\Delta' + \sin\phi_1\right) + \sigma_e \frac{d}{d\phi_1}\right]p_2 + 3b'\, p_3 + \sigma_e\sqrt{4}\frac{d}{d\phi_1}p_4 = 0$$

$$\vdots \qquad \vdots \qquad \vdots \qquad \vdots$$

8.2.3 Truncation of the Infinite Series

A finite number of the 2π-periodic coefficients p_k can be computed numerically. In practice, System 8.2-21 must be truncated to $k_0 + 1$ equations (include equations numbered k, $0 \leq k \leq k_0$ in the numerical algorithm). From the first $k_0 + 1$ equations, retain only those terms involving p_k, $0 \leq k \leq k_0$. Since a scalar multiple of p_0 is used to approximate the desired marginal density p_1 (see Equation 8.2-18), integer k_0 should be selected large enough so that the computed p_0 changes little when k_0 is incremented. Fortunately, the algorithm converges rapidly for the RC loop filter case considered here, and only a few of the periodic coefficients need to be calculated.

NUMERICAL METHODS/NOISE ANALYSIS IN NONLINEAR PLL MODEL 303

This procedure produces a finite system that involves $dp_k/d\phi_1$, $0 \le k \le k_0$, in an implicit manner. If $k_0 \ge 1$ is odd, then this finite and implicit system can be used to write an explicit, "normal" system involving $dp_k/d\phi_1$, $0 \le k \le k_0$ (i.e., these derivatives can be expressed explicitly in terms of the dependent variables). If $k_0 \ge 2$ is even, then the results of the truncation procedure can be used to write $dp_k/d\phi_1$, $0 \le k \le k_0 - 1$, explicitly in terms of the dependent variables.

Two examples serve to clarify the above-mentioned truncation procedure. First, suppose that $k_0 = 1$ so that the first two equations of Equation 8.2-21 are considered. These two equations can be used to write

$$\frac{d}{d\phi_1} p_1 = 0$$

$$\frac{d}{d\phi_1} p_0 = -\frac{b'}{\sigma_e^2}\left(-\omega_\Delta' + \sin\phi_1\right)p_0 - \frac{b'}{\sigma_e} p_1, \quad (8.2\text{-}22)$$

and these equations can be solved for periodic p_0 and p_1. Next, consider the case $k_0 = 2$. The first three equations of Equation 8.2-21 can be truncated to

$$\sigma_e \frac{d}{d\phi_1} p_1 = 0$$

$$\left[\frac{b'}{\sigma_e}\left(-\omega_\Delta' + \sin\phi_1\right) + \sigma_e \frac{d}{d\phi_1}\right] p_0 + b' p_1 + \sigma_e \sqrt{2} \frac{d}{d\phi_1} p_2 = 0 \quad (8.2\text{-}23)$$

$$\sqrt{2}\left[\frac{b'}{\sigma_e}\left(-\omega_\Delta' + \sin\phi_1\right) + \sigma_e \frac{d}{d\phi_1}\right] p_1 + 2b' p_2 = 0.$$

Now, these three equations can be used to write the explicit equations

$$\frac{d}{d\phi_1} p_1 = 0$$

$$\frac{d}{d\phi_1} p_0 = -\frac{b'}{\sigma_e^2}\left(-\omega_\Delta' + \sin\phi_1\right)p_0 - \frac{b'}{\sigma_e} p_1 + \sigma_e^{-1}\left(\cos\phi_1\right)p_1, \quad (8.2\text{-}24)$$

and these can be solved for periodic p_0 and p_1.

The process of truncating Equation 8.2-21 in the manner described above produces a system of differential equations that has a 2π-periodic solution defined up to a scalar multiplicative constant. According to Equation 8.2-18, the coefficient $p_0(\phi_1)$ is proportional to the desired marginal density $p_1(\phi_1)$. Hence, a reasonable approach is to scale the computed p_0 so that it satisfies

$$\int_{-\pi}^{\pi} p_0(\phi_1) d\phi_1 = 1, \qquad (8.2\text{-}25)$$

and then use this scaled p_0 as an approximation of the desired marginal density $p_1(\phi_1)$. This procedure is used to produce the results discussed in the remainder of this section.

8.2.4 Computing the Periodic Solution and Approximating $p_1(\phi_1)$

A 2π-periodic solution must be computed for the truncated system. There are several practical numerical methods for accomplishing this. The periodic solution can be expanded into a trigonometric or exponential Fourier series, and an algorithm can be developed for computing the Fourier coefficients. Alternatively, classical numerical methods for initial value problems can be used to integrate the truncated system of differential equations over the interval $-\pi \le \phi_1 \le \pi$. For this method to work, the correct initial conditions must be found (i.e., solved for) to ensure that a periodic solution results from the integration. This second approach makes more extensive use of the computer than does the Fourier series method, and it is the procedure utilized here.

Regardless of the value of integer k_0, it is apparent from Equation 8.2-21 that coefficient p_1 is a constant. It is easily shown that $p_1 = 0$ if and only if normalized detuning $\omega_\Delta' = 0$ (see Section 11.5 of Risken).[36] In Section 7.3.3, a closed-form solution is given for Equation 8.2-1 with zero detuning. Hence, it is assumed here that $\omega_\Delta' \ne 0$, and $p_1 \equiv 1$ is selected for convenience. Once computed, the coefficients must be scaled so that Constraint 8.2-25 is satisfied. Under these conditions, the computed coefficient p_0 approximates the desired marginal density $p_1(\phi_1)$.

Consider the case where k_0 is odd, and the differential equations in the variables p_k, $0 \le k \le k_0$ must be integrated (a similar case occurs for even k_0). Since $p_1 \equiv 1$, let $p_{0\varnothing} \equiv p_0(0)$, $p_{2\varnothing} \equiv p_2(0)$, ..., $p_{k_0\varnothing} \equiv p_{k_0}(0)$ denote the k_0 unknown initial conditions that must be solved for. As discussed in Section 8.1, for $k = 0, 2, 3, \ldots, k_0$, let $p_k^{(f)}(\phi_1; p_{0\varnothing}, p_{2\varnothing}, \ldots, p_{k_0\varnothing})$ denote the results of a forward integration over the phase interval $[0, \pi]$. In a similar fashion, let $p_k^{(b)}(\phi_1; p_{0\varnothing}, p_{2\varnothing}, \ldots, p_{k_0\varnothing})$ denote the backward integration over the interval $[0, -\pi]$. In order to obtain a periodic solution, the k_0 unknown initial conditions can be obtained by solving the k_0 algebraic equations

$$g_k(p_{0\varnothing}, p_{2\varnothing}, p_{3\varnothing}, \ldots, p_{k_0\varnothing})$$
$$\equiv p_k^{(f)}(\pi; p_{0\varnothing}, p_{2\varnothing}, p_{3\varnothing}, \ldots, p_{k_0\varnothing}) \qquad (8.2\text{-}26)$$
$$- p_k^{(b)}(-\pi; p_{0\varnothing}, p_{2\varnothing}, p_{3\varnothing}, \ldots, p_{k_0\varnothing}),$$

$k = 0, 2, 3, \ldots, k_0$.

The numerical procedure outlined above was implemented by using standard routines running on an inexpensive personal computer. As discussed in Section 8.1, the routine DNEQNF (or equivalent) can be used to solve the system of algebraic equations generated by Equation 8.2-26, and DIVPRK (or equivalent) can be used to accomplish the required numerical integration. For small values of ρ_2 and k_0 (recall that $\rho_2 = 4A^2/GN_0$ is the in-loop SNR parameter for the RC loop filter case), convergence of DNEQNF is obtained easily; almost any reasonable estimate can be used for the unknown initial conditions. A succession of computer "runs" can be performed, and the output from a successful "run" can be used to form the estimates for the next trial with a larger value of ρ_2 and/or k_0.

The loop parameters $b' = 2$ and $\omega_\Delta' = 0.25$ were used in all of the numerical analyses described in this section. In the linear model, this value of b' yields a very desirable damping factor of $1/\sqrt{2}$. The value $\omega_\Delta' = 0.25$ ensures that the steady-state density is centered around a value of phase error $\phi_0 \approx 0.25$ radians. Let SNR denote the value of in-loop signal-to-noise ratio ρ_2 in decibels (SNR $\equiv 10\text{Log}\rho_2$); in terms of this quantity, Equation 8.2-3 can be used to write

$$\sigma_e^2 = \frac{b'}{10^{\text{SNR}/10}}. \qquad (8.2\text{-}27)$$

Results are provided below for values of SNR equal to 3 and 6 dB.

Figure 8.1 depicts numerical results for a value of in-loop SNR equal to 3 dB. The results shown correspond to the cases $k_0 = 1$ through $k_0 = 3$. Note the fast convergence of the algorithm; very little difference could be detected between the p_0 computed for the cases $k_0 = 2$ and $k_0 = 3$. As outlined above, all of the plots are normalized so that they satisfy Equation 8.2-25, and each plot represents an approximation of the marginal density $p_1(\phi_1)$.

Figure 8.2 portrays results for an SNR of 6 dB. Again, results are given for values of $k_0 = 1$ to $k_0 = 3$. Compared to the 3-dB case discussed previously, the algorithm converges faster for an SNR of 6 dB. The computed p_0 changed very little when $k_0 = 1$ ($k_0 = (2)$ was increased to $k_0 = 2$ ($k_0 = 3$). As expected, the 6-dB case produces a phase error density that is more sharply peaked than the 3-dB case.

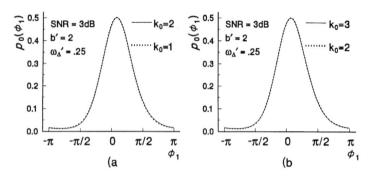

FIGURE 8.1
Approximations to marginal density $p_1(\phi_1)$ for the RC loop filter case with SNR = 3 dB.

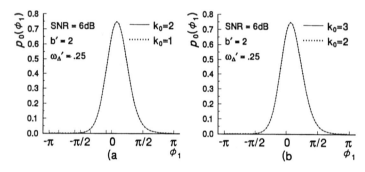

FIGURE 8.2
Approximations to marginal density $p_1(\phi_1)$ for the RC loop filter case with SNR = 6 dB.

8.2.5 Practical Limitations of the Algorithm

As can be seen from the previous example, the algorithm for expanding a steady-state density into a complete orthonormal set can be practical, and it can yield good results with modest computational requirements. However, it does have practical limitations. First, the computational complexity of the algorithm grows rapidly with PLL order, but it should be practical for most second-order PLLs. Also, the truncated system of differential equations has a periodic solution which becomes more difficult to compute with increases in SNR and k_0. With increasing values of SNR, the system of periodic differential equations can have some multipliers which grow quickly, and this phenomenon can cause computational problems (see Yakubovich and Starzhinskii[45] for an extensive discussion of differential equations with periodic coefficients). In most cases, the truncated periodic system becomes "stiff" for large values of SNR; this system attribute makes difficult the task of computing the required periodic solution.

8.3 APPROXIMATING $p_1(\phi_1)$ FOR A PLL CONTAINING A PERFECT INTEGRATOR LOOP FILTER

In the previously considered example, the process of expanding $p_1(\phi_1,e)$ into a complete orthonormal series is made relatively easy by the simple nature of the PLL containing an RC lowpass loop filter. A similar analysis is practical for the PLL containing a perfect integrator in its loop filter; this claim is supported by the results outlined in this section. For this example, symmetry plays an important role. Note that loop detuning ω_Δ' is not important since it does not appear in the Fokker–Planck equation that describes the PLL considered in this section. In the presentation of this example, some details are omitted since they are similar to those discussed in the previous section.

The Fokker–Planck equation for the PLL containing a perfect integrator in its loop filter (i.e., $F(s) = 1 + \alpha/s$) can be obtained easily from the results provided in Section 7.3. To write this equation, use the parameters $a_0 = 1$, $a_1' = \alpha'$, and $b' = 0$ in Equations 7.3-12 through 7.3-15 to obtain

$$L_{FP}\, p_1(\phi_1, e) = 0, \qquad (8.3\text{-}1)$$

where

$$L_{FP} \equiv -\frac{\partial}{\partial \phi_1}\left[-e + \sin\phi_1 + \rho_2^{-1}\left(\frac{\partial}{\partial \phi_1} + 2\alpha'\frac{\partial}{\partial e}\right)\right]$$

$$-\frac{\partial}{\partial e}\left[\alpha'\sin\phi_1 + \rho_2^{-1}(\alpha')^2\frac{\partial}{\partial e}\right]. \qquad (8.3\text{-}2)$$

In the remainder of this section, the steady-state joint density that satisfies Equation 8.3-1 is expanded according to Equation 8.2-16. In this expansion, the quantity σ_e^2 is the variance of variable e in the linear model of the PLL under consideration. By using the linear model outlined in Chapter 3, it is readily shown that

$$\sigma_e^2 = (\alpha')^2 \frac{GN_0}{4A^2}$$
$$= (\alpha')^2\, \rho_2^{-1}, \qquad (8.3\text{-}3)$$

where $\rho_2 = 4A^2/GN_0$.

8.3.1 Transformation of the Fokker–Planck Operator

Unlike Equation 8.2-2, the Fokker–Planck Operator 8.3-2 cannot be simplified by a transformation of the form described in Section 8.2.1. That is, after such a transformation, it cannot be subdivided into a part (such as Equation 8.2-10) that has as eigenfunctions the exponentially weighted Hermite polynomials ψ_k, $k \geq 0$. Nevertheless, in this section, the Transformation 8.2-15 is applied to Equation 8.3-2. By transforming the Fokker–Planck operator, an expansion of the Form 8.2-16 can be used for the joint density, and the advantage implied by Equation 8.2-18 can be attained. That is, in the expansion of the joint density, the first computed coefficient $p_0(\phi_1)$ is proportional to the desired marginal density $p_1(\phi_1)$.

The Fokker–Planck Operator 8.3-2 can be transformed by using the techniques detailed in Section 8.2. This transformation produces the operator

$$\hat{L}_{FP} \equiv \exp\left[e^2/4\sigma_e^2\right] L_{FP} \exp\left[-e^2/4\sigma_e^2\right]$$

$$= -\frac{\partial}{\partial \phi_1}\left[-\rho_2^{-\frac{1}{2}} \alpha'(b+b^+) + \sin\phi_1 + \rho_2^{-1}\frac{\partial}{\partial \phi_1} - 2\rho_2^{-\frac{1}{2}} b^+\right] \quad (8.3\text{-}4)$$

$$+ b^+\left[\rho_2^{\frac{1}{2}} \sin\phi_1 - b^+\right],$$

where operators b^+ and b are given by Equations 8.2-11 and 8.2-12, respectively, with σ_e^2 given by Equation 8.3-3.

8.3.2 Development of the Coupled System of Differential Equations

As shown by Equation 8.2-17, $\hat{L}_{FP}$ operates on a function that is represented as an infinite-dimensional series. An infinite coupled system of differential equations results if $\hat{L}_{FP}$ is allowed to operate directly on the terms in the series. This system is developed in this section.

As given by Equation 8.2-17, the transformed Operator 8.3-4 must satisfy

$$\hat{L}_{FP}\left[\sum_{k=0}^{\infty} p_k \psi_k\right] = \sum_{k=0}^{\infty} Q_k \psi_k \quad (8.3\text{-}5)$$

$$= 0,$$

where

NUMERICAL METHODS/NOISE ANALYSIS IN NONLINEAR PLL MODEL

$$Q_k \equiv \frac{d}{d\phi_1}\left[\rho_2^{-\frac{1}{2}}\alpha'\left(\sqrt{k+1}\,p_{k+1} + \sqrt{k}\,p_{k-1}\right)\right.$$

$$\left. -\left(\sin\phi_1 + \rho_2^{-1}\frac{d}{d\phi_1}\right)p_k + 2\sqrt{k}\rho_2^{-\frac{1}{2}}p_{k-1}\right] \quad (8.3\text{-}6)$$

$$+\sqrt{k}\rho_2^{\frac{1}{2}}\sin\phi_1\,p_{k-1} - \sqrt{k(k-1)}\,p_{k-2}$$

For each $k \geq 0$, set Q_k to zero, define function h_k as the bracketed term in Equation 8.3-6, and obtain the infinite-dimensional, first-order system

$$\frac{d}{d\phi_1}p_k + \rho_2\sin\phi_1\,p_k - \sqrt{k}\rho_2^{\frac{1}{2}}(2+\alpha')p_{k-1}$$

$$-\sqrt{k+1}\,\rho_2^{\frac{1}{2}}\alpha'\,p_{k+1} = -\rho_2 h_k \quad (8.3\text{-}7)$$

$$\frac{d}{d\phi_1}h_k + \sqrt{k}\rho_2^{\frac{1}{2}}\sin\phi_1\,p_{k-1} - \sqrt{k(k-1)}\,p_{k-2} = 0$$

The new dependent variable h_k is introduced so that the system of second-order equations can be written as a system of first-order equations. For example, the first few equations in this infinite-dimensional, first-order system can be written as

$k = 0$: $\quad \dfrac{d}{d\phi_1}p_0 = -\rho_2 \sin\phi_1\,p_0 + \rho_2^{\frac{1}{2}}\alpha'p_1 - \rho_2 h_0$

$\quad\quad\quad \dfrac{d}{d\phi_1}h_0 = 0$

$k = 1$: $\quad \dfrac{d}{d\phi_1}p_1 = -\rho_2 \sin\phi_1\,p_1 + \rho_2^{\frac{1}{2}}(2+\alpha')p_0 + \sqrt{2}\rho_2^{\frac{1}{2}}\alpha'p_2 - \rho_2 h_1$

$\quad\quad\quad \dfrac{d}{d\phi_1}h_1 = -\rho_2^{\frac{1}{2}}\sin\phi_1\,p_0 \quad\quad (8.3\text{-}8)$

$k = 2$: $\quad \dfrac{d}{d\phi_1}p_2 = -\rho_2 \sin\phi_1\,p_2 + \sqrt{2}\rho_2^{\frac{1}{2}}(2+\alpha')p_1 + \sqrt{3}\rho_2^{\frac{1}{2}}\alpha'p_3 - \rho_2 h_2$

$\quad\quad\quad \dfrac{d}{d\phi_1}h_2 = -\sqrt{2}\rho_2^{\frac{1}{2}}\sin\phi_1\,p_1 + \sqrt{2}p_0$

$\quad\vdots \quad\quad\quad\quad\quad \vdots \quad\quad\quad\quad\quad \vdots \quad\quad\quad\quad\quad \vdots$

As discussed in Section 8.1, for computational purposes, this system must be truncated to the first $2(k_0 + 1)$ equations involving the dependent variables p_k and h_k, $0 \le k \le k_0$ (p_{k+1} is dropped from the last equation). This produces a first-order system of dimension $2k_0 + 2$. Integer k_0 must be selected large enough so that further increases in its value produce insignificant change in the computed p_0, an approximation (after scaling) of the marginal density p_1.

8.3.3 Computing an Approximation to the Marginal Density $p_1(\phi_1)$

Symmetry can be used to advantage when computing a periodic solution of the truncated system. From inspection of Equation 8.3-8, note that the functions p_k, k an even integer, and the functions h_k, k an odd integer, are even functions of ϕ_1. Likewise, the functions p_k, k an odd integer, and h_k, k an even integer are odd functions of ϕ_1. Hence, the numerical integration need only be performed over the interval $[0, \pi]$. Also, in computing a periodic solution, symmetry and solution continuity implies that the odd 2π-periodic functions must vanish at $\phi_1 = 0$ and $\phi_1 = \pi$; this leads to the initial conditions

$$p_{k\emptyset} \equiv p_k(0) = 0, \qquad \text{k odd}$$
$$h_{k\emptyset} \equiv h_k(0) = 0, \qquad \text{k even} \tag{8.3-9}$$

and the boundary conditions

$$p_k(\pi) = 0, \qquad \text{k odd}$$
$$h_k(\pi) = 0, \qquad \text{k even.} \tag{8.3-10}$$

Finally, the facts $dh_0/d\phi_1 = 0$ (see Equation 8.3-8) and $h_0(0) = 0$ imply that $h_0(\phi_1) \equiv 0$.

Once all of the initial conditions are known, the desired periodic solution can be computed by a simple numerical integration over the interval $[0, \pi]$. Now, all that remains is the determination of the initial conditions $p_{k\emptyset}$, k an even integer, and $h_{k\emptyset}$, k an odd integer. Since Equation 8.3-8 defines the periodic solution up to a multiplicative scale factor, the initial condition $p_{0\emptyset} \equiv p_0(0)$ can be set arbitrarily (set $p_{0\emptyset} \equiv 1$), and the computed $p_0(\phi_1)$ can be scaled to a proper density function.

The remaining k_0 unknown initial conditions are denoted as $p_{2\emptyset}, p_{4\emptyset}, \dots, p_{k'\emptyset}, h_{1\emptyset}, h_{3\emptyset}, \dots, h_{k''\emptyset}$, where k' and k'' are the largest even and odd, respectively, integers less than or equal to k_0. These unknowns can be computed by solving the k_0 algebraic equations

$$g_k(p_{2\varnothing}, p_{4\varnothing}, \ldots, p_{k'\varnothing}; h_{1\varnothing}, h_{3\varnothing}, \ldots, h_{k''\varnothing})$$

$$= p_k^{(f)}(\pi; p_{2\varnothing}, \ldots, p_{k'\varnothing}; h_{1\varnothing}, \ldots, h_{k''\varnothing}) = 0, \quad k \text{ odd} \quad (8.3\text{-}11)$$

$$= h_k^{(f)}(\pi; p_{2\varnothing}, \ldots, p_{k'\varnothing}; h_{1\varnothing}, \ldots, h_{k''\varnothing}) = 0, \quad k \text{ even},$$

$1 \le k \le k_0$, that follow from the boundary Conditions 8.3-10. On the right-hand side of Equation 8.3-11, the values $p_k^{(f)}$ and $h_k^{(f)}$ are obtained from a forward numerical integration of Equation 8.3-8 over the phase interval $[0, \pi]$.

For several values of k_0, the routine DNEQNF was used to solve Equation 8.3-11, and DIVPRK was used to perform the numerical integration (see Section 8.1 for a discussion of these routines). These computations were completed for $\alpha' = 0.5$ and values of in-loop SNR of 3 and 6 dB. As shown by Equation 3.2-4, this value for α' produces a damping factor of $1/\sqrt{2}$, a highly desirable value. Also, as given by Equation 7.3-71, the quantity ρ represents the value of in-loop SNR for the perfect integrator loop filter case. In terms of in-loop SNR in decibels (SNR $\equiv 10\text{Log}\rho$), the quantity ρ_2 in Equation 8.3-8 is given by

$$\rho_2 = \rho(1+\alpha')$$
$$= 10^{\text{SNR}/10}(1+\alpha'). \quad (8.3\text{-}12)$$

Figure 8.3 depicts results computed for an in-loop SNR value of 3 dB. The results shown correspond to the cases $k_0 = 1$ through $k_0 = 3$. The algorithm converges quickly; little difference could be detected between p_0 computed for the cases $k_0 = 2$ and $k_0 = 3$ (although not shown, very little difference in p_0 was observed for the cases $k_0 = 3$ and $k_0 = 4$). All of the plots are normalized so that they satisfy Equation 8.2-25 and approximate the marginal density $p_1(\phi_1)$.

Figure 8.4 depicts a comparison between two approximations of $p_1(\phi_1)$ for an in-loop SNR of 3 dB. The solid-line plot represents results obtained by using $k_0 = 3$ in the algorithm described in this section. The dashed-line plot represents Equation 7.3-70, an approximation based on the conditional expectation method. The poor agreement between the plots illustrates the main limitation of Equation 7.3-70 and the conditional expectation method (as it is implemented in Section 7.3). Recall that the development of Equation 7.3-70 uses the linear PLL theory to approximate the conditional expectation, and this approximation is not accurate for the low SNR case. From a practical standpoint, a value of in-loop SNR of 3 dB is low; apparently, it is too low for Equation 7.3-70 to yield accurate results.

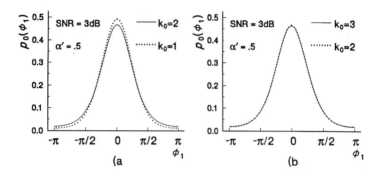

FIGURE 8.3
Approximations to the marginal density $p_1(\phi_1)$ for the perfect integrator loop filter case with SNR = 3 dB.

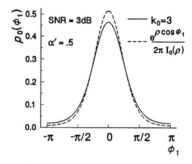

FIGURE 8.4
Comparison of series expansion and conditional expectation methods for SNR = 3 dB.

A value of 6 dB was used for the in-loop SNR parameter when the results depicted by Figure 8.5 were computed. These plots show results for the cases $k_0 = 1$ through $k_0 = 3$. As can be seen by comparing Figures 8.3 and 8.5, algorithm convergence is faster for the SNR = 6 dB case (on Figure 8.5, there is less difference between the plots than there is on Figure 8.3). As expected, higher values of SNR produce results which are more sharply peaked about the mean value of the phase error density. For a value of in-loop SNR of 6 dB, Figure 8.6 depicts a comparison between the algorithm developed in this section and the conditional expectation method described in Section 7.3. The solid-line plot represents results obtained by using $k_0 = 3$ in the algorithm described here. Equation 7.3-70 was used to construct the dashed-line plot depicted on Figure 8.6. This time, the two algorithms produce results which are in close agreement. Apparently, the conditional expectation method works well for an in-loop SNR of 6 dB or more.

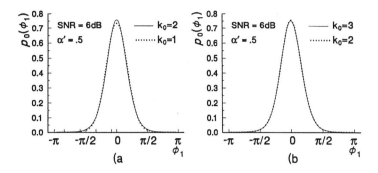

FIGURE 8.5
Approximations to the marginal density $p_1(\phi_1)$ for the perfect integrator loop filter case with SNR = 6 dB.

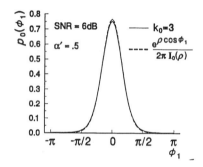

FIGURE 8.6
Comparison of series expansion and conditional expectation methods for SNR = 6 dB.

8.4 MODELING CYCLE SLIPS AS STATE TRANSITIONS OF A MARKOV JUMP PROCESS

A quantitative analysis of the cycle slip phenomenon is required in some PLL applications. As shown in Section 7.2, such an analysis can be given for the first-order PLL with a constant-frequency reference embedded in wideband Gaussian noise. For this simple case, a closed-form result can be given for the rate at which the loop slips cycles. However, as shown in Section 7.3, in second- and higher-order PLLs, a general quantitative analysis of the cycle slip phenomenon is a very daunting task, and closed-form formulas for cycle slip rates cannot be determined. In fact, in commonly used second- and higher-order PLLs, the numerical determination of cycle slip rates is no simple matter.

Because of the implied difficulties, it is natural to restructure the cycle slip problem into a simpler one that can be solved for some meaningful results. Fortunately, under conditions that are valid for many practical PLLs, the problem can be modeled in terms of a Markov jump process with a finite number of states, and some transition rates can be obtained with a reasonable amount of effort when the model is applied to many common first- and second-order PLLs. The numerical method for obtaining these results is the subject of this section, and it is applied to numerical examples in Sections 8.5 and 8.6. The goal of this and the next two sections is to provide enough information so that the reader can apply the method to simple PLLs.

8.4.1 Modeling the Cycle Slip Problem as a Finite State Markov Jump Process

In modeling the cycle slip problem, the first requirement to be imposed concerns the phase error model: the m-attractor model must be used in this section. This model is introduced in Section 7.1.3. To obtain information on the cycle slip rates, user-specified integer m must be equal to or larger than two. Generally speaking, as m increases, more information can be obtained from an analysis of the model; the reason for this is simple. The m-attractor phase error model utilizes m stable states, and phase error is interpreted in a modulo-$2\pi m$ fashion. By using the model, it is possible to distinguish between state transitions that skip up to $m-1$ domains. However, a transition that skipped m or more domains is interpreted incorrectly as a transition that skipped fewer than m domains.

In the development of the model used in this section, a second requirement is imposed on the state vector that describes the PLL. The PLL state is modeled as a vector-valued random process that approaches a steady-state stationary limit as time approaches infinity. In the steady state, the PLL state vector is a random process described by a unique, time-invariant probability density $p_m(\phi_m, \vec{Y})$. In many applications this will be the case if the PLL state is attracted by only stable equilibrium points in the state space.

A third modeling assumption is required that concerns the SNR within the loop. This noise level must be "moderate" as described in Section 7.1.4. For relatively large periods of time, the state vector must remain in a domain of attraction surrounding a stable state. Now, each of the m stable states is surrounded by a similar domain of attraction, and the domains R_i, $1 \leq i \leq m$, are pair-wise disjoint (they are widely separated in most applications). Due to noise, the PLL state can transition between different domains. However, for any given domain, the average occupancy time is large relative to the timescale associated with intradomain state movement. Also, the average amount of time the state spends outside of all domains is insignificant compared to the average amount of time the state is in a domain.

In essence, two timescales can be identified that are relevant to the model. The *fast timescale* is associated with state motion within a domain, and it is on the order of $1/B_L$ seconds. In practical PLL applications, the average domain occupancy time is substantially larger than $1/B_L$, so that transitions between distinct domains occur on the *slow timescale*. Now, transitions between distinct domains correspond to cycle slips in the PLL. Hence, under moderate noise conditions, cycle slips occur on the slow timescale, and they are rare events.

For the PLL state, there is an important consequence of a relatively large average domain occupancy time. Once in a domain, the state tends to "lose its memory" before it makes a transition out of the domain. That is, the sequence of domains occupied by the state *prior* to its arrival in any domain R_k is not relevant in determining the sequence of domains to be occupied once the state leaves R_k. Any probabilistic description of future domain occupancy, when conditioned on present and past domain occupancy, does not depend on past domain occupancy. Essentially, domain occupancy is *modeled* as a discrete-state, continuous-time random process that obeys the Markov property.

The dual timescale and "loss of memory" features are exploited here to simplify further the m-attractor phase error model. Basically, they allow the m-attractor model to be "coarse grained" into a simpler model; only some major features of the original model are retained. First, coarse graining of a temporal nature results by ignoring the fast-timescale state motion within the domains (the "state jitter" inside of a domain is ignored). The new model describes only slow timescale information on transitions between the domains. Second, coarse graining of a spatial nature results by ignoring the continuum of values that the state vector normally assumes. Instead, the state vector is specified by an integer index; at any point t in time, the state is specified by an integer-valued variable $J(t)$, $1 \leq J(t) \leq m$. Equivalently, at any point in time, the state vector is described by the domain R_k, $1 \leq k \leq m$, that it currently occupies.

On this m-attractor, "coarse-grained" state space model, the PLL state is modeled as a Markov jump process. Consider any number of arbitrary times $0 \leq t_0 \leq t_1 \leq t_2 \leq \ldots \leq t_n \leq t_{n+1}$ and arbitrary integers $1 \leq i_0, i_1, \ldots, i_n, i_{n+1} \leq m$. Then the Markov property implies that

$$P\big[J(t_{n+1}) = i_{n+1} \mid J(t_n) = i_n, J(t_{n-1}) = i_{n-1}, \ldots, J(t_0) = i_0\big]$$
$$= P\big[J(t_{n+1}) = i_{n+1} \mid J(t_n) = i_n\big].$$

(8.4-1)

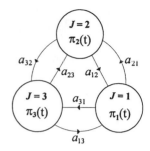

FIGURE 8.7
State transition rate diagram for m = 3.

That is, the probabilistic description of the state at time t_{n+1} depends only on the state at time t_n; not important are the values of the state at times $t_{n-1}, t_{n-2}, \ldots, t_0$ (i.e., there is no state memory). Furthermore, the amount of time spent in the current state is not important in determining the next state (i.e., there is no age memory).

Quantitative characterizations of this Markov jump process model can be given that lead to numerical algorithms relevant to the PLL cycle slip phenomenon. First, a set of *transition rates* a_{ik}, $1 \le i, k \le m$, describes the frequencies of transitions between the m states. This set of transition rates can be expressed in the form of a *transition rate matrix* $\mathbf{A} = \{a_{ik}\}_{1 \le i, k \le m}$. Many applications of the theory are characterized well by this set of rates. When both nonlinearity and noise have to be included in a PLL analysis, this set of transition rates best describes the phenomenon of PLL cycle slips. The transition rates are discussed further in Sections 8.4.2 and 8.4.3. Secondly, the time-varying *occupancy probabilities*,

$$\pi_k(t) \equiv P[J(t) = k], \qquad (8.4\text{-}2)$$

$1 \le k \le m$, describe the propensity for occupancy of the m states. These probabilities can be used to form the vector

$$\bar{\Pi}(t) \equiv [\pi_1(t) \; \pi_2(t) \; \cdots \; \pi_m(t)]^T. \qquad (8.4\text{-}3)$$

These occupancy probabilities are discussed in Section 8.4.4. Finally, the transition rates and occupancy probabilities can be illustrated by a *state transition rate diagram*. Figure 8.7 depicts a typical diagram for the case of m = 3.

8.4.2 The Transition Rate Matrix

Noise-induced transitions can occur between the domains of the m-attractor, "coarse-grained" model. That is, noise can induce the state to jump from $J = k$ to $J = i$, where k and i are dissimilar integers that satisfy $1 \leq i, k \leq m$. Such transitions occur at random values of time in the time continuum. The quantitative characterization of this jump behavior is the subject of this section.

For $t_d \geq 0$, let $p_{ik}(t_d)$, $1 \leq i, k \leq m$, denote the *transition functions*

$$p_{ik}(t_d) \equiv P[J(t+t_d) = i \mid J(t) = k] \qquad (8.4\text{-}4)$$

between the m states of the model. It is assumed that the Markov jump process is homogeneous so that the transition functions do not depend on absolute time t. However, the transition functions depend on t_d, a time difference. They satisfy the initial conditions

$$p_{ik}(0) = \begin{cases} 1 \text{ if } i = k \\ 0 \text{ if } i \neq k. \end{cases} \qquad (8.4\text{-}5)$$

Furthermore, they satisfy the continuous-time Chapman–Kolmogorov equation

$$p_{ik}(t_d + \Delta t_d) = \sum_{j=1}^{m} p_{ij}(\Delta t_d) p_{jk}(t_d), \qquad (8.4\text{-}6)$$

where $\Delta t_d \geq 0$ is an increment of t_d. This result is the discrete state-space counterpart of Equation 6.2-12. Finally, the functions described by Equation 8.4-4 can be used to form an $m \times m$ *transition function matrix*

$$P(t_d) \equiv \{p_{ik}(t_d)\}_{1 \leq i, k \leq m}, \qquad (8.4\text{-}7)$$

where $P(0) = I$ is the $m \times m$ identity matrix.

For t_d and increment Δt_d, the transition function matrix satisfies

$$P(t_d + \Delta t_d) = P(\Delta t_d) P(t_d), \qquad (8.4\text{-}8)$$

the matrix counterpart of Equation 8.4-6. From both sides of this last result, subtract $P(t_d)$ to obtain

$$P(t_d + \Delta t_d) - P(t_d) = [P(\Delta t_d) - I] P(t_d). \qquad (8.4\text{-}9)$$

Now, divide both sides of this last result by Δt_d, and take the formal limit as Δt_d approaches zero from the right to obtain the $m \times m$ matrix equation

$$\frac{d}{dt_d} P(t_d) = A P(t_d), \qquad (8.4\text{-}10)$$

where

$$A \equiv \lim_{\Delta t_d \to 0^+} \frac{P(\Delta t_d) - I}{\Delta t_d} \qquad (8.4\text{-}11)$$

is the constant $m \times m$ *transition rate matrix*. The elements a_{ik}, $1 \le i, k \le m$, of matrix A are called the *transition rates*, and they play an important role in the analysis of cycle slips in PLLs. Finally, Equations 8.4-5 and 8.4-10 imply that the transition function matrix can be expressed as

$$P(t_d) = \exp[A t_d] \qquad (8.4\text{-}12)$$

for $t_d \ge 0$.

8.4.3 Physical Interpretation and Properties of the Transition Rates

For $1 \le i, k \le m$, the quantity a_{ik} is the instantaneous rate of transition from state $J = k$ to state $J = i$. As implied by Equations 8.4-4 and 8.4-11, it is a conditional jump rate since its specification involves a probability that is conditioned on a prejump state of $J = k$. In this section these ideas are elaborated upon in an attempt to convey insight into the physical meaning of the transition rates.

First, consider the case $J = k = i$. From Equation 8.4-10, note that

$$\frac{dp_{ii}(t_d)}{dt_d} = \sum_{k=1}^{m} a_{ik} p_{ki}(t_d) \qquad (8.4\text{-}13)$$

for each i, $1 \le i \le m$. Now, use Equation 8.4-5 and evaluate this last equation at $t_d = 0^+$ to obtain

$$a_{ii} = \frac{dp_{ii}(t_d)}{dt_d}\bigg|_{t_d = 0^+}. \qquad (8.4\text{-}14)$$

However, this result can be written as

$$-a_{ii} = \frac{d}{dt_d}\left[1-p_{ii}(t_d)\right]\bigg|_{t_d=0^+} = \lim_{\Delta t_d \to 0^+} \frac{1-p_{ii}(\Delta t_d)}{\Delta t_d}. \qquad (8.4\text{-}15)$$

Given that the state is $J = i$ at some t, the quantity $1 - p_{ii}(\Delta t_d)$ is the probability that the state leaves $J = i$ during the next interval of length Δt_d. By using a relative frequency interpretation of probability, note that this quantity represents the fraction of Δt_d-length intervals during which such a state transition occurs. But, there are $1/\Delta t_d$ such intervals per unit length of time. Hence, Equation 8.4-15 suggests that $-a_{ii}$ can be interpreted as the *instantaneous rate* at which a state transition out of $J = i$ takes place. A dimensional analysis of Equation 8.4-15 leads to the fact that a_{ii} has units of seconds^{-1}.

A similar interpretation can be given for the case $J = k \neq i$. Note that Equation 8.4-11 leads to

$$a_{ik} = \frac{dp_{ik}(t_d)}{dt_d}\bigg|_{t_d=0^+}, \qquad (8.4\text{-}16)$$

and this equation can be interpreted in terms of a rate. That is, given that the state is $J = k$ at time t, the nonnegative quantity a_{ik} is the *instantaneous rate* at which the state transitions out of $J = k$ and into $J = i$.

As suggested by the discussion of Equations 8.4-15 and 8.4-16, matrix **A** has negative diagonal terms and nonnegative off-diagonal terms. In fact, each column of **A** must sum to zero. To see this, note that Equation 8.4-4 implies that

$$\sum_{i=1}^{m} p_{ik}(t_d) = 1 \qquad (8.4\text{-}17)$$

and

$$\sum_{i=1}^{m} \frac{d}{dt_d} p_{ik}(t_d) = 0 \qquad (8.4\text{-}18)$$

for each k, $1 \leq k \leq m$. However, the element in the ith row and kth column of Equation 8.4-10 can be used to write Equation 8.4-18 as

$$\sum_{i=1}^{m}\left(\sum_{r=1}^{m} a_{ir} P_{rk}(t_d)\right) = \sum_{r=1}^{m}\left(\sum_{i=1}^{m} a_{ir}\right) P_{rk}(t_d) \qquad (8.4\text{-}19)$$

$$= 0,$$

and this leads to the requirement that each column of **A** sum to zero so that

$$-a_{rr} = \sum_{\substack{i=1 \\ i \neq r}}^{m} a_{ir} \qquad (8.4\text{-}20)$$

for each r, $1 \leq r \leq m$.

As a result of Equation 8.4-20, the trace of matrix **A** can be expressed as the negative sum of all transition rates. That is, the negative sum of the diagonal of matrix **A** can be expressed as

$$-\text{Trace } \mathbf{A} = \sum_{r=1}^{m} \sum_{\substack{i=1 \\ i \neq r}}^{m} a_{ir}. \qquad (8.4\text{-}21)$$

This important result is used in Section 8.4.7.

8.4.4 The Dynamics of State Occupancy

The probabilities associated with state occupancy are denoted as shown by Equation 8.4-2, and they can be written in terms of a vector as shown by Equation 8.4-3. As functions of time, these occupancy probabilities can be expressed in terms of their initial values and the transition rate matrix **A**. Furthermore, the occupancy probabilities and the matrix **A** can be used to develop expressions for the unconditional transition rates. These developments are covered in this section.

For $t \geq 0$ and $1 \leq k \leq m$, the state occupancy probabilities can be expressed as

$$\pi_k(t) = P[J(t) = k]$$

$$= \sum_{i=1}^{m} P[J(t) = k | J(0) = i] P[J(0) = i] \qquad (8.4\text{-}22)$$

$$= \sum_{i=1}^{m} P_{ki}(t) \pi_i(0),$$

NUMERICAL METHODS/NOISE ANALYSIS IN NONLINEAR PLL MODEL   321

where the p_{ki} are the transition functions described by Equation 8.4-4. Hence, the state occupancy vector can be expressed as

$$\vec{\Pi}(t) = P(t)\vec{\Pi}(0), \qquad (8.4\text{-}23)$$

where the matrix P is defined by Equation 8.4-7 and $\vec{\Pi}(0)$ represents the vector of initial occupancy probabilities.

By using the state occupancy probabilities, unconditional transition rates between the states can be obtained. Recall that Equation 8.4-11 defines conditional rates; for $1 \leq i, k \leq m$, the quantity a_{ik} represents the transition rate from $J = k$ into $J = i$. However, the system is in $J = k$ only a fraction of the time, and the *unconditional transition rate* from $J = k$ into $J = i$ is $a_{ik}\pi_k(t)$.

A simple differential equation can be obtained for the occupancy probabilities. This equation results by using Equation 8.4-10 in the derivative of Equation 8.4-23; simple algebra leads to

$$\frac{d}{dt}\vec{\Pi}(t) = AP(t)\vec{\Pi}(0) \qquad (8.4\text{-}24)$$

$$= A\vec{\Pi}(t).$$

Finally, this last result implies that the occupancy vector has the form

$$\vec{\Pi}(t) = \exp[At]\vec{\Pi}(0). \qquad (8.4\text{-}25)$$

The right-hand side of Equation 8.4-24 can be given an interpretation of a discrete (in a spatial sense) probability current vector. To see this, note that for each i, $1 \leq i \leq m$, the probability currents flowing into and out of state $J = i$ can be expressed as

$$\left(\text{Total Probability Current Flowing } \textit{Into State } J = i\right)$$

$$= \sum_{\substack{k=1 \\ k \neq i}}^{m} a_{ik}\pi_k(t)$$

$$\left(\text{Total Probability Current Flowing } \textit{Out Of State } J = i\right) \qquad (8.4\text{-}26)$$

$$= \sum_{\substack{k=1 \\ k \neq i}}^{m} a_{ki}\pi_i(t).$$

The net probability current entering a state is the difference between the current flowing into and out of the state. Finally, a net probability current flowing into a state causes a change in the state occupancy probability. With the aid of Equation 8.4-20, this relationship is expressed as

$$\frac{d}{dt}\pi_i(t) = \sum_{\substack{k=1 \\ k \neq i}}^{m} a_{ik}\pi_k(t) - \left(\sum_{\substack{k=1 \\ k \neq i}}^{m} a_{ki}\right)\pi_i(t)$$

$$= a_{ii}\pi_i(t) + \sum_{\substack{k=1 \\ k \neq i}}^{m} a_{ik}\pi_k(t)$$

(8.4-27)

for each i, $1 \leq i \leq m$. However, this last result is a restatement of Equation 8.4-24. Hence, the right-hand side of Equation 8.4-24 can be interpreted as a probability current vector.

For the PLL application under consideration, the occupancy probabilities have steady-state limits as t approaches infinity. That is, in the steady state, every system state has flowing into it a net probability current of zero. From the steady-state limit of Equation 8.4-27, this requirement leads to

$$a_{ii}\pi_i(\infty) + \sum_{\substack{k=1 \\ k \neq i}}^{m} a_{ik}\pi_k(\infty) = 0 \qquad (8.4\text{-}28)$$

for each i, $1 \leq i \leq m$.

As shown by the discussion presented in this section, the transition rate matrix **A** plays an important role in characterizing the dynamics of the Markov jump process. In many practical PLL applications where noise and nonlinearity must be considered, calculation of the transition rates is the most important method for characterizing the cycle slip phenomenon. As shown in Section 8.4.7, under moderate noise conditions, the transition rates can be approximated by an algorithm that employs the m smallest (in magnitude) eigenvalues of the Fokker–Planck operator. Also, the transition rate algorithm requires certain integrals of the eigenfunctions associated with the small eigenvalues. Because of their usefulness, the eigenvalues and eigenfunctions of L_{FP} are discussed next.

8.4.5 Eigenvalues and Eigenfunctions of the Fokker–Planck Operator

The general n-dimensional Fokker–Planck operator L_{FP} is developed in Section 6.3, and it is given by Equation 6.3-15. In Section 7.2, the operator is developed for the first-order PLL, and this result is given by Equation 7.2-18. Equations 8.2-2 and 8.3-2 describe L_{FP} for common second-order PLLs; the former (latter) equation applies when an RC lowpass network (a perfect integrator) is used as the PLL loop filter. For first- and second-order PLLs, these realizations of L_{FP} are of interest in this section.

As discussed in Section 7.1, the m-attractor phase error model interprets phase in a modulo-$2\pi m$ sense, and ϕ_m denotes the modulo-$2\pi m$ phase error variable. The Fokker–Planck operators under consideration employ ϕ_m as their phase variable. They operate on a vector space of functions that are $2\pi m$-periodic in phase and satisfy natural boundary conditions in their nonphase variables. Furthermore, by assumption, the operators generate a unique stationary probability density; a $p_m(\phi_m, \vec{Y})$ is assumed to exist that satisfies $L_{FP}[p_m] = 0$.

The Fokker–Planck operator L_{FP} has eigenvalues and eigenfunctions denoted here as λ_k and $e_k(\phi_m, \vec{Y})$, respectively. For convenience, both of these quantities are indexed on the nonnegative integers. The eigenvalues are constants, and the eigenfunctions are nonzero functions of $(\phi_m, \vec{Y})$, the PLL state. In general, the λ_k and e_k are complex valued. Note that operator L_{FP} reproduces e_k up to the constant $-\lambda_k$; that is, the eigenvalues and eigenfunctions satisfy

$$L_{FP}\, e_k = -\lambda_k\, e_k. \tag{8.4-29}$$

The eigenfunctions must satisfy boundary Conditions 7.1-13 and 7.1-14; they satisfy the same boundary conditions that are imposed on $p_m(\phi_m, \vec{Y}, t)$. Finally, by definition, eigenvalue $\lambda_0 \equiv 0$; hence, eigenfunction $e_0(\phi_m, \vec{Y})$ is proportional to the steady-state density function $p_m(\phi_m, \vec{Y})$.

An eigenfunction is unique up to a nonzero multiplicative scale factor. After multiplication by any nonzero scalar, it remains an eigenfunction. In what follows, the eigenfunction e_0 is assumed to satisfy

$$e_0(\phi_m, \vec{Y}) = p_m(\phi_m, \vec{Y}). \tag{8.4-30}$$

To the remaining eigenfunctions e_k, $k \geq 1$, nonzero scale factors can be assigned by using any convenient method. Finally, the integral of the eigenfunctions over the state space can be found. By considering the eigenvalues of the formal adjoint to L_{FP}, Ryter[64] showed that

$$\underbrace{\int_{-\infty}^{\infty}\cdots\int_{-\infty}^{\infty}}_{n-1 \text{ times}}\int_{-m\pi}^{m\pi}e_k(\vec{\phi}_m,\vec{Y})d\vec{\phi}_m d\vec{Y} = 1 \quad \text{for } k = 0 \qquad (8.4\text{-}31)$$

$$= 0 \quad \text{for } k \geq 1.$$

The eigenfunctions are assumed to form a complete basis for the above-mentioned vector space. Let $p_m(\vec{\phi}_m,\vec{Y},t)$ denote a solution of the Fokker–Planck equation, and suppose that it satisfies Conditions 7.1-12 through 7.1-14. Then, it can be represented as a generalized Fourier series expansion of the form

$$p_m(\vec{\phi}_m,\vec{Y},t) = \sum_k c_k(t) e_k(\vec{\phi}_m,\vec{Y}), \qquad (8.4\text{-}32)$$

where the $c_k(t)$ are a solution-dependent set of time-varying weight functions.

The structure of the time-varying coefficients $c_k(t)$ is simple. Substitute Equation 8.4-32 into the Fokker–Planck equation $\partial p/\partial t = L_{FP}p$, and use Equation 8.4-29 to obtain

$$\sum_k \left[\frac{d}{dt}c_k(t)\right]e_k(\vec{\phi}_m,\vec{Y}) = \sum_k c_k(t)(-\lambda_k)e_k(\vec{\phi}_m,\vec{Y}). \qquad (8.4\text{-}33)$$

This last result leads to the requirements

$$\frac{d}{dt}c_k(t) = -\lambda_k c_k(t) \qquad (8.4\text{-}34)$$

and

$$c_k(t) = c_k(0)\exp[-\lambda_k t], \qquad (8.4\text{-}35)$$

where the $c_k(0)$ are constants. Finally, these observations lead to the conclusion that Equation 8.4-32 can be written as

$$p_m(\vec{\phi}_m,\vec{Y},t) = \sum_k c_k(0)\exp[-\lambda_k t]e_k(\vec{\phi}_m,\vec{Y}). \qquad (8.4\text{-}36)$$

This result shows that the eigenfunctions control the general shape of the density in state space, and the eigenvalues control the temporal decay of the modes that make up the density.

As stated in Section 8.4.1, a unique steady-state density $p_m(\phi_m,\vec{Y})$ is assumed to exist, and all densities $p_m(\phi_m,\vec{Y},t)$ (that satisfy Equations 7.1-12 through 7.1-14) converge to it as time approaches infinity. Hence, all nonzero eigenvalues have to satisfy

$$\text{Re}[\lambda] > 0. \tag{8.4-37}$$

These eigenvalues are indexed on the nonnegative integers. For the present, this indexing scheme can be implemented in any manner where $\text{Re}[\lambda_k]$ is a nondecreasing sequence for $k = 0, 1, 2, 3, \ldots$. In Section 8.4.10, the first $m - 1$ nonzero eigenvalues are reordered to aid the development of an algorithm for computing transition rates. Finally, Equations 8.4-30 and 8.4-36 imply

$$c_0(0) = 1. \tag{8.4-38}$$

The reader should keep in mind a limitation of the notation that is used in this chapter (and the PLL literature). The set of eigenvalues λ_k, $k \geq 0$, is dependent on the value of m in the m-attractor phase error model and the Fokker–Planck operator. However, this dependence is not indicated by the notation used to express the eigenvalues.

8.4.6 Eigenvalues and Eigenfunctions Under Moderate Noise Conditions

For the moderate noise case and the m-attractor model, the eigenvalues and eigenfunctions of L_{FP} have important properties that can be exploited. Some of these properties are summarized here; in Section 8.4.7 they are put to use in the development of an algorithm for calculating the transition rates a_{ik}, $1 \leq i, k \leq m$, discussed briefly at the end of Section 8.4.1. The interested reader should consult Ryter[64] for a more extensive discussion of the implications of moderate noise on the eigenfunctions and eigenvalues.

Based on magnitude, the eigenvalues λ_k, $k \geq 0$, are segregated into two groups. The first m eigenvalues $\lambda_0, \lambda_1, \lambda_2, \ldots \lambda_{m-1}$ are in the first group (recall that $\lambda_0 = 0$), and the remaining λ_k, $k \geq m$, are in the second group. Under moderate noise conditions, the eigenvalues in the first group have a distinctly smaller magnitude than the eigenvalues in the second group. For practical PLLs operating with usable values of in-loop SNR, eigenvalues from the second group are often one, two, or more orders of magnitude larger than eigenvalues from the first group. In Section 8.4.7 this characteristic of the eigenvalues plays a major role in the development of an algorithm for approximating the transition rates.

Associated with the m small eigenvalues in the first group are eigenfunctions that have useful properties. When the noise is moderate, the eigenfunctions e_k, $0 \le k \le m - 1$, are concentrated onto well-separated domains R_k, $1 \le k \le m$. These domains are pair-wise disjoint, and they are centered at the stable states of the m-attractor model. The first m eigenfunctions are approximately zero outside of this collection of domains. However, on the system of domains, these eigenfunctions have important similarities that are discussed in the remainder of this section.

First, as discussed in the previous section, the eigenfunctions e_k, $0 \le k \le m - 1$, are $2\pi m$-periodic in the phase variable. In addition, eigenfunction e_0 is 2π-periodic in phase; on any two distinct domains, this eigenfunction is identical in shape, scale, and overall appearance. This claim follows from Equations 7.1-16 and 8.4-30. In general, the remaining e_k, $1 \le k \le m - 1$, are not 2π-periodic in phase. However, when restricted to any single domain R_i, $1 \le i \le m$, each eigenfunction e_k, $1 \le k \le m - 1$, can be approximated by a complex-valued scalar q_{ki} multiple of e_0. That is, for integers k, $1 \le k \le m - 1$, there exist complex-valued constants q_{ki}, $1 \le i \le m$, such that the eigenfunction e_k can be approximated as

$$e_k(\phi_m, \vec{Y}) \approx e_0(\phi_m, \vec{Y}) \sum_{i=1}^{m} q_{ki} U_{R_i}(\phi_m, \vec{Y}), \qquad (8.4\text{-}39)$$

where the U_{R_i}, $1 \le i \le m$, are state-space windowing functions that satisfy

$$\begin{aligned} U_{R_i}(\phi_m, \vec{Y}) &= 1, \quad (\phi_m, \vec{Y}) \in R_i \\ &= 0, \quad (\phi_m, \vec{Y}) \notin R_i . \end{aligned} \qquad (8.4\text{-}40)$$

The bistable case (m = 2) is simple. Of interest in this case are the small eigenvalues $\lambda_0 = 0$ and λ_1 that correspond to eigenfunctions e_0 and e_1, respectively. Now, eigenfunction e_0 is 2π-periodic in phase, and it has an identical shape and scale over R_1 and R_2. This fact, and Equation 8.4-39, implies that eigenfunction e_1, when evaluated over R_1, can be approximated as a scalar multiple of e_1 over R_2. However, over the state space, the integral of e_1 must vanish, as shown by Equation 8.4-31. This argument leads to the result

$$e_1(\phi_2, \vec{Y}) \approx +qe_0(\phi_2, \vec{Y}) \quad (\phi_2, \vec{Y}) \in R_1$$
$$\approx -qe_0(\phi_2, \vec{Y}) \quad (\phi_2, \vec{Y}) \in R_2, \tag{8.4-41}$$

where q is a factor dependent on the scale assigned to e_1 (nonzero q can be assigned in any convenient manner). Hence, eigenfunction e_1 can be approximated in terms of e_0 when m = 2 (as discussed in Section 8.5, the solid-line graph on Figure 8.11 depicts e_1 for the first-order PLL).

The qualitative nature of the eigenvalues and eigenfunctions outlined in this section can be used to write a slow timescale version of Equation 8.4-36 for the moderate noise case. Probability density functions are assumed to satisfy Conditions 7.1-12 through 7.1-14 and approach a unique steady-state density as time approaches infinity. For large time and under moderate noise conditions, the magnitude difference in eigenvalues implies that these densities can be approximated as

$$P_m(\phi_m, \vec{Y}, t) \approx \sum_{k=0}^{m-1} c_k(0) \exp[-\lambda_k t] e_k(\phi_m, \vec{Y}), \tag{8.4-42}$$

where the $c_k(0)$, $1 \le k \le m-1$, are appropriately chosen constants. However, in view of Equation 8.4-39, it is possible to write a slow timescale approximation as

$$P_m(\phi_m, \vec{Y}, t) \approx e_0(\phi_m, \vec{Y}) \sum_{i=1}^{m} d_i(t) U_{R_i}(\phi_m, \vec{Y}), \tag{8.4-43}$$

where $d_i(t)$, $1 \le i \le m$, are time-dependent scaling functions that satisfy

$$d_i(t) > 0$$
$$d_i(\infty) = 1 \tag{8.4-44}$$

for all t, and the windowing functions U_{R_i} are given by Equation 8.4-40.

8.4.7 Calculation of the Transition Rates a_{ij}

As discussed in Section 8.4.2, the quantity a_{ik}, $1 \le i, k \le m$, describes the transition rate from R_k to R_i. This rate can be calculated by the algorithm described in this section. This algorithm requires the numerical calculation of the m − 1 smallest (in magnitude) nonzero eigenvalues of the Fokker–Planck operator; also required are integrals of the eigenfunctions that correspond to the small eigenvalues. Calculation of these integrals is the most difficult part of the algorithm. For this reason, Section 8.4.8 describes how to eliminate this difficult requirement by imposing a cyclic structure on the basic "coarse-grained" Markov jump model.

By integrating Density 8.4-42 over the domain R_i, $1 \le i \le m$, an approximation can be obtained that describes the time-varying probability that R_i is occupied. This result is

$$\pi_i(t) \approx \sum_{k=0}^{m-1} c_k(0) \exp(-\lambda_k t) e_{ki}, \qquad (8.4\text{-}45)$$

where

$$\pi_i(t) \equiv \underbrace{\int \cdots \int}_{\text{over } R_i} P_m(\vec{\phi}_m, \vec{Y}, t) d\vec{\phi}_m d\vec{Y} \qquad (8.4\text{-}46)$$

denotes the probability that R_i is occupied, and

$$e_{ki} \equiv \underbrace{\int \cdots \int}_{\text{over } R_i} e_k(\vec{\phi}_m, \vec{Y}) d\vec{\phi}_m d\vec{Y} \qquad (8.4\text{-}47)$$

is the integral of eigenfunction e_k over domain R_i.

By using vectors and matrices, an m-dimensional version of Equation 8.4-45 can be written. First, define the time-varying occupancy vector

$$\vec{\Pi}(t) \equiv [\pi_1(t) \ \pi_2(t) \ \cdots \ \pi_m(t)]^T. \qquad (8.4\text{-}48)$$

Furthermore, for each k, $0 \le k \le m - 1$, the e_{ki}, $1 \le i \le m$, described by Equation 8.4-47 can be used to write the constant vector

$$\vec{e}_k \equiv [e_{k1} \ e_{k2} \ \cdots \ e_{km}]^T, \qquad (8.4\text{-}49)$$

and these vectors can be used as columns of the m × m matrix

$$E \equiv [\vec{e}_0 \vdots \vec{e}_1 \vdots \cdots \vdots \vec{e}_{m-1}]. \qquad (8.4\text{-}50)$$

In the definition of Equation 8.4-50, the elements e_{ki} have indices that are reversed from the normal matrix convention (as is the case in the archived literature on this subject). Finally, an m-dimensional version of Equation 8.4-45 can be written as

$$\vec{\Pi}(t) \approx E\vec{c}(t), \tag{8.4-51}$$

where

$$\vec{c}(t) \equiv \begin{bmatrix} c_0(0) \\ c_1(0)\exp(-\lambda_1 t) \\ c_2(0)\exp(-\lambda_2 t) \\ \vdots \\ c_{m-1}(0)\exp(-\lambda_{m-1} t) \end{bmatrix} \tag{8.4-52}$$

denotes an m × 1 time-varying vector.

An equation for the time rate of change of the occupancy probabilities can be obtained. Simply differentiate Equation 8.4-51, and use Equation 8.4-52 to obtain

$$\frac{d}{dt}\vec{\Pi}(t) \approx E\frac{d}{dt}\vec{c}(t)$$
$$= -E\Lambda\vec{c}(t), \tag{8.4-53}$$

where m × m diagonal matrix Λ has the first m eigenvalues on its diagonal ($\lambda_0 = 0$ is in the (1, 1) position, and λ_{m-1} is in the (m, m) position). Now, the first m eigenfunctions of the Fokker–Planck operator are linearly independent (as argued in Section 8.5 and 8.6, these m eigenfunctions lie in m orthogonal subspaces, one eigenfunction per subspace), and they satisfy Equation 8.4-39. As a result of these facts, the m columns of Equation 8.4-50 are linearly independent, and the matrix E is nonsingular. Hence, Equation 8.4-51 can be used to eliminate vector $\vec{c}(t)$ from Equation 8.4-53 and write

$$\frac{d}{dt}\vec{\Pi}(t) \approx -E\Lambda E^{-1}\vec{\Pi}. \tag{8.4-54}$$

Important observations can be deduced from the fact that vector $\vec{\Pi}$ satisfies a differential equation of the Form 8.4-54. First, at any specific time, the rate of change of the occupancy Probabilities 8.4-46 can be approximated by the occupancy probabilities themselves. That is, domain occupancy can be *modeled* as a Markov process as discussed in Section 8.4.1. The second, and more important, observation follows by comparing Equation 8.4-54 with Equation 8.4-24. Under moderate noise conditions, the m × m transition rate matrix **A** appears on the right-hand side of Equation 8.4-54. In the numerical algorithm for approximating the transition rates, the matrix that assumes the role of the transition rate matrix can be computed as

$$\mathbf{A} = -\mathbf{E}\Lambda\mathbf{E}^{-1}. \qquad (8.4\text{-}55)$$

Hence, transition rates in the m state jump model can be computed by a numerical algorithm based on Equation 8.4-55, a matrix determined by the first m eigenvalues and eigenfunctions of the Fokker–Planck operator.

Equation 8.4-55 implies that $-\lambda_k$ and $\vec{e}_k$ are the eigenvalues and eigenvectors, respectively, of matrix **A**. This claim follows from elementary linear algebra, and it can be seen by comparing columns of

$$\begin{aligned}\mathbf{AE} &\equiv \left[\mathbf{A}\vec{e}_0 \,\vdots\, \mathbf{A}\vec{e}_1 \,\vdots\, \cdots \,\vdots\, \mathbf{A}\vec{e}_{m-1}\right] \\ &= -\mathbf{E}\Lambda \\ &= -\left[\vec{0} \,\vdots\, \lambda_1\vec{e}_1 \,\vdots\, \cdots \,\vdots\, \lambda_{m-1}\vec{e}_{m-1}\right],\end{aligned} \qquad (8.4\text{-}56)$$

a result that is obtained from Equation 8.4-55. Hence, the $-\lambda_k$ and $\vec{e}_k$ determine completely the matrix **A**.

The unspecified scales of eigenvectors $\vec{e}_k$, $1 \le k \le m-1$, (the Requirement 8.4-30 specifies the scale of $\vec{e}_0$) do not influence the matrix **A**. To see this, note that a rescaling of the eigenvectors (i.e., columns of **E**) in Equation 8.4-50 amounts to replacing **E** by **ET**, where **T** is a diagonal matrix that contains the arbitrary nonzero scale factors. But simple manipulations lead to

$$\begin{aligned}-\mathbf{ET}\Lambda(\mathbf{ET})^{-1} &= -\mathbf{E}(\mathbf{T}\Lambda\mathbf{T}^{-1})\mathbf{E}^{-1} \\ &= \mathbf{E}\Lambda\mathbf{E}^{-1} \qquad (8.4\text{-}57) \\ &= \mathbf{A},\end{aligned}$$

and the conclusion that the eigenvector scale factors do not influence the transition rates described by matrix **A**.

NUMERICAL METHODS/NOISE ANALYSIS IN NONLINEAR PLL MODEL

Recall from Equation 8.4-21 that the trace of matrix **A** can be expressed as the negative sum of all transition rates. Now, a matrix has a trace that is the sum of its eigenvalues, and the trace and eigenvalues are invariant under a similarity transformation of the Form 8.4-55. As a result of these observations, an important relationship between the significant eigenvalues and the sum of the transition rates is (see Equation 8.4-21)

$$\sum_{k=1}^{m-1} \lambda_k = -\text{Trace } \mathbf{A} = \sum_{r=1}^{m} \left(\sum_{\substack{i=1 \\ i \neq r}}^{m} a_{ir} \right). \qquad (8.4\text{-}58)$$

As an example, consider the bistable case where m = 2. First, note that $\lambda_0 = 0$; this result is true for all values of m. Now, Equation 8.4-58 yields the eigenvalue relationship

$$\lambda_1 = a_{12} + a_{21}. \qquad (8.4\text{-}59)$$

However, Equations 8.4-28 and 8.4-20 yield the steady-state $(d\pi_i(\infty)/dt = 0, i = 1, 2)$ result

$$a_{12}\pi_2(\infty) - a_{21}\pi_1(\infty) = 0 \qquad (8.4\text{-}60)$$

(i.e., in the steady state, a net probability current of zero enters domain R_1). Combined with the requirement that $\pi_1(t) + \pi_2(t) = 1$, the last two equations yield

$$\begin{aligned} a_{12} &= \lambda_1 \pi_1(\infty) \\ a_{21} &= \lambda_1 \pi_2(\infty) \end{aligned} \qquad (8.4\text{-}61)$$

for the transition rates in the general case. As discussed in Section 7.1.3 (see Figure 7.6), the steady-state density $p_2(\infty)$ has an identical shape and scale on the domains R_1 and R_2. Hence, the steady-state domain occupancy probabilities $\pi_1(\infty)$ and $\pi_2(\infty)$ are equal, and Equation 8.4-61 implies that

$$a_{12} = a_{21} = \frac{\lambda_1}{2} \qquad (8.4\text{-}62)$$

for the case m = 2.

8.4.8 An m-Attractor Markov Jump Model with a Cyclic Structure

As shown by Equation 8.4-55, the quantities λ_k and $\vec{e}_k$, $0 \le k \le m-1$, can be used to compute the transition rate matrix **A**. However, as shown by Equations 8.4-47 and 8.4-49, computation of $\vec{e}_k$ requires the integration of eigenfunction $e_k(\phi_m, \vec{Y})$ over each of the m domains. In general, this integration requirement is not met easily. Hence, it is natural to try and restructure the problem in order to avoid integrating eigenfunctions.

Fortunately, this can be accomplished by imposing a cyclic structure on the underlying Markov jump model. As described in this section, this new model characterizes transition events by the number of domains that are skipped and not by the starting (or ending) state index. This simplification is predicated on the assumption that there is little real difference between one state/domain pair and the next. Certainly, this assumption is valid in the PLL model considered here since state/domain pairs differ only in the phase variable, and the phase difference between any two state/domain pairs is a multiple of 2π.

In the reformulated jump model, it is convenient to retain the previously introduced notation for referencing the states and domains. As before, the m states are referenced by an integer between one and m, and the integer-valued variable $J(t)$, $1 \le J \le m$, denotes the current state index at time t. Also, the domains are denoted as R_j, $1 \le j \le m$, and domain R_j corresponds to the state with index $J = j$. The nature of a transition event is new in the reformulated model.

When a transition event occurs, the starting (or ending) state index is of minor importance. The reason for this is simple: in most respects, the domains of the m-attractor model are indistinguishable. What is important (the feature that distinguishes one transition event from another) is the number k of domains that are changed. For example, the value k = 1 corresponds to a jump between adjacent domains, and k = 2 corresponds to a jump of two domains (bypassing completely the intermediate domain). Hence, an obvious simplification results by ignoring state indices and characterizing transition events by the size of the jump.

An additional simplification must be made to obtain a tractable numerical algorithm for the transition rates. In this section, jump direction is ignored, and it is assumed that a positive integer k denotes the size of a jump. Furthermore, a jump of size k takes the model from *any* initial $J = i$, $1 \le i \le m$, to the final $J = 1 + \text{mod}_m(i - k - 1)$, where $0 \le \text{mod}_m(i - k - 1) \le m - 1$. Appendix 8.4.1 describes the *least positive residue function* $\text{mod}_m(i)$; for every integer i, $\text{mod}_m(i)$ produces an integer between 0 and m – 1. This additional structure causes the m-attractor, "coarse-grained" model to depict *all* transitions in the negative direction; when i = 1 is reached, the model wraps around to i = m on the next single slip (i.e., k = 1). For this reason, the structure imposed here gives the Markov jump model a cyclic structure with a period of m.

There are only m − 1 transitions that can be resolved unambiguously, and they correspond to k = 1 through k = m − 1. Simply stated, the m-attractor, "coarse-grained", cyclic Markov jump model interprets all transitions as being jumps of from 1 to m − 1 states. For this reason, the model utilizes a set of nonnegative real numbers a_k, $1 \leq k \leq m - 1$, to describe the rates of the m − 1 unambiguous transitions. From *each* of the m states, unambiguous transitions of size 1, 2, ..., m − 1 occur at rates of $a_1, a_2, ..., a_{m-1}$, respectively.

For the case m = 4, Figure 8.8 depicts the reformulated "coarse-grained", m-attractor cyclic jump model. On the figure, note that the states appear as an interconnected system of indexed circles. Furthermore, a value of k is shown alongside of each path. For the path to be taken, the "mod_4 size" of the jump must be equal to the value of k shown (i.e., $k = mod_4(n)$, where n is the actual jump size). A jump of size k takes the model from *any* state $J = i$, $1 \leq i \leq 4$, to the state $J = 1 + mod_4(i - k - 1)$. Note that only downward transitions through the state indices occur; when i = 1 is reached, the model wraps around to i = 4 on the next domain slip. From each state, unambiguous transitions of size k = 1, 2, and 3 are possible. However, transitions of size equal to, or larger than, four are ambiguous; they are missed completely (as when the jump size is an integer multiple of 4), or they are interpreted incorrectly as a transition of either 1, 2, or 3 states. Finally, transition rates are shown on the diagram; they are assigned to the unambiguous transitions.

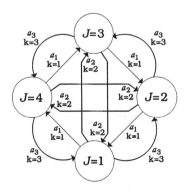

FIGURE 8.8
Transition diagram for a Markov jump model with a cyclic structure (m = 4 states).

8.4.9 Relationship of the Cyclic Jump Model to an Unrestricted Jump Model

Any complete statistical description of the cycle slip phenomenon has to consider the unrestricted phase error model, where $-\infty < \phi < \infty$ (the unrestricted phase error model is introduced in Section 7.1). When "coarse-grained" as discussed in Section 8.4.1, this unrestricted phase error model becomes the *unrestricted jump model* with an infinite number of distinctly indexed state/domain pairs. A transition from R_j to R_k corresponds to a change in phase of $2\pi(k-j)$ radians, so the unrestricted model incorporates a bidirectional jump capability.

In the unrestricted jump model, the transition rates are denoted as r_j, $-\infty < j < \infty$. The magnitude of integer j denotes the number of slipped domains, and the sign of j denotes slip direction. A positive (negative) value for j denotes a negative (positive) slip direction, and a decrease (increase) in the actual value of phase error. For example, a transition involving a decrease of two domains (i.e., a decrease of 4π radians in phase) contributes to the rate r_2. This rule may seem backwards at first glance; however, it is based on the convention established for the transition rates of the m-attractor cyclic jump model (where a downward transition through the domain indices contributes to a_k for some positive k). In the remainder of this section, this unrestricted jump model is compared to the m-attractor cyclic jump model. Also, the relationship is developed between the two sets of rates r_j, $-\infty < j < \infty$ and a_k, $1 \leq k \leq m-1$.

When used to describe cycle slips in the PLL, the m-attractor cyclic jump model imposes limitations that are not encountered when the unrestricted jump model is used. First, for any integer v, a transition of vm domains (i.e., a cycle slip of vm cycles) is identified correctly by the unrestricted jump model; however, such a transition is missed completely by the m-attractor cyclic jump model. Secondly, cycle slip direction is (is not) relevant in the unrestricted (m-attractor cyclic) jump model. For example, in the unrestricted jump model, an advance of two domains (i.e., an advance of 4π in phase) is distinguishable from a decrease of two domains (i.e., a decrease of 4π in phase), and both of these transitions are interpreted correctly. However, as can be seen from Figure 8.8, they are counted as the same k = 2 event in the 4-attractor cyclic model. Finally, the unrestricted jump model always interprets PLL cycle slip length correctly; however, in the m-attractor cyclic jump model, cycle slip length may be reported incorrectly. For example, an advance of one domain is interpreted correctly as a slip of length one by the unrestricted jump model; however, as can be seen from Figure 8.8, it is counted as a slip of length k = 3 in the 4-attractor cyclic jump model.

CHART 8.1

Attributes of the Coarse-Grained, m-Attractor, Cyclic Model as Applied to the PLL Cycle Slip Phenomenon

---
- The cyclic model misses cycle slip bursts of νm periods, ν an integer.
- Cycle slip direction may not be distinguishable. For $1 \leq k \leq m - 1$, r_k (r_{k-m}) corresponds to a slip in the negative (positive) direction, but both r_k and r_{k-m} contribute to the same a_k.
- Ambiguities in cycle slip lengths may occur. For $1 \leq k \leq m - 1$, r_k (r_{k-m}) corresponds to a slip of length k cycles (k − m cycles). Both of these slips are interpreted by the cyclic model as k cycle slips, and both contribute to a_k.
- The cyclic model serves as the basis of a practical algorithm for computing cycle slip information for low-order PLLs.

Chart 8.1 summarizes these limitations. As communicated by the chart and the discussion given in the previous paragraph, the m-attractor cyclic jump model has limitations that must be kept in mind when the model is used in PLL applications. However, in one very important respect this model has a big advantage. As shown in Section 8.4.10, a general numerical algorithm exists for approximating the transition rates a_k, $1 \leq k \leq m - 1$. As is demonstrated in Sections 8.5 and 8.6, the a_k generated by the algorithm can be used to produce useful approximations for the significant r_k in low-order PLLs.

Transition rates for the m-attractor cyclic jump model can be related to the rates that describe the unrestricted jump model. The cyclic jump model considers k-size jumps in a modulo-m sense; because of this, it is easily seen that

$$a_k = \sum_{n=-\infty}^{\infty} r_{k+nm}, \qquad (8.4\text{-}63)$$

for $1 \leq k \leq m - 1$. That is, in the unrestricted jump model, transitions that differ in size from k by a multiple of m count towards the same transition of size k in the m-attractor cyclic jump model.

In many applications, the a_k can be computed numerically (the subject of Section 8.4.10), but the r_k are desired. Hence, of interest are conditions under which Equation 8.4-63 can be used to solve for the significant r_k. In general, an "inversion" of Equation 8.4-63 can be accomplished in only an approximate fashion; in attempting this feat, the key issues are the value of integer m and the number of significant r_k. Basically, to obtain all of the significant r_k (this may not be required in a given application), a relatively small upper limit on jump size must exist. That is, the r_k must be insignificant outside of some (hopefully small) range of k. As can be seen from Equation 8.4-63, integer m should

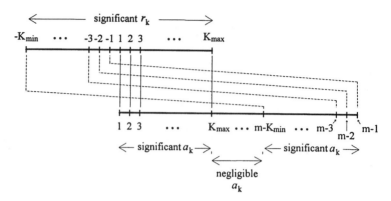

FIGURE 8.9
Use $m > K_{min} + K_{max}$ and obtain $r_k \approx a_k$, $1 \leq k \leq K_{max}$, and $r_{-k} \approx a_{m-k}$, $-K_{min} \leq -k \leq -1$.

be chosen, if possible, so that each a_k can be approximated by only one r_k.

As a first case, suppose the maximum cycle slip lengths in both directions are known *a priori*. Specifically, assume positive integers K_{min} and K_{max} are known such that $r_k \approx 0$ for both positive $k > K_{max}$ and negative $k < -K_{min}$. Then, by using Equation 8.4-63, it is possible to determine the significant r_k from the a_k, $1 \leq k \leq m - 1$, if integer m is chosen greater than $K_{min} + K_{max}$. In fact, for these values of m, the quantity $r_k \approx a_k$, $1 \leq k \leq K_{max}$, describes a slip in the negative direction, and the quantity $r_{-k} \approx a_{m-k}$, $-K_{min} \leq -k \leq -1$, describes a slip in the positive direction.

Figure 8.9 illustrates this relationship. On two separate, indexed, horizontal line segments, the figure symbolically represents the significant r_k, $-K_{min} \leq k \leq K_{max}$, and the computed a_k, $1 \leq k \leq m - 1$. Due to the fact that $m > K_{min} + K_{max}$, for $1 \leq k \leq K_{max}$, Equation 8.4-63 establishes a one-to-one correspondence between the *significant* r_k and the a_k. That is, for $1 \leq k \leq K_{max}$, the significant r_k can be approximated by the a_k. On Figure 8.9, this is indicated by the vertical solid lines that connect the r_k and the a_k for $1 \leq k \leq K_{max}$. In a similar fashion, the a_k, $m - K_{min} \leq k \leq m - 1$, correspond in a one-to-one manner to the r_j, $-K_{min} \leq j \leq -1$. On Figure 8.9, sloping dashed lines depict this correspondence. Finally, values of k in the range $K_{max} < k < m - K_{min}$ are associated with negligible values of a_k. Hence, when $m > K_{min} + K_{max}$, it is possible to approximate the significant r_k in terms of the computed a_k.

Unfortunately, in most applications, integer values for K_{min} and K_{max} are not known *a priori*. First-order systems are an exception to this general rule; in these systems, integers $K_{min} = K_{max} = 1$ since the continuous sample functions of the system cannot bypass, or jump over, a stable equilibrium point. In the more general case, a "trial-and-error" procedure must be employed if all of the significant r_k are desired. For increasing values of m, sets of a_k, $1 \le k \le m-1$, should be computed until the data show the trend illustrated by Figure 8.9. That is, the significant a_k should fall into two sets that are divided by a span of relatively insignificant a_k. Then, the first set contains the significant a_k for which $r_k \approx a_k$, rates for slips in the direction of decreasing ϕ. The second set contains the significant a_{m-k} for which $r_{-k} \approx a_{m-k}$, rates for slips in the direction of increasing ϕ. For example, the value $m = 4$ might be tried first. If a_2 is not insignificant compared to a_1 and a_3, then m should be increased, and a new set of rates should be computed.

8.4.10 Computing the Slip Rates in the "Coarse-Grained", m-Attractor, Cyclic State Model

Section 8.4.7 describes an algorithm for calculating the transition rates for the Markov jump model without the cyclic structure introduced in Section 8.4.8. Recall that the algorithm requires numerically integrating eigenfunctions over domains of attraction. Because of this requirement, the algorithm is difficult to implement in most practical applications. Hence, in the previous section, the underlying jump model is given a cyclic structure, and the promise is made that this additional structure will simplify calculation of a modified set of transition rates by eliminating the requirement of eigenfunction integration. The present section makes good on this promise; here, the algorithm of Section 8.4.7 is modified to utilize the cyclic structure discussed in Section 8.4.8.

The cyclic structure introduced in Section 8.4.8 imposes a cyclic structure on the transition rate matrix $\mathbf{A}$. First, consider the case $m = 4$, and write $\mathbf{A}$ for the model that is illustrated by Figure 8.8. By pulling the off-diagonal elements directly from Figure 8.8, the transition rate matrix can be written as

$$\mathbf{A} = \begin{bmatrix} a_0 & a_1 & a_2 & a_3 \\ a_3 & a_0 & a_1 & a_2 \\ a_2 & a_3 & a_0 & a_1 \\ a_1 & a_2 & a_3 & a_0 \end{bmatrix}, \quad (8.4\text{-}64)$$

where the common diagonal element

$$a_0 = -[a_1 + a_2 + a_3] \quad (8.4\text{-}65)$$

follows from Equation 8.4-20. From Equation 8.4-64, note that the first row is a sequential listing of the transition rates, and each successive row is obtained by applying a circular right shift to the previous row.

The generalization of Equation 8.4-64 follows easily. By applying to the rows the circular right-shift procedure, the m × m **A** for the general case of m stable states is given by

$$\mathbf{A} = \begin{bmatrix} a_0 & a_1 & a_2 & a_3 & \cdots & a_{m-1} \\ a_{m-1} & a_0 & a_1 & a_2 & \cdots & a_{m-2} \\ a_{m-2} & a_{m-1} & a_0 & a_1 & \cdots & a_{m-3} \\ \vdots & \vdots & \vdots & \vdots & & \vdots \\ a_1 & a_2 & a_3 & a_4 & \cdots & a_0 \end{bmatrix}, \quad (8.4\text{-}66)$$

where the common diagonal element is obtained from

$$a_0 = -\sum_{k=1}^{m-1} a_k. \quad (8.4\text{-}67)$$

In matrix **A**, note that the (i, j)th entry is a_k, where $k = \text{mod}_m(j - i)$.

The eigenvalues and eigenvectors of Equation 8.4-66 have a simple form. Let $\vec{e}_k$, $1 \le k \le m$, denote the eigenvectors of this matrix. Then the cyclic structure of **A** implies that

$$\vec{e}_k = \begin{bmatrix} 1 & \kappa^k & \kappa^{2k} & \cdots & \kappa^{(m-1)k} \end{bmatrix}^T \quad (8.4\text{-}68)$$

$$\kappa \equiv \exp[j2\pi/m]. \quad (8.4\text{-}69)$$

Since $\kappa^m = 1$, the components of each eigenvector are roots of unity. The eigenvalues of **A** can be expressed in terms of the rates as

$$-\lambda_k = \sum_{i=0}^{m-1} \kappa^{ki} a_i, \quad (8.4\text{-}70)$$

for $0 \le k \le m - 1$. These results for the eigenvectors and eigenvalues can be verified directly by showing $\mathbf{A}\vec{e}_k = -\lambda_k \vec{e}_k$.

It is possible to relate the eigenvalues computed for a given value of m to the eigenvalues computed for $m' \equiv im$, where i is a positive integer. For the case $m' \equiv im$, denote the "new" nonzero eigenvalues

NUMERICAL METHODS/NOISE ANALYSIS IN NONLINEAR PLL MODEL 339

as λ_k', $1 \leq k \leq m' - 1$. If m is increased by the multiplicative factor i to form m', then the "new" λ_{ik}' (subscript ik denotes the product of i and k) is equal to the "old" eigenvalue (computed for m) λ_k, $1 \leq k \leq m - 1$. To see this, use Equations 8.4-63 and 8.4-70 to write

$$-\lambda_k = \sum_{j=0}^{m-1} \kappa^{kj} \sum_{n=-\infty}^{\infty} r_{j+nm}, \qquad (8.4\text{-}71)$$

where $\kappa = \exp[2\pi j/m]$. Now, consider a version of this last expression written for the case $m' \equiv im$. For each k, $1 \leq k \leq m - 1$, it is possible to write

$$-\lambda_{ik}' = \sum_{j=0}^{m'-1} \left(\kappa^{1/i}\right)^{ikj} \sum_{n=-\infty}^{\infty} r_{j+nm'}$$

$$= \sum_{n=-\infty}^{\infty} \sum_{j=0}^{im-1} \kappa^{kj} r_{j+nim}. \qquad (8.4\text{-}72)$$

On the inner summation, as j ranges from $j = 0$ to $j = im - 1$, the factor κ^{kj} runs through a batch of i periods, each period of length m. The outer sum uses an index that selects different batches of rates. Hence, it is possible to rewrite the double sum so that the inner sum ranges from $j = 0$ to $j = m - 1$ if the batch size is reduced to one period. This observation leads to

$$\sum_{n=-\infty}^{\infty} \sum_{j=0}^{im-1} \kappa^{kj} r_{j+nim} = \sum_{n=-\infty}^{\infty} \sum_{j=0}^{m-1} \kappa^{kj} r_{j+nm} \qquad (8.4\text{-}73)$$

$$= -\lambda_k.$$

Finally, a comparison of Equations 8.4-72 and 8.4-73 leads to the desired result $\lambda_{ik}' = \lambda_k$, $1 \leq k \leq m - 1$.

As claimed previously, the rates can be expressed in terms of the first $m - 1$ nonzero eigenvalues. To see this, multiply Equation 8.4-70 by κ^{-kj}, sum the product from $k = 0$ to $k = m - 1$, and use the intermediate result

$$\sum_{k=0}^{m-1} \kappa^{k(i-j)} = m \quad \text{for } i-j = 0$$
$$= 0 \quad \text{for } i-j \neq 0 \qquad (8.4\text{-}74)$$

to obtain

$$-\sum_{k=0}^{m-1} \lambda_k \kappa^{-kj} = \sum_{k=0}^{m-1}\sum_{i=0}^{m-1} \kappa^{k(i-j)} a_i$$

$$= \sum_{i=0}^{m-1} a_i \sum_{k=0}^{m-1} \kappa^{k(i-j)} \qquad (8.4\text{-}75)$$

$$= m a_j.$$

Finally, Equation 8.4-75 leads to the desired result

$$a_j = -\frac{1}{m}\sum_{k=1}^{m-1} \lambda_k \kappa^{-kj}, \qquad (8.4\text{-}76)$$

where $0 \le j \le m - 1$. This last equation expresses the transition rates in terms of the first $m - 1$ nonzero eigenvalues.

Equation 8.4-76 can be used to calculate the transition rates by using the eigenvalues, which must be computed first (eigenvalue calculation is covered in Sections 8.5 and 8.6). However, Equation 8.4-76 must be used carefully; when applying this result, the appropriate numbering of the λ_k must be observed. For the cases $m = 3$ and $m = 4$, numbering of the eigenvalues can be accomplished by using the requirement that

$$\lambda_{m-k} = \lambda_k^*, \quad 1 \le k \le m-1, \qquad (8.4\text{-}77)$$

a result that follows from Equation 8.4-70 (λ_k^* denotes the complex conjugate of λ_k). However, for $m \ge 5$, Equation 8.4-77 does not provide enough guidance to number the eigenvalues for use in Equation 8.4-76. For example, for the case $m = 5$, Equation 8.4-77 indicates that the four nonzero eigenvalues occur as two pairs of complex conjugate numbers; eigenvalue numbering cannot be inferred from this information. For $m \ge 5$, Equation 8.4-68 must be used to establish eigenvalue numbering. That is, eigenvalue numbering is established by inspection of the eigenvectors. As shown by Equation 8.4-68, the kth eigenvalue corresponds to an eigenvector with adjacent components that vary by the factor κ^k (this method of eigenvalue numbering differs from the method used in Section 8.4.5). In Sections 8.5 and 8.6, this fact is used in the numerical algorithm for computing the eigenvalues.

NUMERICAL METHODS/NOISE ANALYSIS IN NONLINEAR PLL MODEL

The mean occurrence of transitions of any kind (i.e., the frequency of transition events, regardless of jump size) can be expressed as the average of the first m − 1 nonzero eigenvalues. To obtain this result, substitute j = 0 into Equation 8.4-76 to obtain

$$a_0 = -\frac{1}{m} \sum_{k=1}^{m-1} \lambda_k. \qquad (8.4\text{-}78)$$

Now, Equations 8.4-78 and 8.4-67 can be combined to define the quantity

$$R \equiv \sum_{k=1}^{m-1} a_k = \frac{1}{m} \sum_{k=1}^{m-1} \lambda_k. \qquad (8.4\text{-}79)$$

As the sum of the a_k, $1 \le k \le m - 1$, R is the rate of occurrence for transitions of any kind. According to Equation 8.4-79, this "frequency of undesirable events" is the average of the m − 1 smallest nonzero eigenvalues.

Transition rate R is a characterization of the cycle slip phenomenon. In a given PLL application, it may be a sufficient descriptor of the severity of the cycle slip problem. In calculating R, a few details should be noted. First, eigenvalue numbering is not important when computing the Average 8.4-79. Second, as implied by Equation 8.4-77, only the real parts of the eigenvalues influence R. Finally, integer m should be incremented upward until no significant increase is noted in the eigenvalue average R; integer m is determined by this method when R is the quantity of interest.

8.5 THE FIRST-ORDER PLL AS A SIMPLE EXAMPLE

The first-order PLL with a constant frequency reference can serve as the basis of a relatively simple example that illustrates some of the ideas discussed in Section 8.4. In addition to showing how the theory can be applied, a first-order example can illustrate the strengths and weaknesses of the theory. In Section 7.2.7, many closed-form results are given that describe exactly the cycle slip phenomenon in the first-order PLL (similar results are not available for higher-order PLLs). These exact results are compared here with numerical results obtained by using the eigenvalue-based theory, and this effort supports the general claim that eigenvalue-based methods can provide accurate results.

Recall that in the first-order PLL under consideration, transitions with a size greater than one domain are not possible. In terms of the unrestricted jump model discussed in Section 8.4.9, only r_1 and r_{-1} are nonzero for the first-order PLL under consideration. As shown in this section, the sum of these rates can be determined by using the eigenvalue-based theory with m = 2, and both of these rates can be determined by using m = 4. For the first-order example considered here, there is little reason for using a value of m greater than four.

For the case m = 2, a_1 is the only transition rate defined for the cyclic model used to describe the first-order loop. An examination of Equation 8.4-63 and Figure 8.9 leads to the conclusion

$$a_1 = r_1 + r_{-1}. \tag{8.5-1}$$

Equations 8.4-69 and 8.4-76 yield $2a_1 = \lambda_1$; this result produces

$$\frac{\lambda_1}{2} = r_1 + r_{-1} \tag{8.5-2}$$

with the aid of Equation 8.5-1. Hence, for the unrestricted jump model, the sum of the nonzero transition rates can be determined by using m = 2.

For the first-order PLL, values of m larger than two must be used in order to compute r_1 and r_{-1}. For these values of m, the transition rates of the cyclic and unconstrained jump models are related by

$$\begin{aligned} a_1 &= r_1 \\ a_2 &= 0 \\ &\vdots \\ a_{m-2} &= 0 \\ a_{m-1} &= r_{-1}, \end{aligned} \tag{8.5-3}$$

a result that follows from Equation 8.4-63 and Figure 8.9. Also, Equations 8.4-70 and 8.4-67 can be combined to express

$$\begin{aligned} \lambda_k &= (a_1 + a_{m-1}) - \kappa^k a_1 - \kappa^{k(m-1)} a_{m-1} \\ &= (1 - \kappa^k) a_1 + (1 - \kappa^{-k}) a_{m-1}, \end{aligned} \tag{8.5-4}$$

where $0 \leq k \leq m - 1$, and $m \geq 3$.

NUMERICAL METHODS/NOISE ANALYSIS IN NONLINEAR PLL MODEL 343

Consider applying these results to the case m = 4. First, Equation 8.5-3 yields

$$a_1 = r_1$$
$$a_2 = 0 \qquad (8.5\text{-}5)$$
$$a_3 = r_{-1}.$$

Also, Equations 8.4-69 and 8.5-4 yield

$$\lambda_1 = [1-j]a_1 + [1+j]a_3$$
$$\lambda_2 = 2(a_1 + a_3) \qquad (8.5\text{-}6)$$
$$\lambda_3 = [1+j]a_1 + [1-j]a_3.$$

Now, the results of the previous two equations can be combined to obtain

$$r_1 = \frac{1}{2}\left[\operatorname{Re}[\lambda_1] - \operatorname{Im}[\lambda_1]\right]$$
$$r_{-1} = \frac{1}{2}\left[\operatorname{Re}[\lambda_1] + \operatorname{Im}[\lambda_1]\right], \qquad (8.5\text{-}7)$$

where $\operatorname{Re}[\lambda_1]$ and $\operatorname{Im}[\lambda_1]$ denote the real and imaginary parts, respectively, of eigenvalue λ_1. For the unrestricted phase error model applied to the first-order PLL, this last result can be used to compute the rates in terms of λ_1 (computed for the case m = 4). Finally, Equation 8.5-6 implies

$$\lambda_1 = \lambda_3^*$$
$$\lambda_2 = 2\operatorname{Re}[\lambda_1] \qquad (8.5\text{-}8)$$

for the case m = 4.

8.5.1 The Eigenvalue Problem Formulated in Terms of a First-Order System of Differential Equations

For the first-order PLL, the eigenvalue problem is formulated in terms of a first-order system of linear differential equations with periodic coefficients. Each eigenvalue and its eigenfunction appears as a parameter and a solution, respectively, of this system of differential equations. In Sections 8.5.2 and 8.5.3, this system is integrated numerically in an algorithm that computes the desired eigenvalue and eigenfunction.

In Section 7.2, the Fokker–Planck operator is derived for a first-order PLL with a noisy reference; this operator is given by Equation 7.2-18. For the m-attractor cyclic model, the eigenvalues and their associated $2\pi m$-periodic eigenfunctions must satisfy

$$L_{FP}[e_k] = -\lambda_k e_k \qquad (8.5\text{-}9)$$

$$L_{FP} \equiv -\frac{\partial}{\partial \phi_m}(\omega_\Delta' - \sin\phi_m) + \rho^{-1}\frac{\partial^2}{\partial \phi_m^2}. \qquad (8.5\text{-}10)$$

This equation uses the modulo-$2\pi m$ phase error variable ϕ_m first introduced in Section 7.1.3.

This eigenvalue problem can be formulated in terms of a first-order system. Define $\dot{e}_k \equiv de_k/d\phi_m$, and use the last two equations to write the system

$$\frac{d}{d\phi_m}\begin{bmatrix} e_k \\ \dot{e}_k \end{bmatrix} = \begin{bmatrix} 0 & 1 \\ -\rho(\lambda_k + \cos\phi_m) & \rho(\omega_\Delta' - \sin\phi_m) \end{bmatrix}\begin{bmatrix} e_k \\ \dot{e}_k \end{bmatrix}. \qquad (8.5\text{-}11)$$

Eigenvalue λ_k appears as an unknown parameter in Equation 8.5-11. In the next two sections this system is used in an algorithm for computing the desired λ_k and e_k. With $e_{k\emptyset} \equiv e_k(0) = 1$, the algorithm seeks values for λ_k and the initial condition $\dot{e}_{k\emptyset} \equiv \dot{e}_k(0)$ so that Equation 8.5-11 can be integrated numerically to obtain the eigenfunction $e_k(\phi_m)$, $0 \leq \phi_m \leq 2\pi$. It accomplishes this by solving algebraic constraints that the unknown λ_k and $\dot{e}_{k\emptyset}$ must satisfy. In Section 8.5.3, these algebraic constraints are developed by using geometric considerations outlined in the next section (see also Section 12.8 of Meyr and Ascheid[8] for a discussion of the geometric foundation of the constraint).

8.5.2 Decomposition of the Vector Space of $2\pi m$-Periodic Functions

Consider the vector space $\mathcal{L}$ of $2\pi m$-periodic, square-integrable (on the phase interval $[0, 2\pi m]$) functions. As is well known from the theory of exponential Fourier series, the set of functions

NUMERICAL METHODS/NOISE ANALYSIS IN NONLINEAR PLL MODEL

$$z_n(\phi_m) \equiv \exp\left(jn\frac{\phi_m}{m}\right), \qquad (8.5\text{-}12)$$

$-\infty < n < \infty$, serves as a basis for this space. In this section, the eigenfunctions of interest are shown to lie in m orthogonal subspaces of $\mathcal{L}$. This characterization of the eigenfunctions is used to form constraints that can be solved numerically for the m smallest eigenvalues.

As given by Equation 8.5-10, the Fokker–Planck operator maps $\mathcal{L}$ into itself. By using simple trigonometry, operator L_{FP} can be shown to map z_n according to

$$L_{FP}[z_n(\phi_m)] = -\frac{n-m}{2m}z_{n-m}(\phi_m) - \left[j\omega_\Delta' \frac{n}{m} + \rho^{-1}\left(\frac{n}{m}\right)^2\right]z_n(\phi_m)$$

$$+ \frac{n+m}{2m}z_{n+m}(\phi_m). \qquad (8.5\text{-}13)$$

Operator L_{FP} maps z_n into a vector with components in the orthogonal directions z_{n-m}, z_n, and z_{n+m}. In the Image 8.5-13, the three components are separated from each other by m basis vectors.

Based on this observation, the vector space $\mathcal{L}$ can be "split up" into m orthogonal subspaces $\mathcal{L}^{(k)}$, $0 \le k \le m - 1$, each of which is invariant under the mapping L_{FP}. First, introduce the notation

$$\mathcal{L} \equiv \{z_r(\phi_m)\}_{r=-\infty}^{r=+\infty}, \qquad (8.5\text{-}14)$$

to represent the fact that $\mathcal{L}$ is the span (i.e., generated by taking all linear combinations) of the vectors z_r, $-\infty < r < \infty$. Now, "split up" $\mathcal{L}$ into the m orthogonal subspaces

$$\mathcal{L}^{(0)} \equiv \{z_{rm+0}(\phi_m)\}_{r=-\infty}^{r=+\infty}$$

$$\mathcal{L}^{(1)} \equiv \{z_{rm+1}(\phi_m)\}_{r=-\infty}^{r=+\infty}$$

$$\mathcal{L}^{(2)} \equiv \{z_{rm+2}(\phi_m)\}_{r=-\infty}^{r=+\infty} \qquad (8.5\text{-}15)$$

$$\vdots \qquad \vdots$$

$$\mathcal{L}^{(m-1)} \equiv \{z_{rm+(m-1)}(\phi_m)\}_{r=-\infty}^{r=+\infty}.$$

Finally, note that $\mathcal{L}$ is the direct sum of the subspaces, and this fact is expressed as

$$\mathcal{L} = \mathcal{L}^{(0)} \oplus \mathcal{L}^{(1)} \oplus \mathcal{L}^{(2)} \oplus \ldots \mathcal{L}^{(m-1)}. \quad (8.5\text{-}16)$$

Furthermore, each subspace is invariant under L_{FP} (each subspace is mapped into itself), and this observation is expressed as

$$L_{FP} : \mathcal{L}^{(k)} \to \mathcal{L}^{(k)}. \quad (8.5\text{-}17)$$

This decomposition of $\mathcal{L}$ is invariant under the mapping induced by L_{FP}.

Equations 8.5-16 and 8.5-17 imply that the eigenfunctions of L_{FP} must lie in the subspaces. For subspace $\mathcal{L}^{(k)}$, $0 \le k \le m-1$, the eigenfunctions are denoted as $e_i^{(k)}$, where subscript $i \ge 0$ serves to index the eigenfunctions within the subspace. Using similar notation, let $\lambda_i^{(k)}$ denote the eigenvalue that corresponds to $e_i^{(k)}$, and assume that these eigenvalues are ordered by magnitude. That is, as indexed on subscript i, the pairs $\lambda_i^{(k)}$, $e_i^{(k)}$ are ordered so that $j > i$ implies

$$\left|\lambda_j^{(k)}\right| > \left|\lambda_i^{(k)}\right|, \quad (8.5\text{-}18)$$

$0 \le k \le m-1$.

In terms of the notation just introduced, the quantity $\lambda_0^{(k)}$ is the smallest (in magnitude) eigenvalue associated with the kth subspace (note that $\lambda_0^{(0)} = 0$). Also, as outlined in the remainder of this section, it is equal to eigenvalue λ_k required for implementing the slip rate algorithm based on the "coarse-grained" cyclic jump model. That is, $\lambda_0^{(k)} = \lambda_k$, $1 \le k \le m-1$, are the eigenvalues that must be used in Equation 8.4-76 to compute the slip rates a_i. Hence, the goal is to calculate the smallest (in magnitude) eigenvalue associated with an eigenfunction in each subspace $\mathcal{L}^{(k)}$, $1 \le k \le m-1$.

The claim $\lambda_0^{(k)} = \lambda_k$, $0 \le k \le m-1$ can be argued easily. Consider subspace $\mathcal{L}^{(k)}$ that is spanned by vectors z_{rm+k}, $-\infty < r < \infty$, each of which satisfies the property

$$z_{rm+k}(\phi_m + 2\pi) = \kappa^k z_{rm+k}(\phi_m), \quad (8.5\text{-}19)$$

where $\kappa \equiv \exp(j2\pi/m)$. Now, all eigenfunctions in $\mathcal{L}^{(k)}$ must have the Property 8.5-19. And, in each $\mathcal{L}^{(k)}$, $0 \le k \le m-1$, there is exactly one eigenfunction associated with an eigenvalue in the set of m smallest (in magnitude) eigenvalues. This eigenfunction is denoted as $e_0^{(k)}$, it is sharply concentrated on the m domains, and it can be scaled to satisfy

$$e_0^{(k)}(\phi_m) \approx \kappa^{ki} e_0(\phi_m), \qquad \phi_m \in R_i, \qquad (8.5\text{-}20)$$

$1 \le i \le m$. This result follows from Equation 8.5-19 (since adjacent domains are 2π radians apart in phase) and the qualitative properties outlined in Section 8.4.6. Now, integrate $e_0^{(k)}$ over each domain, and compute vector $\vec{e}_k$ given by Equation 8.4-49. Up to a scale factor, the computed $\vec{e}_k$ has the Form 8.4-68. Hence, for $0 \le k \le m - 1$, eigenvalue $\lambda_0^{(k)}$ is equal to λ_k. To compute the rates a_k, $1 \le k \le m - 1$, Equation 8.4-76 can be used. This equation uses the nonzero eigenvalues $\lambda_0^{(1)}$, $\lambda_0^{(2)}$, ..., $\lambda_0^{(m-1)}$; for $1 \le k \le m - 1$, $\lambda_0^{(k)}$ is the smallest (in magnitude) eigenvalue associated with an eigenfunction in subspace $\mathscr{L}^{(k)}$.

8.5.3 An Algorithm for Computing the $\lambda_0^{(k)}$

Each basis vector z_{rm+k}, $-\infty < r < \infty$, satisfies Equation 8.5-19, and eigenfunction $e_0^{(k)}$ is composed of these basis vectors. Hence, the eigenfunction has Property 8.5-19. Starting at $\phi_m = 0$ (defined as the point where domain R_1 is centered), the eigenfunction changes in scale by the factor κ^k by the time $\phi_m = 2\pi$ (where domain R_2 is centered) is reached. This feature establishes the boundary conditions

$$\begin{aligned} e_0^{(k)}(2\pi) &= \kappa^k \, e_0^{(k)}(0) \\ \dot{e}_0^{(k)}(2\pi) &= \kappa^k \, \dot{e}_0^{(k)}(0). \end{aligned} \qquad (8.5\text{-}21)$$

Because of the scaling feature exhibited by κ^k in Equation 8.5-21, the quantity $\kappa^k = \exp[j2\pi k/m]$ is known as a *multiplier* of Equation 8.5-11 (see Section 2.1 of Yakubovich and Starzhinskii[45] for an introduction to multipliers). Equation 8.5-21 serves to impose constraints on Equation 8.5-11, the system of linear differential equations that $e_0^{(k)}$ satisfies. These constraints can be solved for $\lambda_k = \lambda_0^{(k)}$ and the initial conditions that must be imposed to obtain eigenfunction $e_0^{(k)}$ by a simple numerical integration.

The development of these constraints is straightforward. Basically, set $e_{k\varnothing} \equiv e_k(0)$ to any convenient constant value, and let λ_k and $\dot{e}_{k\varnothing} \equiv \dot{e}_k(0)$ be unknowns that can be solved for by an iterative procedure. Start by using estimates of these unknowns, and integrate Equation 8.5-11 over a 2π interval. After the integration, use the computed end points to evaluate the error incurred when trying to achieve equality in Equation 8.5-21. Use this error to update the estimates of unknown λ_k and $\dot{e}_{k\varnothing}$. Then restart this iterative process with the new estimates of the unknowns. After iterating to convergence, the value of λ_k is the desired eigenvalue $\lambda_0^{(k)}$. In the remainder of this section the implementation of this algorithm is discussed in detail.

As discussed in the previous paragraph, algebraic equations satisfied by λ_k and $\dot{e}_{k\varnothing}$ can be developed by using Equation 8.5-21. As initial conditions at $\phi_m = 0$, use $e_{k\varnothing} \equiv e_k(0) = 1$ and an estimate of the pair $(\dot{e}_{k\varnothing}, \lambda_k)$, and numerically integrate Equation 8.5-11 in a *forward* direction (i.e., increasing values of ϕ_m) over the interval $0 \le \phi_m \le \pi$ to produce the vector $[e_k^{(f)}(\phi_m; \dot{e}_{k\varnothing}, \lambda_k) \; \dot{e}_k^{(f)}(\phi_m; \dot{e}_{k\varnothing}, \lambda_k)]^T$ (superscript (f) denotes a forward integration). Note that

$$\left[e_k^{(f)}(\phi_m; \dot{e}_{k\varnothing}, \lambda_k) \; \dot{e}_k^{(f)}(\phi_m; \dot{e}_{k\varnothing}, \lambda_k)\right]^T \bigg|_{\phi_m = 0} = \begin{bmatrix} 1 & \dot{e}_{k\varnothing} \end{bmatrix}^T. \quad (8.5\text{-}22)$$

Now, starting at $\phi_m = 2\pi$, integrate Equation 8.5-11 in a *backward* direction (i.e., decreasing ϕ_m) over the interval $2\pi \ge \phi_m \ge \pi$. In this second integration, use λ_k, $e_k(2\pi) = \kappa^k$ and $\dot{e}_k(2\pi) = \kappa^k \dot{e}_{k\varnothing}$ (λ_k and $\dot{e}_{k\varnothing}$ are the same values used in the forward integration) as initial conditions at the starting point $\phi_m = 2\pi$, and backward integrate Equation 8.5-11 to produce the vector $[e_k^{(b)}(\phi_m; \dot{e}_{k\varnothing}, \lambda_k) \; \dot{e}_k^{(b)}(\phi_m; \dot{e}_{k\varnothing}, \lambda_k)]^T$ (superscript (b) denotes that a backwards integration is required). Note that

$$\left[e_k^{(b)}(\phi_m; \dot{e}_{k\varnothing}, \lambda_k) \; \dot{e}_k^{(b)}(\phi_m; \dot{e}_{k\varnothing}, \lambda_k)\right]^T \bigg|_{\phi_m = 2\pi} = \kappa^k \begin{bmatrix} 1 & \dot{e}_{k\varnothing} \end{bmatrix}^T. \quad (8.5\text{-}23)$$

Unknowns $\dot{e}_{k\varnothing}$ and λ_k must be selected so that the results of the two integrations match up at the midpoint $\phi_m = \pi$. That is, unknowns $\dot{e}_{k\varnothing}$ and λ_k must be selected to satisfy the algebraic equations

$$\begin{aligned} g_1(\dot{e}_{k\varnothing}, \lambda_k) &\equiv e_k^{(f)}(\pi; \dot{e}_{k\varnothing}, \lambda_k) - e_k^{(b)}(\pi; \dot{e}_{k\varnothing}, \lambda_k) = 0 \\ g_2(\dot{e}_{k\varnothing}, \lambda_k) &\equiv \dot{e}_k^{(f)}(\pi; \dot{e}_{k\varnothing}, \lambda_k) - \dot{e}_k^{(b)}(\pi; \dot{e}_{k\varnothing}, \lambda_k) = 0. \end{aligned} \quad (8.5\text{-}24)$$

Equation 8.5-24 must be solved numerically, and details of a computer program for accomplishing this are discussed in Appendix 8.5.1. The main program passes initial, user-supplied estimates for Re[λ_k], Im[λ_k], Re[$\dot{e}_{k\varnothing}$], and Im[$\dot{e}_{k\varnothing}$] to the root-finding subroutine DNEQNF (described in Section 8.1). Then, DNEQNF passes these initial estimates to the user-supplied routine FCN that integrates the real and imaginary parts of Equation 8.5-11. The results of this integration are used to compute Equation 8.5-24, and FCN passes values for g_1 and g_2 back to DNEQNF where they are used to update estimates of the unknowns.

The root-finding subroutine iterates this procedure until it converges (calling FCN many times); each time FCN is called, it is supplied with new and more accurate estimates for $\text{Re}[\lambda_k]$, $\text{Im}[\lambda_k]$, $\text{Re}[\dot{e}_{k\varnothing}]$, and $\text{Im}[\dot{e}_{k\varnothing}]$. In general, root-finding subroutines based on the well-known Newton–Raphson technique (or a derivative of this technique) seem to work best, and routines can be used that compute a finite-difference approximation to the Jacobian.

For small values of signal-to-noise ratio ρ and detuning ω_Δ', the user-supplied initial estimates for $\dot{e}_{k\varnothing}$ and λ_k are not extremely critical, and convergence of DNEQNF is easy to obtain (however, the problems outlined in Section 8.2.5 have counterparts here). Also, in each case (for each k), the root $\lambda_k \equiv \lambda_0^{(k)}$ that must be computed is the root with the smallest magnitude, and it is well separated (typically, by one or more orders of magnitude) from the larger roots of Equation 8.5-24. While computing the values of $\lambda_0^{(k)}$ given in the next section, the algorithm never converged to the wrong root (this potential problem is simple to identify by inspecting the computed eigenfunction).

8.5.4 Numerical Results

For the cases m = 2 and m = 4, both eigenvalues and eigenfunctions were computed by using the algorithm and ideas outlined in the previous section. Values of ρ between 1 and 8 dB were used to obtain these results. For m = 2, results are given for both of the permissible values of k; note that k = 0 (k = 1) corresponds to a multiplier of $\kappa^k = 1$ ($\kappa^k = -1$). For the cases with m = 4, the three significant, nonzero eigenvalues can be computed from what is given. The eigenfunction plots show symmetry since they were computed for the case $\omega_\Delta' = 0$; however, they become skewed for nonzero values of detuning. In what follows, these results are discussed in detail; where possible, they are compared with exact, closed-form results given in Section 7.2.

Table 8.1 contains results for m = 2, k = 1, and $\omega_\Delta' = 0$. For this case of zero detuning, the loop has no favored slip direction, and the eigenfunction $e_0^{(1)}(\phi_2)$ is even valued. Hence, it was no surprise when the algorithm produced values for $\dot{e}_{1\varnothing}$ that are less than 10^{-15} in magnitude. Equation 8.5-2 relates the slip rates r_1 and r_{-1} to eigenvalue λ_1. However, for the case $\omega_\Delta' = 0$ under consideration, the slip rates are equal, and $r_1 = r_{-1} \approx \lambda_1/4$ for moderate noise. However, according to Equation 7.2-55 with $\omega_\Delta' = 0$, the exact, gain-normalized slip rate (under all levels of noise) is given by

$$\frac{N_+}{G} = \frac{N_-}{G} = \frac{1}{\left(2\pi I_0(\rho)\right)^2 \rho}. \tag{8.5-25}$$

TABLE 8.1

Numerical Data for the Case m = 2, k = 1, and $\omega_\Delta' = 0$

ρ (in dB)	$\lambda_1 = \lambda_0^{(1)}$	$\dot{e}_{1\varnothing} = \dot{e}_0^{(1)}(0)$	N_+/G (Eqn. 8.5-25)	Error% (Eqn. 8.5-26)
1 dB	0.4605×10^{-1}	≈0	0.9740×10^{-2}	18.2%
2 dB	0.2350×10^{-1}	≈0	0.5317×10^{-2}	10.5%
3 dB	0.1034×10^{-1}	≈0	0.2459×10^{-2}	5.1%
4 dB	0.3735×10^{-2}	≈0	0.9149×10^{-3}	2.1%
5 dB	0.1039×10^{-2}	≈0	0.2580×10^{-3}	0.68%
6 dB	0.2061×10^{-3}	≈0	0.5147×10^{-4}	0.16%
7 dB	0.2668×10^{-4}	≈0	0.6668×10^{-5}	0.07%
8 dB	0.2017×10^{-5}	≈0	0.5041×10^{-6}	0.18%

This exact result is compared with $r_1 = r_{-1} \approx \lambda_1/4$, and the percent error

$$\text{Error\%} \equiv \frac{100}{(N_+/G)} \left| \frac{\lambda_1}{4} - \frac{N_+}{G} \right| \qquad (8.5\text{-}26)$$

is given in Table 8.1. At $\rho = 1$ dB, the percent error is 18.2%; note that the percent error decreases as ρ increases. At $\rho = 1$ dB, the poor results suggest that the so-called "moderate noise assumption" (that the transition rates can be approximated by the m − 1 smallest, nonzero eigenvalues of the Fokker–Planck operator) is questionable. However, from a practical applications standpoint, this is not a limitation of the algorithm; in most communication system and instrumentation applications, the value of SNR ρ must be significantly larger than 1 dB for the PLL to be useful.

The numerical results given in Table 8.2 correspond to the case m = 2, k = 1, and $\omega_\Delta' = \sin 10°$. Since ω_Δ' is positive, cycle slips in the forward direction are more likely. For m = 2, individual slip rates cannot be determined from the computed data. Instead, according to Equation 8.5-2 the sum $r_1 + r_{-1}$ is approximated by one half of eigenvalue λ_1. However, by using Equation 7.2-53 an exact expression for the sum of the normalized slip rates is given by

$$\frac{N_+}{G} + \frac{N_-}{G} = \mathfrak{I}_{ss} \coth(\pi \rho \omega_\Delta'), \qquad (8.5\text{-}27)$$

where $\mathfrak{I}_{ss}$ is given by Equation 7.2-36. Data from this exact result appears in Table 8.2. When using $r_1 + r_{-1} \approx \lambda_1/2$, the absolute error is

$$\text{Error\%} \equiv \frac{100}{(N_+ + N_-)/G} \left| \frac{\lambda_1}{2} - \frac{N_+ + N_-}{G} \right|, \qquad (8.5\text{-}28)$$

NUMERICAL METHODS/NOISE ANALYSIS IN NONLINEAR PLL MODEL

TABLE 8.2

Numerical Data for the Case $m = 2$, $k = 1$, and $\omega_\Delta' = \sin(10°)$

ρ (in dB)	$\lambda_1 = \lambda_0^{(1)}$	$\dot{e}_{1\emptyset} = \dot{e}_0^{(1)}(0)$	$(N_+ + N_-)/G$ Eqn. (8.5-27)	Error% Eqn. (8.5-28)
1 dB	0.5637×10^{-1}	0.2374	0.2303×10^{-1}	22.4%
2 dB	0.3169×10^{-1}	0.2877	0.1388×10^{-1}	14.1%
3 dB	0.1612×10^{-1}	0.3540	0.7462×10^{-2}	8.0%
4 dB	0.7188×10^{-2}	0.4402	0.3458×10^{-2}	3.9%
5 dB	0.2689×10^{-2}	0.5509	0.1323×10^{-2}	1.6%
6 dB	0.7963×10^{-3}	0.6920	0.3961×10^{-3}	0.50%
7 dB	0.1737×10^{-3}	0.8705	0.8676×10^{-4}	0.09%
8 dB	0.2556×10^{-4}	1.096	0.1278×10^{-4}	0.02%

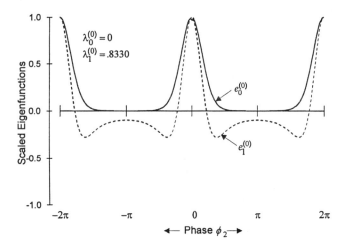

FIGURE 8.10
Scaled eigenfunctions $e_0^{(0)}$ and $e_1^{(0)}$ of subspace $\mathcal{L}^{(0)}$ for the case $m = 2$. For a first-order PLL with $\rho = 6$ dB and $\omega_\Delta' = 0$.

and values computed with this formula appear in Table 8.2. Again, the error is significant for small values of ρ (where the "moderate noise assumption" is questionable), and it decreases as ρ increases.

For the case $m = 2$, $\rho = 6$ dB, and $\omega_\Delta' = 0$, Figure 8.10 shows plots of scaled (so that both have the same unity peak value) eigenfunctions $e_0^{(0)}(\phi_2)$ and $e_1^{(0)}(\phi_2)$ as a solid and as a dashed curve, respectively. Recall that both of these lie in subspace $\mathcal{L}^{(0)}$ described by Equation 8.5-15. As expected, eigenfunction $e_0^{(0)}$ is highly concentrated on domains R_1 and R_2, centered at $\phi_2 = 0$ and $\phi_2 = \pm 2\pi$, respectively. As discussed in Section 8.4, after appropriate scaling, the plot of $e_0^{(0)}$ becomes $p_2(\phi_2)$, the steady-state density that describes phase error ϕ_2. Also, $e_0^{(0)}$ corresponds to the eigenvalue $\lambda_0 \equiv \lambda_0^{(0)} = 0$. In subspace $\mathcal{L}^{(0)}$, $e_1^{(0)}$ is the "next" (in

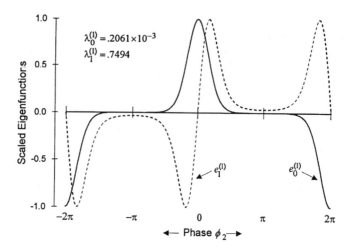

FIGURE 8.11
Scaled eigenfunctions $e_0^{(1)}$ and $e_1^{(1)}$ of subspace $\mathscr{L}^{(1)}$ for the case m = 2. For a first-order PLL with $\rho = 6$ dB and $\omega_\Delta' = 0$.

the sense given by Equation 8.5-18) eigenfunction. It is *not* concentrated on the domains; on each domain, it *cannot* be approximated as a scalar multiple of $e_0^{(0)}$. Finally, the eigenvalue $\lambda_0^{(0)} = 0.8330$ corresponds to $e_1^{(0)}$; this computed eigenvalue is not used in the slip rate algorithm.

For the case m = 2, $\rho = 6$ dB, and $\omega_\Delta' = 0$, Figure 8.11 shows plots of scaled eigenfunctions $e_0^{(1)}(\phi_2)$ and $e_1^{(1)}(\phi_2)$ as a solid and as a dashed curve, respectively. Both of these lie in subspace $\mathscr{L}^{(1)}$ described by Equation 8.5-15. As predicted by the theory discussed in Section 8.4, eigenfunction $e_0^{(1)}$ is highly concentrated on domains R_1 and R_2. On both domains it has the same general shape as $e_0^{(0)}$. However, as predicted by Equation 8.4-41, the sign of the scale factor changes from one domain to the next. Also, eigenvalue $\lambda_1 \equiv \lambda_0^{(1)} = 0.2061 \times 10^{-3}$ corresponds to $e_0^{(1)}$, and it is the smallest eigenvalue corresponding to an eigenfunction in $\mathscr{L}^{(1)}$ (this eigenvalue is required in the transition rate algorithm discussed in Section 8.4). In subspace $\mathscr{L}^{(1)}$, $e_1^{(1)}$ is the "next" (in the sense given by Equation 8.5-18) eigenfunction. On each domain, it *cannot* be approximated as a scalar multiple of $e_0^{(0)}$. Finally, the eigenvalue $\lambda_1^{(1)} = 0.7494$ corresponds to $e_1^{(1)}$. Note that eigenvalue $\lambda_1^{(1)}$ is three orders of magnitude larger than $\lambda_0^{(1)}$; such a large magnitude difference is predicted by the theory outlined in Section 8.4. Finally, eigenvalue $\lambda_1^{(1)}$ is not used in the slip rate algorithm.

Table 8.3 contains numerical results that correspond to the case m = 4, k = 1, and $\omega_\Delta' = \sin 10°$. To generate the data contained in this table, the multiplier $\kappa^k = \exp[j\pi/2] = j$ was used to integrate Equation 8.5-11

TABLE 8.3

Numerical Data for the Case m = 4, k = 1, and $\omega_\Delta' = \sin(10°)$

ρ (in dB)	$\lambda_1 = \lambda_0^{(1)}$	$\dot{e}_{1\emptyset} = \dot{e}_0^{(1)}(0)$	N_+/G Eqn. (8.5-29)	Error% Eqn. (8.5-30)
1 dB	$(0.2417 + 0.1639j) \times 10^{-1}$	$0.1973 + 0.4570 \times 10^{-1}j$	0.1838×10^{-1}	10.3%
2 dB	$(0.1424 + 0.1095j) \times 10^{-1}$	$0.2599 + 0.3037 \times 10^{-1}j$	0.1179×10^{-1}	6.8%
3 dB	$(0.7543 + 0.6394j) \times 10^{-2}$	$0.3368 + 0.1820 \times 10^{-1}j$	0.6702×10^{-2}	4.0%
4 dB	$(0.3470 + 0.3155j) \times 10^{-2}$	$0.4310 + 0.9497 \times 10^{-2}j$	0.3248×10^{-2}	2.0%
5 dB	$(0.1324 + 0.1261j) \times 10^{-2}$	$0.5469 + 0.4132 \times 10^{-2}j$	0.1282×10^{-2}	0.8%
6 dB	$(0.3962 + 0.3880j) \times 10^{-3}$	$0.6905 + 0.1419 \times 10^{-2}j$	0.3910×10^{-3}	0.3%
7 dB	$(0.8676 + 0.8614j) \times 10^{-4}$	$0.8701 + 0.3592 \times 10^{-3}j$	0.8640×10^{-4}	0.05%
8 dB	$(0.1278 + 0.1276j) \times 10^{-4}$	$1.096 + 0.6165 \times 10^{-4}j$	0.1277×10^{-4}	0.04%

and solve Equation 8.5-24. Both λ_1 and $e_1(\phi_2)$ are complex valued for this case. Since ω_Δ' is positive, cycle slips in a forward direction (increasing ϕ_2) are favored, and r_{-1} is dominant over r_1. In terms of λ_1 and the moderate noise theory, the forward slip rate is given by $r_{-1} = (\text{Re}[\lambda_1] + \text{Im}[\lambda_1])/2$, the second formula in Equation 8.5-7. However, the forward slip rate can be determined exactly (for all noise levels) by the methods outlined in Section 7.2; adapt Equation 7.2-51 to the case under consideration, and write the exact normalized forward slip rate as

$$\frac{N_+}{G} = \frac{\mathfrak{S}_{ss}}{1-\exp[-2\pi\rho\omega_\Delta']}, \qquad (8.5\text{-}29)$$

where $\mathfrak{S}_{ss}$ is given by Equation 7.2-36. This formula was used to compute the exact slip rate data that appear in Table 8.3. When using r_{-1} computed by the eigenvalue-based theory, the absolute error is

$$\text{Error\%} \equiv \frac{100}{N_+/G}\left|\frac{\text{Re}[\lambda_1]+\text{Im}[\lambda_1]}{2} - \frac{N_+}{G}\right|, \qquad (8.5\text{-}30)$$

and values computed with this formula appear in Table 8.3.

For the first-order PLL, the eigenvalue-based method only approximates cycle slip rates that can be computed exactly by using the formulas provided in Section 7.2. However, the first-order PLL example of this section has value since it shows how to use the eigenvalue-based method, and it may help the reader better understand the general theory covered in Section 8.4. This section should help prepare the reader for the challenge of applying the eigenvalue-based method to second- and higher-order PLLs for which exact cycle slip rate formulas do not exist.

8.6 EIGENVALUE METHOD APPLIED TO A SECOND-ORDER SYSTEM

In this section, the eigenvalue method is applied to a simple second-order system. The system considered here is a simple extension of what is covered in Section 8.5; it consists of a first-order PLL driven by a reference source that is frequency modulated (FM) by band-limited Gaussian noise. As in Section 8.5, the reference signal is embedded in additive white Gaussian noise (in addition to being frequency modulated). Besides being relatively simple to analyze, this example was chosen so that the computed results could be compared with similar results from the PLL literature (see Section 12.9 of Meyr and Ascheid).[8]

8.6.1 Development of the System Model and Fokker–Planck Equation

The message source y that is used to frequency modulate the reference is obtained by passing white Gaussian noise through a single-pole filter. It is described by the formal representation

$$\frac{dy}{dt} = -DGy + \sqrt{2DG}\, \sigma_y\, \eta_2, \qquad (8.6\text{-}1)$$

where $D > 0$, $\sigma_y > 0$. Also, η_2 represents zero-mean, Gaussian white noise with a double-sided spectral density of 1 watt/Hz. As can be shown, message y is band-limited, zero-mean Gaussian noise with a variance (AC power) of σ_y^2.

In the first-order PLL model, the relative frequency of the reference is $d\theta_1/dt = -Gy$ (note that loop detuning ω_Δ is assumed to be zero). By using the model development in Section 7.2.1 (use Equation 7.2-3 with ω_Δ' replaced by $-y$), the second-order model that describes the message source and PLL is described formally as

$$\frac{d\phi}{d\tau} = -y - \sin\phi - A^{-1}\eta$$

$$\frac{dy}{d\tau} = -Dy + \sqrt{2D}\, \sigma_y \left(\frac{\eta_2}{\sqrt{G}}\right), \qquad (8.6\text{-}2)$$

where $\tau \equiv Gt$. As shown by Equation 7.2-7, in Equation 8.6-2, noise sources $\eta(\tau)$ and $\eta_2(\tau)$ have $(N_0/2)\delta(\tau/G)$ and $\delta(\tau/G)$, respectively, as autocorrelation functions. Also, it is assumed that η_2 and η are independent. Finally, parameter D represents the bandwidth of y divided by the bandwidth of the PLL linear model; in this section, D is referred to as the *bandwidth ratio*.

The Fokker–Planck operator for this second-order system can be developed by using the methods outlined in Sections 7.2 and 7.3. In terms of the modulo-$2\pi m$ phase variable ϕ_m, this operator can be written as

$$L_{FP} \equiv \frac{\partial}{\partial \phi_m}\left[y + \sin\phi_m + \rho^{-1}\frac{\partial}{\partial \phi_m}\right] + D\frac{\partial}{\partial y}\left[y + \sigma_y^2 \frac{\partial}{\partial y}\right], \qquad (8.6\text{-}3)$$

where ρ is the in-loop SNR parameter that describes the effects of additive noise η acting alone (without random modulation y). The small eigenvalues of Equation 8.6-3 can be computed by using methods that are similar to those used in Sections 8.2 through 8.5.

8.6.2 Development of the Coupled System of Differential Equations

In this section, eigenfunctions of Equation 8.6-3 are represented as weighted series of Hermite polynomials. That is, each eigenfunction is represented as

$$e_k(\phi_m, y) = \exp\left[-y^2/4\sigma_y^2\right]\sum_{n=0}^{\infty} p_n(\phi_m)\psi_n(y;\sigma_y), \qquad (8.6\text{-}4)$$

where each $p_n(\phi_m)$, $n \geq 0$, is a $2\pi m$-periodic, complex-valued function. For each e_k, there is a different set of functions $\{p_n, n \geq 0\}$ that must be computed (implicit is the fact that p_n is indexed on k). In the eigenfunction expansion, note that standard deviation σ_y is used in the composition of the weighted Hermite polynomials.

In the Representation 8.6-4, the series is weighted by an exponential term that is proportional to ψ_0. The reason for this is spelled out in Sections 8.2 and 8.3. The periodic $p_n(\phi_m)$ and the eigenvalues λ_k must satisfy

$$\hat{L}_{FP}\left[\sum_{n=0}^{\infty} p_n(\phi_m)\psi_n(y;\sigma_y)\right] = -\lambda_k \sum_{n=0}^{\infty} p_n(\phi_m)\psi_n(y;\sigma_y), \qquad (8.6\text{-}5)$$

where

$$\hat{L}_{FP} \equiv \exp\left[y^2/4\sigma_y^2\right] L_{FP} \exp\left[-y^2/4\sigma_y^2\right]$$

$$= \frac{\partial}{\partial \phi_m}\left[\sigma_y(b+b^+)+\sin\phi_m + \rho^{-1}\frac{\partial}{\partial \phi_m}\right] - Db^+b. \qquad (8.6\text{-}6)$$

Note that linear differential operators b and b^+ have been used in Equation 8.6-6. These operators are defined by Equations 8.1.1-11 and 8.1.1-12, where constant c_2 is replaced by σ_y. Operators b and b^+ have important properties that are given by Equations 8.1.1-15 through 8.1.1-17.

The left-hand side of Equation 8.6-5 must be expressed as a weighted sum of Hermite polynomials. By using the approach introduced in Sections 8.2 and 8.3, this expansion has the form

$$\hat{L}_{FP}\left[\sum_{n=0}^{\infty} p_n \psi_n\right] = \sum_{n=0}^{\infty} Q_n \psi_n, \qquad (8.6\text{-}7)$$

where

$$Q_n \equiv \frac{\partial}{\partial \phi_m}\left[\sqrt{n+1}\,\sigma_y p_{n+1} + \left(\sin\phi_m + \rho^{-1}\frac{\partial}{\partial \phi_m}\right) p_n + \sqrt{n}\,\sigma_y p_{n-1}\right] \qquad (8.6\text{-}8)$$

$$-nDp_n.$$

Finally, the representation given by Equation 8.6-7 can be used with Equation 8.6-5 to write

$$Q_n + \lambda_k p_n = 0 \qquad (8.6\text{-}9)$$

for $n \geq 0$.

As indexed on n, Equation 8.6-9 represents a system of coupled, second-order linear differential equations with periodic coefficients. It can be used to write a system of first-order equations by introducing an auxiliary set of variables h_n. By following the procedure outlined in Section 8.3.2, Equation 8.6-9 can be used to write

$$\sqrt{n+1}\,\sigma_y p_{n+1} + \left(\sin\phi_m + \rho^{-1}\frac{\partial}{\partial\phi_m}\right)p_n + \sqrt{n}\,\sigma_y p_{n-1} = h_n \tag{8.6-10}$$

$$\frac{\partial}{\partial\phi_m} h_n + (\lambda_k - nD)p_n = 0$$

for $n \geq 0$. Finally, this last result leads to the explicit system

$$n = 0: \quad \frac{\partial}{\partial\phi_m} p_0 = -\rho\left[\sigma_y p_1 + \sin\phi_m\, p_0 - h_0\right]$$

$$\frac{\partial}{\partial\phi_m} h_0 = -\lambda_k p_0$$

$$n = 1: \quad \frac{\partial}{\partial\phi_m} p_1 = -\rho\left[\sqrt{2}\,\sigma_y p_2 + \sin\phi_m\, p_1 + \sigma_y p_0 - h_1\right]$$

$$\frac{\partial}{\partial\phi_m} h_1 = -(\lambda_k - D)p_1 \tag{8.6-11}$$

$$n = 2: \quad \frac{\partial}{\partial\phi_m} p_2 = -\rho\left[\sqrt{3}\,\sigma_y p_3 + \sin\phi_m\, p_2 + \sqrt{2}\,\sigma_y p_1 - h_2\right]$$

$$\frac{\partial}{\partial\phi_m} h_2 = -(\lambda_k - 2D)p_2$$

$$\vdots \qquad \vdots \qquad \vdots$$

For computational purposes, this infinite-dimensional system must be truncated to include equations through $n = n_0$ (i.e., include terms ψ_n, $0 \leq n \leq n_0$, in the expansion of e_k). Once truncated to the first $2n_0 + 2$ equations, the periodic solution of the system and λ_k must be computed; a procedure to accomplish this is discussed in the next section.

8.6.3 Algorithm for Computing the p_n, h_n, and λ_k

The ideas and notation introduced in Section 8.5 can be used here. For each fixed integer k, $0 \leq k \leq m - 1$, there exists a sequence of eigenvalues $\lambda_i^{(k)}$, i in an index set I of nonnegative integers. For each k and i, eigenvalue $\lambda_i^{(k)}$ is associated with an eigenfunction $e_i^{(k)}$ with the Representation 8.6-4, where the $2\pi m$-periodic weight functions p_n, $n \geq 0$, are in subspace $\mathcal{L}^{(k)}$ defined by Equation 8.5-15. Also, according to magnitude, the eigenvalues $\lambda_i^{(k)}$ are indexed on the nonnegative integers $i \in I$ (the ordering established by Equation 8.5-18 is used), and $\lambda_0^{(k)}$ is the smallest (also $\lambda_0 \equiv \lambda_0^{(0)} = 0$). As discussed in Section 8.5, the small eigenvalues $\lambda_k \equiv \lambda_0^{(k)}$, $1 \leq k \leq m-1$, are of interest since they can be used with Equation 8.4-76 to compute the slip rates. In the Expansion 8.6-4 of $e_i^{(k)}$, the $2\pi m$-periodic weight functions p_n, $n \geq 0$, depend on indices k and i; when it is helpful to denote this, the notation $\{p_n\}_i^{(k)}$, $0 \leq n < \infty$ is used.

When $\lambda_i^{(k)}$ is used on the right-hand side of Equation 8.6-11, this infinite system has a $2\pi m$-periodic solution with pairs $\{p_n, h_n\}_i^{(k)}$, $0 \leq n < \infty$, in the set

$$S^{(k)} \equiv \left[\{p, h\} : p \in \mathcal{L}^{(k)}, h \in \mathcal{L}^{(k)} \right]. \qquad (8.6\text{-}12)$$

For each k, $1 \leq k \leq m-1$, the algorithm discussed in this section can approximate the small eigenvalue $\lambda_0^{(k)}$ and the pairs $\{p_n, h_n\}_0^{(k)} \in S^{(k)}$, $0 \leq n \leq n_0$. Integer n_0 sets the number of terms in the approximation, and it should be incremented upward until convergence is noted in the computed approximation of $\lambda_0^{(k)}$.

For each k, $1 \leq k \leq m-1$, numerically approximating the small eigenvalue $\lambda_0^{(k)}$ and the relevant $2\pi m$-periodic pairs $\{p_n, h_n\}_0^{(k)} \in S^{(k)}$, $0 \leq n \leq n_0$, involves determining unknown initial conditions. Like every example discussed up to this point, these unknowns are formulated as roots of an algebraic system solved by iteration using DNEQNF (or equivalent). Since eigenfunctions are determined up to a freely chosen multiplicative scale factor, the first initial condition can be selected as $p_{0\emptyset} \equiv p_0(0) = 1$. Now, given System 8.6-11 truncated to the first $2n_0 + 2$ differential equations, the number of unknown initial conditions that must be found is $2n_0 + 1$, and they are denoted as $p_{n\emptyset} \equiv p_n(0)$, $1 \leq n \leq n_0$, and $h_{n\emptyset} \equiv h_n(0)$, $0 \leq n \leq n_0$ ($\lambda_0^{(k)}$ raises the total number of unknowns to $2n_0 + 2$). In general, all of these unknowns can be complex valued.

By using the symmetry that exists for the case $\omega_\Delta' = 0$ under consideration, the number of unknown initial conditions can be cut in half (to $n_0 + 1$ unknowns). For $n \geq 0$, inspection of Equations 8.6-10 and 8.6-11 reveals that functions p_{2n} and h_{2n+1} are even valued, and p_{2n+1} and h_{2n} are odd valued. Hence, for $n \geq 0$, the initial conditions

$$\begin{aligned} p_{2n+1}(0) &= 0 \\ h_{2n}(0) &= 0 \end{aligned} \qquad (8.6\text{-}13)$$

should be used to reduce the computational requirements. Let n_{ev} (n_{od}) denote the largest even (odd) integer less than or equal to n_0. For the

NUMERICAL METHODS/NOISE ANALYSIS IN NONLINEAR PLL MODEL 359

case $\omega_\Delta' = 0$, the unknown initial conditions are $p_{2\varnothing}, p_{4\varnothing}, \ldots, p_{n_{ev}\varnothing}, h_{1\varnothing}, h_{3\varnothing}, \ldots, h_{n_{od}\varnothing}$. On the other hand, if $\omega_\Delta' \neq 0$ (and System 8.6-11 is modified to reflect this condition), the full set of $2n_0 + 2$ unknowns must be considered ($p_{0\varnothing} = 1$ for both cases).

The algebraic equations that these unknowns satisfy can be formulated by using the procedure established in Section 8.5 for the first-order PLL. For $0 \leq n \leq n_0$, let $p_n^{(f)}$ and $h_n^{(f)}$ denote the numerical solution computed in a forward direction over the interval $[0, \pi]$. Of course, these depend on the unknown initial conditions and eigenvalue so that

$$p_n^{(f)}\left(\phi_m; p_{2\varnothing}, \ldots, p_{n_{ev}\varnothing}, h_{1\varnothing}, \ldots, h_{n_{od}\varnothing}, \lambda_k\right)\bigg|_{\phi_m = 0}$$
$$= p_{n\varnothing}, \quad n = 0, 2, 4, \cdots, n_{ev}$$
$$= 0 \quad n = 1, 3, 5, \cdots, n_{od}$$

(8.6-14)

$$h_n^{(f)}\left(\phi_m; p_{2\varnothing}, \ldots, p_{n_{ev}\varnothing}, h_{1\varnothing}, \ldots, h_{n_{od}\varnothing}, \lambda_k\right)\bigg|_{\phi_m = 0}$$
$$= h_{n\varnothing}, \quad n = 1, 3, 5, \cdots, n_{od}$$
$$= 0 \quad n = 0, 2, 4, \cdots, n_{ev}$$

for the case $\omega_\Delta' = 0$ under consideration. In a similar manner, let $p_n^{(b)}$ and $h_n^{(b)}$ denote the numerical solution computed in a reverse direction over the interval $[2\pi, \pi]$. For this backward integration, the initial conditions are

$$p_n^{(b)}\left(\phi_m; p_{2\varnothing}, \ldots, p_{n_{ev}\varnothing}, h_{1\varnothing}, \ldots, h_{n_{od}\varnothing}, \lambda_k\right)\bigg|_{\phi_m = 2\pi}$$
$$= \kappa^k p_{n\varnothing}, \quad n = 0, 2, 4, \cdots, n_{ev}$$
$$= 0 \quad n = 1, 3, 5, \cdots, n_{od}$$

(8.6-15)

$$h_n^{(b)}\left(\phi_m; p_{2\varnothing}, \ldots, p_{n_{ev}\varnothing}, h_{1\varnothing}, \ldots, h_{n_{od}\varnothing}, \lambda_k\right)\bigg|_{\phi_m = 2\pi}$$
$$= \kappa^k h_{n\varnothing}, \quad n = 1, 3, 5, \cdots, n_{od}$$
$$= 0 \quad n = 0, 2, 4, \cdots, n_{ev}.$$

This last set of initial conditions results from the fact that, over adjacent 2π intervals, the basis functions of $\mathcal{L}^{(k)}$ (and the p_n and h_n in $\mathcal{L}^{(k)}$) must change in scale by the factor κ^k (see Equation 8.5-20), where κ is given by Equation 8.4-69. Equivalently, the scalar κ^k is a multiplier of the truncated system of first-order differential equations that is integrated to implement the algorithm. Finally, the unknowns must be selected so that the forward and backward integrations match at their common end point $\phi_m = \pi$. That is, the unknowns must satisfy the algebraic system

$$\begin{aligned} g_n\left(p_{2\emptyset}, \ldots, p_{n_{ev}\emptyset}, h_{1\emptyset}, \ldots, h_{n_{od}\emptyset}, \lambda_k\right) \\ \equiv p_n^{(f)}\left(\pi; p_{2\emptyset}, \ldots, p_{n_{ev}\emptyset}, h_{1\emptyset}, \ldots, h_{n_{od}\emptyset}, \lambda_k\right) \\ - p_n^{(b)}\left(\pi; p_{2\emptyset}, \ldots, p_{n_{ev}\emptyset}, h_{1\emptyset}, \ldots, h_{n_{od}\emptyset}, \lambda_k\right) \\ = 0, \end{aligned} \qquad (8.6\text{-}16)$$

$n = 0, 2, 4, \ldots, n_{ev}$, and

$$\begin{aligned} g_n\left(p_{2\emptyset}, \ldots, p_{n_{ev}\emptyset}, h_{1\emptyset}, \ldots, h_{n_{od}\emptyset}, \lambda_k\right) \\ \equiv h_n^{(f)}\left(\pi; p_{2\emptyset}, \ldots, p_{n_{ev}\emptyset}, h_{1\emptyset}, \ldots, h_{n_{od}\emptyset}, \lambda_k\right) \\ - h_n^{(b)}\left(\pi; p_{2\emptyset}, \ldots, p_{n_{ev}\emptyset}, h_{1\emptyset}, \ldots, h_{n_{od}\emptyset}, \lambda_k\right) \\ = 0, \end{aligned} \qquad (8.6\text{-}17)$$

$n = 1, 3, 5, \ldots, n_{od}$. The system comprised of Equations 8.6-16 and 8.6-17 must be solved numerically. Some typical numerical results are given in the next section.

8.6.4 Numerical Results

The results given in Tables 8.4 through 8.6 demonstrate numerical convergence of the algorithm. Bandwidth ratio $D = 1/2$ and cyclic model order $m = 2$ were used, and the tables contain computed approximations of eigenvalue $\lambda_0^{(1)}$, the smallest eigenvalue corresponding to an eigenfunction with its pairs $\{p_n, h_n\}_0^{(1)}$, $0 \leq n < \infty$, in $S^{(1)}$. This case is relatively easy to compute since the multiplier $\kappa^1 = -1$, and the eigenvalue and eigenfunction are real valued. Values of σ_y^2 between -60 and -5 dB were used, and these values are listed in the left-hand column of each table. In the tables, the column headings indicate n_0, the highest-order Hermite polynomial used in the approximation of $e_0^{(1)}$. At least for $\sigma_y^2 \leq -10$ dB and $\rho \leq 6$ dB, the data suggest that convergence is reached by $n_0 = 5$. Also, to obtain convergence, it shows that more terms

TABLE 8.4

Computed approximations $\lambda_1 \equiv \lambda_0^{(1)}$ for the case m = 2. The value of σ_y^2 increases from row to row. The number of terms used increases from column to column. The values ρ = 2 dB and D = 1/2 were used in all calculations.

σ_y^2	$n_0 = 1$	$n_0 = 2$	$n_0 = 3$	$n_0 = 4$	$n_0 = 5$
−60 dB	0.23499×10^{-1}	0.23499×10^{-1}	0.23499×10^{-1}	0.23499×10^{-1}	0.23499×10^{-1}
−50 dB	0.23501×10^{-1}	0.23501×10^{-1}	0.23501×10^{-1}	0.23501×10^{-1}	0.23501×10^{-1}
−40 dB	0.23513×10^{-1}	0.23513×10^{-1}	0.23513×10^{-1}	0.23513×10^{-1}	0.23513×10^{-1}
−30 dB	0.23632×10^{-1}	0.23632×10^{-1}	0.23632×10^{-1}	0.23632×10^{-1}	0.23632×10^{-1}
−20 dB	0.24829×10^{-1}	0.24832×10^{-1}	0.24832×10^{-1}	0.24832×10^{-1}	0.24832×10^{-1}
−10 dB	0.37480×10^{-1}	0.37704×10^{-1}	0.37649×10^{-1}	0.37652×10^{-1}	0.37652×10^{-1}
−5 db	0.73135×10^{-1}	0.72593×10^{-1}	0.71495×10^{-1}	0.71733×10^{-1}	0.71705×10^{-1}

TABLE 8.5

Computed approximations to $\lambda_1 \equiv \lambda_0^{(1)}$ for the case m = 2. The value of σ_y^2 increases from row to row. The number of terms used increases from column to column. The values ρ = 4 dB and D = 1/2 were used in all calculations.

σ_y^2	$n_0 = 1$	$n_0 = 2$	$n_0 = 3$	$n_0 = 4$	$n_0 = 5$
−60 dB	0.37351×10^{-2}	0.37351×10^{-2}	0.37351×10^{-2}	0.37351×10^{-2}	0.37351×10^{-2}
−50 dB	0.37356×10^{-2}	0.37356×10^{-2}	0.37356×10^{-2}	0.37356×10^{-2}	0.37356×10^{-2}
−40 dB	0.37404×10^{-2}	0.37404×10^{-2}	0.37404×10^{-2}	0.37404×10^{-2}	0.37404×10^{-2}
−30 dB	0.37886×10^{-2}	0.37887×10^{-2}	0.37887×10^{-2}	0.37887×10^{-2}	0.37887×10^{-2}
−20 dB	0.42791×10^{-2}	0.42896×10^{-2}	0.42895×10^{-2}	0.42895×10^{-2}	0.42895×10^{-2}
−10 dB	0.10104×10^{-1}	0.11068×10^{-1}	0.11012×10^{-1}	0.11009×10^{-1}	0.11009×10^{-1}
−5 db	0.30978×10^{-1}	0.37655×10^{-1}	0.35620×10^{-1}	0.35803×10^{-1}	0.35837×10^{-1}

TABLE 8.6

Computed approximations to $\lambda_1 \equiv \lambda_0^{(1)}$ for the case m = 2. The value of σ_y^2 increases from row to row. The number of terms used increases from column to column. The values ρ = 6 dB and D = 1/2 were used in all calculations.

σ_y^2	$n_0 = 1$	$n_0 = 2$	$n_0 = 3$	$n_0 = 4$	$n_0 = 5$
−60 dB	0.20615×10^{-3}	0.20615×10^{-3}	0.20615×10^{-3}	0.20615×10^{-3}	0.20615×10^{-3}
−50 dB	0.20622×10^{-3}	0.20622×10^{-3}	0.20622×10^{-3}	0.20622×10^{-3}	0.20622×10^{-3}
−40 dB	0.20692×10^{-3}	0.20692×10^{-3}	0.20692×10^{-3}	0.20692×10^{-3}	0.20692×10^{-3}
−30 dB	0.21390×10^{-3}	0.21395×10^{-3}	0.21395×10^{-3}	0.21395×10^{-3}	0.21395×10^{-3}
−20 dB	0.28838×10^{-3}	0.29403×10^{-3}	0.29415×10^{-3}	0.29415×10^{-3}	0.29415×10^{-3}
−10 dB	0.15692×10^{-2}	0.23732×10^{-2}	0.24401×10^{-2}	0.24261×10^{-2}	0.24261×10^{-2}
−5 db	0.10060×10^{-1}	0.19913×10^{-1}	0.18746×10^{-1}	0.18327×10^{-1}	0.18526×10^{-1}

are required as σ_y^2 and ρ increase (σ_y^2, much more than ρ, influences the number of required terms).

Figure 8.12 shows plots of computed $\lambda_0^{(1)}$ as a function of σ_y^2 for various values of SNR ρ. Bandwidth ratio D = 1/2, cyclic model order m = 2, and $n_0 = 5$ were used to compute the data. The plot shows the

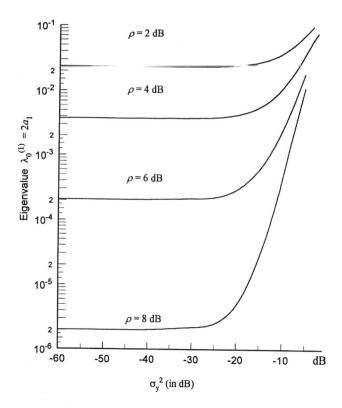

FIGURE 8.12
Eigenvalue $\lambda_0^{(1)} = 2a_1$ as a function of the modulation variance σ_y^2. Parameter ρ is the signal-to-noise ratio, and $D = 0.5$ is the bandwidth ratio.

existence of a threshold that occurs as σ_y^2 increases. Below threshold, the additive reference noise (i.e., η in Equation 8.6-2) dominates as the main cause of domain transitions, and the plots change little with increasing σ_y^2. On the other hand, above threshold, the random frequency modulation on the reference (i.e., y in Equation 8.6-2) becomes the main culprit responsible for cycle slips. Above threshold, the value of $\lambda_0^{(1)}$ increases significantly with σ_y^2.

Table 8.7 contains computed approximations to $\lambda_0^{(1)}$ for the case $D = 1/2$ and cyclic model order m = 4. For this case, the multiplier is $\kappa^1 = j$, and eigenfunction $e_0^{(1)}$ is complex valued. Hence, the truncated system of differential equations must be split into real and imaginary parts that must be integrated simultaneously. However, since detuning $\omega_\Delta' = 0$ (an assumption in this section), the loop has no favored slip direction. Because of this, Equations 8.4-63 and 8.4-70 predict real-valued $\lambda_0^{(k)}$, $1 \le k \le 3$. Also, Equation 8.4-77 yields $\lambda_0^{(1)} = \lambda_0^{(3)}$, so that Table 8.7 gives both

TABLE 8.7

Computed approximations to $\lambda_1 \equiv \lambda_0^{(1)}$ for the case m = 4. The value of σ_y^2 increases from row to row. The number of terms used differs from column to column. The values $\rho = 4$ dB and $D = 1/2$ were used in all calculations.

σ_y^2	$n_0 = 2$	$n_0 = 3$	$n_0 = 4$	$n_0 = 5$
−60 dB	0.18484×10^{-2}	0.18484×10^{-2}	0.18484×10^{-2}	0.18484×10^{-2}
−50 dB	0.18487×10^{-2}	0.18487×10^{-2}	0.18487×10^{-2}	0.18487×10^{-2}
−40 dB	0.18510×10^{-2}	0.18510×10^{-2}	0.18510×10^{-2}	0.18510×10^{-2}
−30 dB	0.18747×10^{-2}	0.18747×10^{-2}	0.18747×10^{-2}	0.18747×10^{-2}
−20 dB	0.21206×10^{-2}	0.21205×10^{-2}	0.21205×10^{-2}	0.21205×10^{-2}
−10 dB	0.54200×10^{-2}	0.53967×10^{-2}	0.53951×10^{-2}	0.53954×10^{-2}

$\lambda_0^{(1)}$ and $\lambda_0^{(3)}$. Finally, as predicted by Equations 8.4-71 through 8.4-73, eigenvalue $\lambda_0^{(2)}$ for the present case m = 4 is equal to $\lambda_0^{(1)}$ for the case m = 2 (i.e., the data contained in Tables 8.4 through 8.6).

For the case m = 4, the eigenvalues just described can be used to calculate the rates a_k, $1 \le k \le 3$, for the second-order system under consideration. With the aid of Equation 8.4-76, it is easy to show that

$$a_1 = a_3 = \frac{1}{4}\lambda_0^{(2)}$$

$$a_2 = \frac{1}{4}\left(2\lambda_0^{(1)} - \lambda_0^{(2)}\right).$$

(8.6-18)

Using the data for $\sigma_y^2 = -60$ dB in Tables 8.4 through 8.7, rate a_1 is about two orders of magnitude larger than a_2. This ratio drops to approximately one order of magnitude as σ_y^2 increases to $\sigma_y^2 = -10$ dB. Over the range of σ_y^2 employed, the result $a_1 \gg a_2$ suggests that simple transitions of one domain dominate the cycle slip behavior of this second-order system.

APPENDIX 8.1.1   HERMITE POLYNOMIALS

Sections 8.1 through 8.3 describe a method for approximating the solution of the Fokker–Planck equation that describes the steady-state probability density function for the state of a PLL driven by a noisy reference. In Sections 8.4 through 8.6, a numerical method is developed and demonstrated that it is useful in analyzing the cycle slip rate in a PLL. These numerical methods employ Hermite polynomials. This appendix introduces Hermite polynomials, and it provides the reader with information that is necessary for an understanding of the applications discussed in Chapter 8. Chapter 5 of Magnus et al.[46] and Section 10.1.4 of Risken[36] can be consulted for further information on Hermite polynomials and their use in orthogonal expansions.

8.1.1.1 Definitions and Elementary Properties

The Hermite polynomials $H_n(y)$, $n \geq 0$ are defined here in terms of derivatives of an exponential function. The first polynomial is defined as $H_0(y) = 1$, and for $n \geq 1$ the remaining polynomials are given as

$$H_n(y) = (-1)^n \exp(y^2) \frac{d^n}{dy^n} \exp(-y^2). \qquad (8.1.1\text{-}1)$$

As an example, the first few polynomials can be computed as

$$\begin{aligned} H_0(y) &= 1 \\ H_1(y) &= 2y \\ H_2(y) &= 4y^2 - 2 \\ H_3(y) &= 8y^3 - 12y. \end{aligned} \qquad (8.1.1\text{-}2)$$

Some elementary properties of the polynomials can be obtained from inspection of Equation 8.1.1-1. First, note that H_n is of degree n, and it has a leading coefficient of 2^n. Furthermore, H_n contains only even (odd) powers of y if n is even (odd) since

$$H_n(-y) = (-1)^n H_n(y). \qquad (8.1.1\text{-}3)$$

The Hermite polynomials are orthogonal with respect to the weight function

$$w(y) \equiv \exp[-y^2]. \qquad (8.1.1\text{-}4)$$

That is, the polynomials satisfy the orthogonality relationship

$$\int_{-\infty}^{\infty} w(y) H_n(y) H_m(y) dy = \begin{array}{ll} 0 & \text{for } n \neq m \\ = \sqrt{\pi}\, 2^n\, n! & \text{for } n = m. \end{array} \qquad (8.1.1\text{-}5)$$

As used in Chapter 8, the weighted Hermite polynomials are defined as

$$\psi_n(y;c_2) \equiv \frac{1}{\sqrt{n!2^n c_2 \sqrt{2\pi}}} \exp\left(\frac{-y^2}{4c_2^2}\right) H_n\left(\frac{y}{\sqrt{2}c_2}\right), \quad (8.1.1\text{-}6)$$

where integer $n \geq 0$, and c_2 denotes a positive constant that is chosen for the specific application under consideration. The weighted Hermite polynomials are real valued, and they are orthonormal since

$$\int_{-\infty}^{\infty} \psi_n(y;c_2) \psi_m(y;c_2) dy = 1 \text{ for } n = m \quad (8.1.1\text{-}7)$$

$$= 0 \text{ for } n \neq m.$$

Finally, note that $\psi_0(y;c_2)$ is proportional to a zero-mean Gaussian function; in Chapter 8, this fact proves helpful in approximating probability density functions which are nearly Gaussian in nature.

These exponentially weighted Hermite polynomials can be used as basis functions in the generalized Fourier series expansion of a function defined on $(-\infty, \infty)$. More precisely, they form a complete orthonormal sequence for the space of square-integrable functions defined on the real line. Let $f(y)$ denote any real-valued function that satisfies

$$\int_{-\infty}^{\infty} f(y)^2 \, dy \leq \infty. \quad (8.1.1\text{-}8)$$

Then the function f can be represented in the generalized Fourier series

$$f(y) = \sum_{k=0}^{\infty} \gamma_k \psi_k(y;c_2), \quad (8.1.1\text{-}9)$$

where the generalized Fourier coefficients are given as

$$\gamma_k = \int_{-\infty}^{\infty} f(y) \psi_k(y;c_2) dy. \quad (8.1.1\text{-}10)$$

Expansions of the Type 8.1.1-9 can be used to represent solutions of some differential equations. In Chapter 8, this idea is exploited to represent solutions of the Fokker–Planck equations for several PLLs.

8.1.1.2 Weighted Hermite Polynomials as Eigenfunctions of a Differential Operator

The numerical methods discussed in Chapter 8 employ the differential operators

$$b \equiv c_2 \frac{\partial}{\partial y} + \frac{1}{2} c_2^{-1} y \qquad (8.1.1\text{-}11)$$

$$b^+ \equiv -c_2 \frac{\partial}{\partial y} + \frac{1}{2} c_2^{-1} y, \qquad (8.1.1\text{-}12)$$

where c_2 is a positive constant. Note that b^+ is the formal adjoint of b (see p. 84 of Coddington and Levinson).[47] Also, note that these definitions can be used to write

$$y = c_2(b + b^+) \qquad (8.1.1\text{-}13)$$

$$\frac{\partial}{\partial y} = \frac{1}{2c_2}(b - b^+). \qquad (8.1.1\text{-}14)$$

This last equation expresses the partial differential operator in terms of the operators b and b^+.

For the examples discussed in Chapter 8, the Fokker–Planck operator is expressed in terms of the operators b and b^+, and it is used to transform a series of weighted Hermite polynomials. As a result of this, how b and b^+ transform the Hermite polynomials is important. Tedious, but straightforward algebraic calculations show that

$$b^+ \psi_n(y;c_2) = \sqrt{n+1}\, \psi_{n+1}(y;c_2) \qquad (8.1.1\text{-}15)$$

$$b\, \psi_n(y;c_2) = \sqrt{n}\, \psi_{n-1}(y;c_2) \qquad (8.1.1\text{-}16)$$

$$b^+ b\, \psi_n(y;c_2) = n\, \psi_n(y;c_2). \qquad (8.1.1\text{-}17)$$

Equation 8.1.1-17 shows that the ψ_n are eigenfunctions of the linear differential operator $b^+ b$. Finally, note that the operators b and b^+ do not commute; that is, note that $b^+ b \neq b b^+$.

APPENDIX 8.4.1   LEAST POSITIVE RESIDUE FUNCTION

The *least positive residue function* is used extensively in Section 8.4. In the literature, there is some confusion regarding the use of this function, especially with integers. Hence, it is prudent to define this function. First, assume that i, j, and m are integers and that m > 0. It is said that integer i *is congruent to* j *modulo* m if positive integer m divides without remainder the quantity i − j. Often, this is denoted as

$$i = j(\mod m). \tag{8.4.1-1}$$

Integer j is said to be a *residue* of i modulo m. If $0 \le j \le m - 1$, then j is the *least positive residue* of i modulo m (integers 0, 1, 2, ..., m − 1 form the *Residue Set*). Finally, the notation

$$j = \mod_m[i] \tag{8.4.1-2}$$

is used to denote the least positive residue of i modulo m. Note that this function satisfies $0 \le \mod_m[i] \le m - 1$.

APPENDIX 8.5.1   COMPUTING EIGENVALUES OF THE FOKKER–PLANCK OPERATOR

Section 8.5.3 discusses an algorithm for computing the significant eigenvalues of the Fokker–Planck operator used in modeling the statistical behavior of the first-order PLL. This appendix describes a simple program for implementing this algorithm. The program listing given here is well commented, and it should help the reader get started in developing similar programs for computing eigenvalues. It served as the starting point for the development of a program for the second-order system discussed in Section 8.6. The program calls routines in the IMSL™ scientific library (see IMSL™),[80] a long-time standard in numerical analysis. Finally, the program was written for Microsoft™ FORTRAN Version 5.1, but it should be easy to port the code to other FORTRAN environments.

In the first 50 lines of code, the program obtains the user-supplied inputs Rho, Wd, Mr, Mj, and the initial guesses for the roots of Equation 8.5-24. Parameter Rho is the in-loop SNR, and it should be supplied in decibels. The quantity Wd is loop detuning, and it should be supplied in degrees. In the absence of noise, the PLL suffers a static phase error of Wd degrees. Parameters Mr and Mj are the real and imaginary parts, respectively, of the complex-valued multiplier κ^k that is used in Equation 8.5-23 to compute values of Equation 8.5-24 (for example, Mr = –1 and Mj = 0 were used to calculate the data that appear in Table 8.1). These parameters are determined by the integer values m (number of attractors in the cyclic state model) and k (index to the significant eigenvalues).

The root finding is accomplished by the IMSL™ routine DNEQNF. This subroutine solves Equation 8.5-24 for its roots λ_k and $\dot{e}_{k\emptyset} \equiv \dot{e}_k(0)$ by using a modified Powell hybrid algorithm; to accomplish this it computes a finite-difference approximation to the Jacobian (a user-supplied Jacobian is not required). This equation solver passes estimates of the roots to a user-supplied routine FCN; these estimated roots are used by FCN to calculate numerical values for the functions g_1 and g_2 defined by Equation 8.5-24. These values of g_1 and g_2 are returned to DNEQNF where they are used to update the root estimates. Ideally, this iterative process continues until convergence.

The subroutine FCN uses the Runge-Kutta–Verner integration routine DIVPRK from the IMSL™ library to integrate Equation 8.5-11, a system with complex-valued dependent variables. DIVPRK calls FCN2 to obtain the real and imaginary parts of the right-hand side of Equation 8.5-11 (hence, DIVPRK solves a fourth-order system of differential equations). FCN uses the results of this numerical integration to calculate numerical vales for g_1 and g_2 which it returns to DNEQNF as described above. The interested reader should consult IMSL's™ documentation for DNEQNF and DIVPRK before executing the code listed below.

```
C   For the first-order PLL eigenvalue problem discussed in Section 8.5.
C   Calculates the Fokker–Planck operator's eigenvalue corresponding to
C   a user-specified multiplier of the equivalent monodromy matrix.
C   Accomplishes this by solving Equation 8.5-17 in the text.

C   Compile using the Microsoft™ FORTRAN Version 5.1
C   compiler for MS-DOS. Link to the IMSL™ Version 2.0 library
C   for the Microsoft™ compiler.

C   With minimal changes, this code can be used in most FORTRAN
C   environments where the IMSL™ library (any recent version) is
C   available.

      implicit double precision (A-H, O-Z)
      double precision Ystart(4), Yend(4)
      double precision Mr, Mj
      common Rho, Wd, Tol, Mr, Mj
      external FCN
```

```
C  Read In-Loop SNR Rho in dB and Convert it to a Usable
C  Value
   write(*,'(" In-loop SNR (in dB Please) = ",\)')
   read(*,'(d15.5)')Rho
   Rho = .1d0*Rho
   Rho = 10.0d0**Rho

C  Read Loop Detuning Angle in Degrees and Convert it to a
C  Usable Value
   write(*,'(" Detuning Angle In Degrees = ",\)')
   read(*,'(d15.5)') Wd
   Wd = Wd*Pi()/180.d0
   Wd = Dsin(Wd)

C  Read the Complex-Valued Multiplier Used in the Algorithm
   write(*,'(" Real Part of Multiplier = ",\)')
   read(*,'(d15.5)') Mr

   write(*,'(" Imaj Part of Multiplier = ",\)')
   read(*,'(d15.5)') Mj

C  Read the Starting Estimate for the Complex-Valued
C  Eigenvalue Lambda
   write(*,'(" Guess For Real Part of Eigenvalue = ",\)')
   read(*,'(d15.5)') Ystart(1)

   write(*,'(" Guess For Imaj Part of Eigenvalue = ",\)')
   read(*,'(d15.5)') Ystart(2)

C  Read the Starting Estimate for the Initial Condition on the
C  Derivative of the Eigenvalue
   write(*,'(" Guess For Real Part of e-dot(0) = ",\)')
   read(*,'(d15.5)') Ystart(3)

   write(*,'(" Guess For Imaj Part of e-dot(0) = ",\)')
   read(*,'(d15.5)') Ystart(4)

C  Read the Stopping/Error Criteria for the IMSL™ Routines
   write(*,'(" ErrrL For DNEQNF = ",\)')
   read(*,'(d15.5)') ErrrL

   write(*,'(" Tol For DIVPRK = ",\)')
   read(*,'(d15.5)') Tol

C  Number of Equations and Maximum Iteration Count
   N = 4
   Itmax = 200
```

```
C     Call IMSL™ Routine to Solve Nonlinear Algebraic System
C     (8.5-17)
      call DNEQNF(FCN, ErrrL, N, Itmax, Ystart, Yend, Fnorm)
      write(*, '(" Norm = ",e10.4)') Fnorm
      write(*,'(" Real Part of Eigenvalue is ",e10.4)') Yend(1)
      write(*,'(" Imaj Part of Eigenvalue is ",e10.4)') Yend(2)
      write(*,'(" Real Part of e-dot(0) Variable is ",e10.4)') Yend(3)
      write(*,'(" Imaj Part of e-dot(0) Variable is ",e10.4)') Yend(4)
      stop
      end
C     - - - - - - - - - - - - - - - - - - - - - - - - - - - - -
C     This Subroutine is Called by the IMSL™ root-finder DNEQNF
C     Passed Estimates of the Roots, this Subroutine calculates
C     numerical values for (8.5-17)

      subroutine FCN(Ystart, Fvec, N)
      implicit double precision (A-H, O-Z)
      double precision Ystart(N), Fvec(N)
      double precision Param(50)
      double precision YF(4), YB(4)
      double precision Mr, Mj
      common Rho, Wd, Tol, Mr, Mj
      common/eval/EvR, EvJ
      external FCN2

      EvR = Ystart(1)
      EvJ = Ystart(2)

      X = 0.d0
      Xend = DCONST('Pi')

      Nu = 4
      Ido = 1

C     Set Initial Conditions for forward integration

      YF(1) = 1.d0
      YF(2) = 0.d0
      YF(3) = Ystart(3)
      YF(4) = Ystart(4)

C     Set Vector PARAM to all default values
      call DSET(50, 0.d0, Param, 1)

C     Integrate in a forward direction over the phase interval 0
C     to Pi
      call Divprk(Ido, Nu, FCN2, X, Xend, Tol, Param, YF)
```

```
C     Release memory. Cannot initialize X and recall Divprk until
C     we do the next call with Ido = 3
      Ido = 3
      call Divprk(Ido, Nu, FCN2, X, Xend, Tol, Param, YF)

C     Integrate Backwards from 2Pi to Pi. Set Limits of Independent
C     Variable
      Xend = DCONST('Pi')
      X = 2.d0*Xend

      Nu = 4
      Ido = 1

C     Set Initial Conditions (at X = 2*Pi) for Backward Integration

      YB(1) = Mr
      YB(2) = Mj
      YB(3) = Mr*Ystart(3) - Mj*Ystart(4)
      YB(4) = Mr*Ystart(4) + Mj*Ystart(3)

C     Set Vector PARAM to all Default Values
      call DSET(50, 0.d0, Param, 1)

C     Integrate in a Backward Direction Over the Phase Interval 2*Pi
C     to Pi
      call Divprk(Ido, Nu, FCN2, X, Xend, Tol, Param, YB)

C     Release Memory. Cannot Initialize X = 0 and Recall Divprk Until
C     We Do the Next Call With Ido = 3
      Ido = 3
      call Divprk(Ido, Nu, FCN2, X, Xend, Tol, Param, YB)

C     Force End Conditions

      Fvec(1) = YF(1) - YB(1)
      Fvec(2) = YF(2) - YB(2)
      Fvec(3) = YF(3) - YB(3)
      Fvec(4) = YF(4) - YB(4)

      return
      end
C     -------------------------------
C     This Subroutine is used by DIVPRK. It computes values for the
C     right-hand side of (8.5-12).

      Subroutine FCN2(Nu, X, Y, Yprime)
      implicit double precision (A-H, O-Z)
```

```
      double precision Y(Nu), Yprime(Nu)
      double precision Mr, Mj
      common Rho, Wd, Tol, Mr, Mj
      common/eval/EvR, EvJ

      tmp1 = Wd – dsin(x)
      tmp2 = EvR – dcos(x)

      Yprime(1) = Y(3)
      Yprime(2) = Y(4)
      Yprime(3) = Rho*(tmp1*Y(3) + tmp2*Y(1) – EvJ*Y(2))
      Yprime(4) = Rho*(tmp1*Y(4) + tmp2*Y(2) + EvJ*Y(1))

      return
      end
```

REFERENCES

1. Lindsey, W. C. and Simon, M. K., Eds., *Phase-Locked Loops and Their Applications*, IEEE Press, New York, 1978.
2. Lindsey, W. C. and Chie, C. M., Eds., *Phase-Locked Loops*, IEEE Press, New York, 1986.
3. Viterbi, A. J., *Principles of Coherent Communication*, McGraw-Hill, New York, 1966.
4. Lindsey, W. C., *Synchronization Systems in Communication and Control*, Prentice-Hall, Englewood Cliffs, NJ, 1972.
5. Lindsey, W. C., *Telecommunication Systems Engineering*, Prentice-Hall, Englewood Cliffs, NJ, 1972.
6. Gardner, F. M., *Phaselock Techniques*, 2nd ed., John Wiley & Sons, New York, 1979.
7. Stiffler, J. J., *Theory of Synchronous Communications*, Prentice-Hall, Englewood Cliffs, NJ, 1971.
8. Meyr, H. and Ascheid, G., *Synchronization in Digital Communications*, Vol. 1, John Wiley & Sons, New York, 1990.
9. Wolaver, D. H., *Phase-Locked Loop Circuit Design*, Prentice-Hall, New York, 1993.
10. Best, R. E., *Phase-Locked Loops*, 2nd ed., McGraw-Hill, New York, 1993.
11. Encinas, J., *Phase Locked Loops*, Chapman & Hall, London, 1993.
12. Blanchard, A., *Phase-Locked Loops: Application to Coherent Receiver Design*, John Wiley & Sons, New York, 1976.
13. Klapper, J. and Frankle, J. T., *Phaselocked and Frequency Feedback Systems*, Academic Press, New York, 1972.
14. Van Trees, H. L., *Detection, Estimation and Modulation Theory*, Part II, John Wiley & Sons, New York, 1971.
15. Rhode, U. L., *Digital PLL Frequency Synthesizers, Theory and Design*, Prentice-Hall, Englewood Cliffs, NJ, 1983.
16. Egan, W. F., *Frequency Synthesis by Phase Lock*, John Wiley & Sons, New York, 1981.
17. Mannessewitch, V., *Frequency Synthesizers; Theory and Design*, 3rd ed., John Wiley & Sons, New York, 1987.
18. Stirling, R. C., *Microwave Frequency Synthesizers*, Prentice-Hall, Englewood Cliffs, NJ, 1987.
19. Jerzy Gorski–Popiel, Ed., *Frequency Synthesis: Techniques and Applications*, IEEE Press, New York, 1975.
20. Papoulis, A., *Probability, Random Variables, and Stochastic Processes*, 3rd ed., McGraw-Hill, New York, 1991.
21. Peebles, P. Z., *Probability, Random Variables, and Random Signal Principles*, 2nd ed., McGraw-Hill, New York, 1987.
22. Abramowitz, M. and Stegun, I., *Handbook of Mathematical Functions*, Applied Mathematics Series, Vol. 55, National Bureau of Standards, June, 1964.
23. Krauss, H. L., Bostian, C. W., and Raab, F. H., *Solid State Radio Engineering*, John Wiley & Sons, New York, 1980.

24. Technical Staff, *The ARRL Handbook For Radio Amateurs*, 71st ed., American Radio Relay League, Newington, CT, 1994.
25. Fink, D. G., Ed., *Electronics Engineers' Handbook*, 3rd ed., McGraw-Hill, New York, 1989.
26. Graeme, J., Tobey, G., and Huelsman, L., *Operational Amplifiers, Design and Applications*, McGraw-Hill, New York, 1971.
27. Parzen, B., *Design of Crystal and Other Harmonic Oscillators*, John Wiley & Sons, New York, 1983.
28. *Phase-Locked Loop Data Book*, 2nd ed., Motorola Semiconductor Products, Phoenix, AZ, August 1973.
29. Barna, A. and Porat, D. I., *Operational Amplifiers*, 2nd ed., John Wiley & Sons, New York, 1989.
30. Perko, L., *Differential Equations and Dynamical Systems*, Springer-Verlag, New York, 1991.
31. Andronov, A. A., Leontovich, E. A., Gordon, I. I., and Maier, A. G., *Theory of Bifurcations of Dynamic Systems on a Plane*, John Wiley & Sons, New York, 1973.
32. Andronov, A. A., Vitt, A. A., and Khaikin, S. E., *Theory of Oscillators*, Dover, New York, 1987.
33. Wax, N., Ed., *Selected Papers on Noise and Stochastic Processes*, Dover, New York, 1954.
34. Arnold, L., *Stochastic Differential Equations: Theory and Applications*, John Wiley & Sons, New York, 1974.
35. Jazwinski, A. H., *Stochastic Processes and Filtering Theory*, Academic Press, New York, 1970.
36. Risken, H., *The Fokker–Planck Equation*, 2nd ed., Springer-Verlag, Berlin, 1989.
37. Stratonovich, R. L., *Topics in the Theory of Random Noise*, Vol. 1, Gordon and Breach, New York, 1963.
38. Gardiner, C. W., *Handbook of Stochastic Methods*, Springer-Verlag, Berlin, 1985.
39. Smith, R. J. and Dorf, R. C., *Circuits, Devices and Systems*, 5th ed., John Wiley & Sons, New York, 1992.
40. Cox, D. R. and Miller, H. D., *The Theory of Stochastic Processes*, Chapman and Hall, London, 1965.
41. Bharucha–Reid, A. T., *Elements of the Theory of Markov Processes and Their Applications*, McGraw-Hill, New York, 1960.
42. Ziemer, R. E. and Tranter, W. H., *Principles of Communications*, 3rd ed., Houghton Mifflin, Boston, 1990.
43. Schuss, Z., *Theory and Application of Stochastic Differential Equations*, John Wiley & Sons, New York, 1980.
44. Horsthemke, W. and Lefever, R., *Noise-Induced Transitions*, Springer-Verlag, Berlin, 1984.
45. Yakubovich, V. A. and Starzhinskii, V. M., *Linear Differential Equations with Periodic Coefficients*, Vol. 1, John Wiley & Sons, New York, 1975.
46. Magnus, W., Oberhettinger, F., and Soni, R. P., *Formulas and Theorems for the Special Functions of Mathematical Physics*, Springer-Verlag, New York, 1966.
47. Coddington, E. A. and Levinson, N., *Theory of Ordinary Differential Equations*, McGraw-Hill, New York, 1955.
48. Cassandra, C. G., *Discrete Event Systems, Modeling and Performance Analysis*, Irwin, Homewood, IL, 1993.
49. Gupta, S. C., Phase-locked loops, *Proc. IEEE*, 63, 2, 1975.
50. Rosenkranz, W., Phase-locked loops with limiter phase detectors in the presence of noise, *IEEE Trans. Commun.*, 30, 10, 1982.
51. Springett, J. C. and Simon, M. K., An analysis of the phase coherent-incoherent output of the bandpass limiter, *IEEE Trans. Commun. Technol.*, 19, 1, 1971.

52. Davenport, W. B., Signal-to-noise ratios in band-pass limiters, *J. Appl. Phys.*, 24, 6, 1953.
53. Gilbert, B., A precise four-quadrant multiplier with subnanosecond response, *IEEE J. Solid-State Circuits*, SC-3, 4, 1968.
54. Endo, T. and Tada, K., Analysis of the pull-in range of phase-locked loops by the Galerkin procedure, *Electron. Commun. Jpn.*, Part 1, 69, 5, 1986 (transl. from *Denshi Tsushin Gakkai Ronbunshi*, 68-B, 2, 1985).
55. Stensby, J. L., False lock and bifurcation in the phase locked loop, *SIAM J. Appl. Math.*, 47, 6, 1987.
56. Stensby, J. L., False lock and bifurcation in costas loops, *SIAM J. Appl. Math.*, 49, 2, 1989.
57. Stensby, J. L., Saddle node bifurcation at a nonhyperbolic limit cycle in a phase locked loop, *J. Franklin Inst.*, 330, 5, 1993.
58. Stensby, J. L., A parametrically driven PLL lock detector, *J. Franklin Inst.*, 329, 2, 1992, pp. 283-290.
59. Stensby, J. L., Lock detection in phase-locked loops, *SIAM J. Appl. Math.*, 52, 5, 1992.
60. Stensby, J. L., On the PLL spectral purity problem, *IEEE Trans. Circuits Syst.*, CAS 30, 4, 1983.
61. Stensby, J. L. and Harb, B., Computing the half-plane pull-in range of second-order PLLs, *Electron. Lett.*, 31, 11, 1995.
62. Greenstein, L. L., Phase-locked loop pull-in frequency, *IEEE Trans. Commun.*, 22, 1974.
63. Richman, D., Color carrier reference phase synchronization accuracy in NTSC color television, *Proc. IRE*, 42, 1954.
64. Ryter, D., On the eigenfunctions of the Fokker–Planck operator and of its adjoint, *Physica*, 142A, 103–121, 1987.
65. Ryter, D., Coarse-graining of diffusion processes with a multistable drift, *Z. Phys. B, Condensed Matter*, 68, 209–211, 1987.
66. Ryter, D. and Meyr, H., Multistable systems with moderate noise: coarse-graining to a Markovian jump process and evaluation of the transition rates, *Physica*, 142A, 122–134, 1987.
67. Ryter, D. and Jordan, P., A way to solve the stationary Fokker–Planck equation for metastable systems, *Phys. Lett.*, 104A, 4, 193–195, 1984.
68. Matkowsky, B. J., Schuss, Z., and Knessl, C., Asymptotic solution of the Kramers–Moyal equation and first-passage time for Markov jump processes, *Phys. Rev. A*, 29, 3359–3369, 1984.
69. Vollmer, H. D. and Risken, H., Eigenvalues and their connection to transition rates for the Brownian motion in an inclined cosine potential, *Z. Phys. B, Condensed Matter*, 52, 259–266, 1983.
70. Bobrovsky, B. Z. and Schuss, Z., A singular perturbation method for the computation of the mean first passage time in a nonlinear filter, *SIAM J. Appl. Math.*, 42, 174, 1982.
71. Viterbi, A. J., Phase-locked loop dynamics in the presence of noise by Fokker–Planck techniques, *Proc. IEEE*, 51, 12, 1737–1753, 1963.
72. Tikhonov, V. I., The effects of noise on phase-lock oscillator operation, *Avtom. Telemekh.*, 22, 9, 1959.
73. Tikhonov, V. I., Phase-lock automatic frequency control application in the presence of noise, *Avtom. Telemekh.*, 23, 3, 1960.
74. Develet, J. A., A threshold criterion for phase-lock demodulator, *Proc. IEEE*, 51, 1963.
75. Develet, J. A., An analytic approximation of phase-lock receiver threshold, *IEEE Trans.*, SET-9, 1963.
76. Charles, F. J. and Lindsey, S. C., Some analytical and experimental phaselocked loop results for low signal-to-noise ratios, *Proc. IEEE*, 54, 1966.

77. Van Trees, H. L., Functional techniques for the analysis of the nonlinear behavior of phase-locked loops, *Proc. IEEE*, 52, 1964.
78. Ascheid, G. and Meyr, H., Cycle slips in phase-locked loops: a tutorial survey, *IEEE Trans. Commun.*, 30, 10, 1982.
79. Tausworthe, R. C., Cycle slipping in phase-locked loops, *IEEE Trans. Commun.*, 15, 1967.
80. Visual Numerics Corporation, *IMSL Math Library*, Visual Numerics, 9990 Richmond Ave., Suite 400, Houston, TX.
81. Garbow, B., Hillstrom, K., and More, J., *MINIPACK Software Documentation*, Argonne National Laboratory, March, 1980 (the MINIPACK subroutine package is available freely on the Internet).

INDEX

A

Adjoint operator, 300, 366
AGC, see Automatic gain control
Amplitude modulation, 6
Analog mixer, see also Phase detector
　double balanced, 93
　singly balanced, 95
Analytic signal, 54
Autocorrelation, 62
Automatic gain control, 116

B

Bandpass filter, 51
　group delay, 55
　input/output, 57
　phase delay, 55
　symmetrical, 54
Bandpass signal, 51
　carrier frequency, 51
　envelop, 51
　lowpass equivalent, 52
　phase, 51
　quadrature components, 51
　random process, 60
Beat note, 124, 138
Bernoulli trials, 180
Bifurcation diagram, 149
　high-gain case, 150
　low-gain case, 151
Bifurcation of limit cycle, 119, 136
　saddle node bifurcation, 142, 144, 150
　separatrix cycle bifurcation, 145
Binary phase shift keying, 8
Bistable cyclic model, 236
Bode plot, 81
Boundary condition, 185, 188, 203
BPSK, see Binary phase shift keying
Brownian motion, see Wiener process

C

Chapman–Kolmogorov equation, 193
　n-dimensional, 217
　one dimensional, 193
Charge pump, 103
Closed loop transfer function, 23
Coarse-grained state model, 315
Coherent detection, 8
Complex envelope, see Lowpass equivalent
Conditional expectation method, 270, 288
Conservation of probability, 184, 235
Continuity equation, 184, 200
Cross correlation, 62
Cycle slip, 131, 231, 256

Cycle slip rate, 231, 257
    coarse-grained model, 316, 333
    first-order PLL, 259
    second-order PLL, 280
Crystal oscillator, 111

D

D flip-flop, 102
Damping factor, 77, 85
Demodulator, 6, 8
DeMoivre–Laplace theorem, 181
Detuning parameter, 25
Diffusion constant, 182, 225
Diffusion current, see Probability current
Diffusion equation, 184
Diffusion process, 192, 198
Digital phase detector, 13, 93
Digital phase-locked loop, 13, 93
Diode ring mixer, 96
Domain of attraction, 231
Doppler shift, 9, 74
Double balanced mixer, see Analog mixer
DPLL, see digital phase-locked loop
Drift coefficient, 200, 225
Drift current, see Probability current

E

Eigenfunctions, 323
    moderate noise case, 325
    subspace of periodic functions, 346
Eigenvalue method, 11, 290
Eigenvalues, 323
    computation of, 347, 357
    moderate noise case, 325

Equilibrium points, 121, 127, 138
Equivalent noise bandwidth, 35
Euclidean norm, 126
Exclusive OR gate detector, 98

F

False lock, 139, 156
Fast timescale, 241, 315
First passage time, 208, 281
    density of, 211
    expected value of, 211
First variation equation, 128, 138
Flip-flop phase detector, 99
Focus, 128, 130, 139
Fokker–Planck equation, 198, 223
    boundary condition, 209
    initial condition, 209
    n-dimensional, 223
    one-dimensional, 198
    series solution, 291
    steady-state, 206
Fokker–Planck operator, 224
    eigenfunctions/eigenvalues, 323
    finite dimensional approximation, 294
    first-order PLL, 247
    n-dimensional, 224
Four-quadrant multiplier, 93
Frequency, 3
    error, 4, 131
    ramp, 74, 78, 87
    rate, 4
Frequency detector, 13, 101
Frequency locked, 131
Frequency modulation, 24, 290, 254
Frequency of undesirable events, 341
Frequency pushing, 160
Frequency synthesis, 7

G

Gambler's ruin, see Random walk
Gaussian white noise, 188
Generalized Fourier series, 291
Generalized stochastic process, 189
Gilbert multiplier, 93

H

Half-plane pull-in frequency, 140, 148
Hang-up, 124
Hard limiter, 37
Hermite polynomials, 292, 363
High-gain case, 140, 142
Hilbert transform, 54, 62
Hold-in range, 122, 140

I

IMSL™, 296, 367
Initial conditions, see Fokker–Planck equation
In-loop SNR, see Signal-to-noise ratio in the loop
Intensity coefficients, 197, 223
   Fokker–Planck equation, 198
   physical meaning of, 199
Intermediate frequency filtering, 46, 156
Itô integral, 193

J

Jump process, 313

K

Kramers–Moyal expansion, 194, 217

L

Large-scale integrated circuits, 91
Least positive residue function, 332, 367
Limiter phase detector, see Phase detector
Lock detector, 115
Lock-in range, 132
Long loop, 9
   false lock, 156
   model, 46
Loop filter, 5
   imperfect integrator, 84, 108
   model, 17
   perfect integrator, 76, 108
   technology, 108
Loop noise bandwidth, 35
Low-gain case, 149
Lowpass equivalent, 48
Lowpass filter, 17
LSI, see Large scale integrated circuits

M

Markov process, 190, 216
   absorption, 187, 204, 214
   diffusion processes, 192
   homogeneous, 192
   jump process, 314
   vector process, 216
m-attractor model, 238
MINPACK, 296
Moderate noise conditions, 230, 241
MOD, see Least positive residue function
Motional arm capacitance, 111
Motional arm inductance, 111
Multiplier for differential equation, 347, 359

Multistable cyclic model, see m-attractor model
Multivibrator, 113

N

Narrowband signals and systems, 51
Natural frequency, 77, 85
Node, 128, 130, 139
Noise equivalent bandwidth, 35
Noise in PLL, 12
  autocorrelation function, 62
  bandpass model, 61
  lowpass equivalent, 63
  symmetrical bandpass process, 65
  time normalized, 244

O

Occupancy probabilities, 316
Open loop transfer function, 26
Ordinary point, 127
Orthogonality principle, see Projection theorem
Overdamped loop, 130

P

Parallel resonance in crystal, 112
Phase acquisition, 123
Phase coherence, 4
Phase detector, 16
  digital, 93
  diode ring, 96
  exclusive OR gate, 98
  four-quadrant multiplier, 93
  gain, 16
  input waveforms types, 92
  limiter phase detector, 37

  memory, 92
  output, 29
  RS flip-flop, 99
  sequential phase/frequency, 101
  threshold, 10, 47, 156
Phase error, 4
  bistable cyclic model, 236
  modulo-$2\pi m$ model, 238
  modulo-2π model, 232
  unrestricted, 4, 230, 334
  variance, 34
Phase-frequency detector, 101
  charge pump, 103
  state diagram, 102
Phase lock, 4
Phase-locked loop, 5
  block diagram, 12
  digital, 93
  first-order, 72
  with imperfect integrator filter, 76
  linear model for, 22
  nonlinear model for, 19
  with perfect integrator filter, 84
  transfer function for, 23
  type I, 85
  type II, 76
Phase-locked receiver, 9
Phase modulation, 25
Phase plane, 121, 128
  structural changes, 129, 133, 139
  trajectory, 129
Phase shift keying, see Binary phase shift keying
PLL, see Phase-locked loop
Potential function, 201, 244, 248, 278
Probability current, 184
  diffusion component, 202, 225
  drift component, 200, 224
  n-dimensional vector, 224

particle flow interpretation, 200
  steady-state, 207, 249
Projection theorem, 276
Pull-in, 6
  first-order PLL, 121-124
  imperfect integrator loop, 139-153
  perfect integrator loop, 128-135
Pull-in range, 122
  approximation, 142
  high-gain loop, 142
  low-gain loop, 150
Pull-in time, 135, 161
Pull-out frequency, 132

Q

Quadrature components, 51, 61
Quasi-stationary behavior, 231
Quiescent frequency, see VCO center frequency

R

Random walk, 178
  absorption, 185
  limit, 181
Reduced Fokker–Planck equation, 269
Reference signal, 6
  angle modulation, 24
  noisy, 27
Reflection principle, 186
Relaxation time constant, 231
Residual frequency detuning, 257, 282
Residues, 37
Residue set, 367
Restoring force, 244
Root locus, 23
Rotated vector fields, 172
RS flip-flop, 99

S

Saddle node bifurcation, see Bifurcation
Saddle point, 128, 130
Saddle-to-saddle separatrix, see Separatrix cycle
Separatrix, 129
Separatrix cycle, 142, 145, 171
  bifurcation from, 147, 151
  computation, 154, 173
  external stability, 147, 152, 174
  structural stability, 146
Series resonance in crystal, 112
Short loop, 5
Signal-to-noise ratio in the loop, 35
Singly balanced mixer, see Analog mixer
Slow timescale, 241, 315
Slow-time variable, 121
Smoluchowski equation, see Chapman–Kolomogorov equation
Squaring loop, 8
Stability, 77, 86
  asymptotic, 127
  equilibrium point, 122, 126, 137
State equation model, 21
State transition rate diagram, 316, 333
Stochastic differential equation, 193
Symmetrical bandpass filter, 54

T

Threshold
  phase detector, 10, 47, 156
  PLL, 231
Time scale separation, 241
Transition density, 191, 202
Transition event, 237

Transition function, 317
Transition function matrix, 317
Transition rate matrix, 318
Transition rates, 290
   m-attractor cyclic model, 333
   relation to eigenvalues, 340
   unconditional, 321
   unrestricted jump model, 334

U

Underdamped loop, 130

V

Varactor diode, 112
VCO, see Voltage-controlled oscillator
VCXO, see Voltage controlled crystal oscillator
Vector space of periodic functions, 344
Voltage-controlled crystal oscillator, 111
Voltage-controlled oscillator, 5
   based on relaxation condition, 113
   based on resonance condition, 111
   center frequency, 18
   gain parameter, 18
   model, 18
   technologies, 111
   tuning range, 111, 113

W

Wiener process, 181